Applied Mathematical Sciences
Volume 126

Springer

New York
Berlin
Heidelberg
Barcelona
Budapest
Hong Kong
London
Milan
Paris
Santa Clara
Singapore
Tokyo

Applied Mathematical Sciences

(continued following index)

Frank C. Hoppensteadt   Eugene M. Izhikevich

Weakly Connected
Neural Networks

With 173 Illustrations

 Springer

Frank C. Hoppensteadt
Eugene M. Izhikevich
Center for Systems Science and
    Engineering
Arizona State University
Tempe, AZ 85287

Editors

J.E. Marsden
Control and Dynamical Systems, 104-44
California Institute of Technology
Pasadena, CA 91125
USA

L. Sirovich
Division of Applied Mathematics
Brown University
Providence, RI 02912
USA

Mathematics Subject Classification (1991): 58F14, 92B20, 92C20

Library of Congress Cataloging-in-Publication Data
Hoppensteadt, F.C.
    Weakly connected neural networks / Frank C. Hoppensteadt, Eugene
M. Izhikevich.
        p.   cm. — (Applied mathematical sciences ; 126)
    Includes bibliographical references and index.
    ISBN 0-387-94948-8 (hardcover : alk. paper)
    1. Neural networks (Computer science)   I. Izhikevich, Eugene M.
II. Title.   III. Series: Applied mathematical sciences (Springer-
Verlag New York Inc) ; v. 126.
QA76.87.H68   1997
006.3′2 — dc21   97-14324

Printed on acid-free paper.

Production managed by Victoria Evarretta; manufacturing supervised by Joe Quatela.
Photocomposed copy prepared from the authors' LaTeX files.
Printed and bound by Maple-Vail Book Manufacturing Group, York, PA.
Printed in the United States of America.

9 8 7 6 5 4 3 2 1

ISBN 0-387-94948-8 Springer-Verlag New York Berlin Heidelberg   SPIN 10557261

To Elizabeth Izhikevich,
who was brought into existence together with this book,

and

To Leslie Hoppensteadt,
a constant source of inspiration.

Preface

This book is devoted to an analysis of general weakly connected neural networks (WCNNs) that can be written in the form

$$\dot{x}_i = f_i(x_i) + \varepsilon g_i(x_1, \ldots, x_n, \varepsilon) , \qquad \varepsilon \ll 1 . \tag{0.1}$$

Here, each $x_i \in \mathbb{R}^m$ is a vector that summarizes all physiological attributes of the ith neuron, n is the number of neurons, f_i describes the dynamics of the ith neuron, and g_i describes the interactions between neurons. The small parameter ε indicates the strength of connections between the neurons.

Weakly connected systems have attracted much attention since the second half of seventeenth century, when Christian Huygens noticed that a pair of pendulum clocks synchronize when they are attached to a lightweight beam instead of a wall. The pair of clocks is among the first weakly connected systems to have been studied.

Systems of the form (0.1) arise in formal perturbation theories developed by Poincaré, Liapunov and Malkin, and in averaging theories developed by Bogoliubov and Mitropolsky.

We treat (0.1) as being an ε-perturbation of the uncoupled system

$$\dot{x}_i = f_i(x_i) , \qquad i = 1, \ldots, n . \tag{0.2}$$

Obviously, (0.2) is not interesting as a model of the brain. A natural question to ask is when the WCNN (0.1) acquires any new features that the uncoupled system (0.2) does not have. The answer depends upon the nature of each equation in (0.2).

A network of resting neurons is described when each equation $\dot{x}_i = f_i(x_i)$ has an equilibrium point – the rest potential. From the Hartman-Grobman theorem it follows that the local behavior of (0.1) and (0.2) is essentially the same (uncoupled and linear) unless one or more neurons are near a bifurcation. The fundamental theorem of WCNN theory, which is a consequence of center manifold reduction, claims that those neurons that are far away from the threshold potential at which the bifurcation occurs can be removed from (0.1) without changing the network's local dynamics. Being near a threshold is a necessary but not a sufficient condition for a neuron to participate in local dynamics of a WCNN (0.1). Additional conditions must be satisfied, such as an adaptation condition. In this case, the WCNN (0.1) has a rich local dynamics; in particular, it can perform pattern memorization and recognition tasks.

A network of periodically spiking neurons is described when each equation $\dot{x}_i = f_i(x_i)$ has a limit cycle attractor, say with a frequency Ω_i. A direct consequence of averaging theory is that the behavior of (0.1) and (0.2) is essentially the same unless there are resonances among the frequencies $\Omega_1, \ldots, \Omega_n$. An important case is that in which the connection functions g_i in WCNN (0.1) have the special form

$$g_i(x_1, \ldots, x_n) = \sum_{j=1}^{n} g_{ij}(x_i, x_j) \, .$$

In this case, the interaction between each pair of oscillators averages to zero unless their frequencies are commensurable (i.e., $a\Omega_i - b\Omega_j = 0$ for some relatively prime integers a and b). Moreover, the interaction is still negligible when the frequencies are barely commensurable (i.e., a and b are large). Another important case is that in which the limit cycle attractors have small amplitudes of order $\sqrt{\varepsilon}$. In this case, the oscillators interact only when they have nearby (ε-close) frequencies. One of the consequence of this fact is that an oscillatory neuron can turn on and off its connections with other neurons simply by adjusting its frequency. This mechanism is similar to the one employed in frequency modulated (FM) radio: The channel of transmission is encoded in the frequency of oscillation, and the information is transmitted via phase deviations (frequency modulations). When a neuron spikes periodically, the frequency of oscillation is the mean firing rate, and the phase deviation is the timing of spikes. Therefore, the firing rate encodes only the channel of communication, whereas the information (the neural code) is encoded in the timing of spikes.

While it is feasible to study weakly connected networks of chaotic neurons, this problem goes beyond the scope of this book.

Weak connections are observed in real neuron networks: Postsynaptic potentials (PSPs) have small amplitude when the postsynaptic neuron is not near the threshold. (When the postsynaptic neuron is sufficiently close to the threshold potential, the PSPs may have large amplitudes or even grow to a full-scale action potential.) We discuss this issue in Chapter 1.

There are two phenomena associated with the generation of action potentials by neurons – neural excitability and the transition from rest to periodic spiking activity. The former is a single response to external perturbations; the latter is a qualitative change in dynamic behavior. The class of neural excitability depends intimately on the type of bifurcation from quiescence to oscillatory activity. We discuss this as well as many other bifurcation issues in Chapter 2.

When a neural network's activity is near a bifurcation point, it is sensitive to even small external perturbations. Future behavior of such a network can be determined by the local events at the bifurcation. This motivates a new neural network type – the nonhyperbolic neural network. We discuss this and two other neural network types in Chapter 3.

One of the most important results in dynamical system theory is the proof that many diverse and complicated systems behave similarly and simply when they operate near critical regimes, such as bifurcations. Often, all such systems can be transformed by a suitable change of variables to another system that is usually simpler – the *canonical model*. Derivation of such canonical models is usually a difficult problem, but once found, the canonical model provides an advantage since its study yields some information about all systems in the critical regime, even those whose equations we do not know. Another advantage is that the question of plausibility of a canonical model is replaced by the question of plausibility of the critical regime. There is no general algorithm for deriving canonical models, though normal form and invariant manifold theories, which we discuss in Chapter 4, are useful.

We derive some canonical models for multiple bifurcations in WCNNs (Chapter 5), singularly perturbed WCNNs (Chapter 6), and weakly connected mappings (Chapter 7). Canonical models for some global bifurcations and for weakly connected oscillatory networks are derived in Chapter 8 and 9, respectively.

The simplicity of the canonical models in comparison with the original WCNN (0.1) is deceptive. They have quite complicated behavior, and most of them have yet to be analyzed. We concentrate our efforts on the canonical model for multiple cusp bifurcations in Chapter 11, where we show that the model operates similarly to classical neural networks. We also study canonical models describing oscillatory phenomena in Chapter 10 and 12. In Chapter 13 we study synaptic organizations and dynamical properties of weakly connected neural oscillators. There we discuss synaptic organizations of networks that can and cannot accomplish certain tasks, such as in- and anti-phase locking, memorization of phase information, etc.

Chapters 5, 6, 7, 8, 10, 11, 12, and 13 were derived in part from the Ph.D. dissertation (Izhikevich 1996) performed under the guidance of F.C.

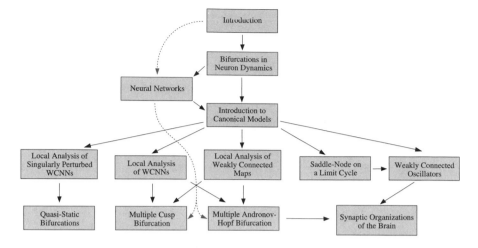

Figure 0.1. Possible routes through this book.

Hoppensteadt. A part of this book (Chapters 5 and 11) received the SIAM award as the best student paper in applied mathematics in 1995.

Acknowledgments

One of the authors (E.I.) acknowledges his parents, Michael and Nina, and all the people who contributed to his growth as a scientist: G.D. Kim, V.N. Pilshikov, E.A. Samarskaya, G.G. Elenin, E.N. Sokolov, A.L. Krulova, A.S. Mikhailov, Yu.L. Klimantovich, G.S. Voronkov, V.V. Shulgovskii, B. Kh. Krivitskii, and G.G. Malinetskii (Moscow State University), L.L. Voronin and A. Kreshevnikov (Brain Institute, Moscow), I.I. Surina (Kurchatov Institute of Atomic Energy, Moscow), A.A. Frolov (Institute of High Nervous Activity, Moscow), R.M. Borisyuk (Russian Biological Center, Puschino), S. Newhouse (Michigan State University). Special thanks to E.I.'s scientific advisors A.S. Mikhailov, G.G. Malinetskii, and F.C. Hoppensteadt. Finally, E.I. thanks his wife, Tatyana (Kazakova), and his daughter, Elizabeth, for their help and patience during preparation of the manuscript.

The authors are grateful to S.M. Baer, B. Ermentrout, T. Erneux, and J. Rinzel for reading various portions of the manuscript and making a number of valuable suggestions. Some topics of the book were polished in lively debates with R. Borisyuk. Assistance of I. Martinez in preparation of various illustrations is greatly appreciated.

June 2, 1997   Frank Hoppensteadt
Scottsdale, Arizona   Eugene Izhikevich

Contents

Part I

Introduction

1
Introduction

In this chapter we give definitions and explanations of basic neurophysiological terminology that we use in the book. We do not intend to provide a comprehensive background on various topics.

We discuss here several types of mathematical models in the neurosciences. Then we present a simplified ordinary language model of a brain, which we use to explain what we mean under the notion of weakness of synaptic connections between neurons. In subsequent chapters we use the additive and Wilson-Cowan neural networks to illustrate bifurcation phenomena.

1.1 Models in Mathematical Biology

Most models in neuroscience can be divided into the following groups:

- *Ordinary Language Models* are used by biologists to explain how the human brain or some of its structures might work. These models are precise where data are known but otherwise are suitably imprecise.

- *Comprehensive Models* are the result of an attempt to take into account all known neurophysiological facts and data. Usually they are cumbersome and are not easily amenable to mathematical analysis. A typical example is the Hodgkin-Huxley (1954) model of a neuron or the Traub-Miles (1991) model of the hippocampus.

- *Empirical Models*, or caricature models, occur when one tries to construct a model reflecting one or more important neurophysiological observations, often without regard to other neurophysiological facts. Typical examples are the McCulloch-Pitts (1943) bistable neuron and Hopfield's network (1982). Although amenable to mathematical analysis, these models have no rigorous connection with real neural systems.

- *Canonical Models* arise when one studies critical regimes, such as bifurcations in brain dynamics. It is often the case that general systems at a critical regime can be transformed by a suitable change of variables to a canonical model that is usually simpler, but that captures the essence of the regime. Such a transformation shifts attention from the plausibility of a model to the plausibility of the critical regime. The major advantage of the canonical model approach is that we can derive canonical models for systems whose equations we do not know. The major drawback is that the canonical models are useful only when the systems operate near the critical regimes. We derive several canonical models in this book.

The division above is artificial, since there are no exact boundaries between the model types. For example, the Hodgkin-Huxley model could be classified as an empirical model because it reflects our knowledge of membrane properties up to the 1950s, which is still incomplete. The canonical models might be considered as empirical models too, because they can be analyzed without resort to computers and/or because they illustrate some basic neurophysiological facts such as excitability, bistability, oscillations. Each of the model types has its own advantages and drawbacks. Neither of them is better or worse than the others. Figure 1.1 indicates some connections between these models.

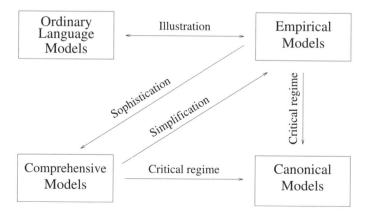

Figure 1.1. Relationship between models in mathematical biology.

Biologists often use ordinary language models since they are precise where data are known and appropriately vague otherwise. These models do not require knowledge of mathematics. Typical examples of such models are given in Section 1.2 where we introduce some basic notions from neurophysiology.

Using comprehensive models could become a trap, since the more neurophysiological facts are taken into consideration during the construction of the model, the more sophisticated and complex the model becomes. As a result, such a model can quickly come to a point beyond reasonable analysis even with the help of a computer. Moreover, the model is still far from being complete.

Most mathematicians and physicists who study brain functions use empirical models. Usually, one has an idea motivated by an ordinary language model and tries to construct a simple formula that captures it. The simpler the model, the better. As a result, one often obtains a system of ordinary differential or difference equations that might have some neurocomputational properties but could be irrelevant to the brain. If we completely understood a brain, we might be able to explain how it works using a simple empirical model. However, we are far away from understanding a brain, and the invention of empirical models capable of performing a useful task is more an art than a science. Some successful models are discussed in the review article by S. Grossberg (1988). We discuss the additive and Wilson-Cowan empirical models of neural networks in Section 1.4

There are examples of canonical models in mathematical neuroscience. The voltage controlled oscillator neuron (VCON) model is canonical for neurons having Class 1 neural excitability, which we discuss in the next chapter. Ermentrout and Kopell (1986) proved that Hill's equation is a canonical model for parabolic bursting. The Bonhoeffer-Van der Pol oscillator is a canonical model for quasi static cusp bifurcation. We discuss these models and derive and analyze others in subsequent chapters.

1.2 Ordinary Language Models

The following ordinary language model of the brain is based on experimental observations, see, e.g., Shepherd (1983) and Kuffler et al. (1984). In Section 1.2.1 we examine some of the claims we make below.

- *A brain has many neurons.* The number of neurons is enormous – estimated to be more than 10^{11}.

  Although neurons differ in size, shape, and function, it is possible to describe typical neuron attributes (see Figure 1.2).

- *Neurons consist of a cell body, a dendritic arborization, and an axon.* Neurons are living cells having metabolisms similar to those found in

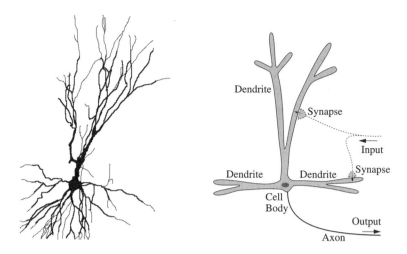

Figure 1.2. *Left:* A pyramidal neuron from the hippocampus of a rat (courtesy of Bruce McNaughton). *Right:* Schematic representation of a typical neuron.

many other cells. For example, the neuron's cell body (soma) has a nucleus, vesicles, mitochondria, etc. Unlike other cells of the body, a neuron has dendrites and an axon. The dendritic tree can be immense, covering vast areas of the brain, and the axons can be longer than one meter.

- *Neurons generate action potentials.* The action potential, which is also called an impulse or voltage spike, is a brief electrophysiological event that occurs because the neuron's membrane has active properties – it is an excitable medium. We discuss this issue in detail in the next chapter. Action potentials usually originate in the axon hillock and propagate away along the axon.

- *Action potentials are basic mechanisms for communication between neurons.* One can think of the action potential as being a signal that is sent by a neuron to other neurons. A neuron receives many signals from other neurons (called convergence) and in turn signals to many others (divergence).

- *Neurons are functionally polarized.* That is, they receive signals via dendrites, process and accumulate them in the soma (integration), and send the response to other neurons via the axon.

- *The junction between an axon of one neuron and a dendritic branch of another neuron is called a synapse.* Synapses can be electrical or chemical. A chemical synapse consists of presynaptic and postsynaptic parts that are separated by the synaptic cleft. When an impulse arrives at an axon terminus, it triggers a chain of chemical and physiological reactions in the presynaptic part, resulting in the release of certain chemicals into the synaptic cleft. The chemicals, which are called *neurotransmitters*, passively diffuse across the synaptic cleft to the postsynaptic region. They react with postsynaptic receptors, producing changes in the postsynaptic membrane potential.

- *Dale's Principle: A neuron is either excitatory or inhibitory.* It is excitatory if the postsynaptic membrane potential increases. This is called *depolarization.* Increasing a membrane's potential facilitates the generation of an action potential in the postsynaptic neuron. If the postsynaptic potential decreases (*hyperpolarization*), the presynaptic neuron is called inhibitory. Hyperpolarization usually impedes generation of an action potential.

This list describes some physiological aspects of a brain, but it does not describe how a brain functions. Indeed, it is known how the neurons form brain structures, such as the neocortex or hippocampus, and a bit is known about how neurons communicate within and outside of these structures. But it is not yet clear how these communications result in recognition, attention, emotions, consciousness, or in any other brain phenomena.

We mention here only those facts that we use in this book and omit many other interesting ones that are not relevant to our modeling.

1.2.1 Qualifications

A brain is a complicated biological entity. The variety of neurons and types of communication between them is so great that it is possible to find counterexamples to all the postulates of the ordinary language model described above. For example,

- *There are many types of cells other than neurons in a brain.* For example, neuroglial and Schwann cells greatly outnumber neurons and make up almost one-half of a brain's volume. Schwann cells form an insulation layer (myelin) around large axons to speed up propagation of action potentials, but the role of neuroglial cells has yet to be clarified. It is believed that they also participate in signal processing.

- *Some neurons do not have dendrites or axons.* Typical examples of such neurons are granule cells in the olfactory bulb, which do not have axons.

- *Some neurons do not generate action potentials,* and there are non-neurons that do generate action potentials. Neurons that do not generate action potentials under normal conditions are, for instance, granule cells in the olfactory bulb. Incidentally, generation of action potentials is a common property of cells that is expressed, for example, in cells from a pumpkin stem, tadpole's skin, or annelid eggs. Action potentials play certain roles in cell division, fertilization, morphogenesis, secretion of hormones, ion transfer, cell volume control, etc., and could be irrelevant to neural signaling.

- *There are many mechanisms for communication between neurons*, and action potentials are only one of them. Neurons can also carry localized, graded potentials. These are membrane potential changes that can spread only short distances, usually from dendritic terminals to the soma. Neurons can also have long-distance chemically mediated interactions through secretion of specific neurohormones into intercellular clefts or into the bloodstream. Thus, communication is possible between distant neurons having no common synapses.

- *Not all neurons are functionally polarized.* Consider, for example, a granule cell of the olfactory bulb: It does not have an axon, but it has dendro-dendritic synapses with mitral cells (see Figure 1.3). That is, the signals (changes of membrane potential) are transferred from dendrites of mitral cells directly to dendrites of granule cells and back to mitral cell dendrites. There could be nontrivial information processing inside dendritic arborization not involving the cell body. Thus, a neuron is not one functional unit. Operation of its units can be relatively autonomous throughout the dendritic tree.

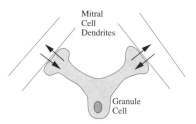

Figure 1.3. Dendro-dendritic (reciprocal) synapses. For simplicity, only two dendrites are depicted.

- *There are dendro-dendritic, axo-axonic, reciprocal, and many other types of synapses.* The role of these has yet to be explained.

- *Dale's principle* as we stated it above is not always true. Actually, there are three definitions of the principle, which we discuss in the next section.

We see that none of the postulates of the brain ordinary language model are entirely accurate. Nevertheless, they are useful metaphors when we discuss brain organizations and functions. When we study general neural networks, we implicitly assume that they consist of typical neurons having dendrites, axons, action potentials, etc. When we study specific brain structures, such as the olfactory bulb or the hippocampus, we explicitly define the type of neurons under consideration since such neurons might not be "typical".

1.2.2 Dale's Principle

Dale's principle has played an important role in the development of neurosciences (Osborne 1983). It is widely used in mathematical neuroscience, since it imposes natural restrictions on the possible dynamics of neural networks. We use it extensively in this book because it gives a useful criterion for various synaptic organizations that we study.

Dale gave the 1934 Dixon Memorial Lecture to the Royal Society of Medicine, in which he discussed chemical transmissions at synapses. He said (Dale 1935),

> When we are dealing with two different endings of the same sensory neurone, the one peripheral and concerned with vasodilatation and the other at a central synapse, can we suppose that the discovery and identification of a chemical transmitter of axon-reflex vasodilatation would furnish a hint as to the nature of the transmission process at a central synapse? The possibility has at least some value as a stimulus to further experiments.

That is, various branches of a neuron have identical chemical transmissions. Sometimes this is called "one neuron – one transmitter", though in fact it is "one neuron – one cocktail of transmitters". Notice that this says nothing about the neuron, i.e., whether it is excitatory or inhibitory. The sign of the synaptic action is determined not only by the neurotransmitter, but also by the postsynaptic receptors. The same neurotransmitter could be excitatory for one receptor and inhibitory for others. For example, acetylcholine has excitatory action when it is released by motoneurons and inhibitory action when released from vagal nerve terminals in the heart.

It is customary to use the following version of Dale's principle: A given neuron has the same physiological action at all its synapses. If all its synapses are excitatory (inhibitory), then the neuron is said to be excitatory (inhibitory). Although this definition is not quite correct, it is a useful paradigm when we discuss anatomical and physiological organizations of various brain structures.

The third version of Dale's principle is mathematical: Suppose $x_i \in \mathbb{R}$ denotes the activity of the ith neuron, $i = 1, \ldots, n$. We assume that increasing (decreasing) x_i corresponds to depolarization (hyperpolarization). Suppose the network of such neurons is governed by a dynamical system of the form

$$\dot{x}_i = f_i(x_1, \ldots, x_n) \, , \qquad i = 1, \ldots, n \, ,$$

where $f_i : \mathbb{R}^n \to \mathbb{R}$ are some smooth functions. Then the mathematical definition of Dale's principle states that all synaptic coefficients

$$s_{ij} = \frac{\partial f_i}{\partial x_j} \, , \qquad i \neq j \, ,$$

from the jth neuron to the other neurons have the same sign. If $s_{ij} \geq 0$ for all i, then the jth neuron is excitatory (inhibitory otherwise). It is easy to see that if $s_{ij} > 0$, then increasing the jth neuron activity x_j facilitates increasing the ith neuron activity x_i. Hence, the jth neuron has an excitatory effect on the ith neuron. From Dale's principle it follows that the effect is the same for any other neuron, provided that the neurons are connected (that is, if the synaptic coefficient is not identically zero).

Remark. The coefficient

$$s_{ii} = \frac{\partial f_i}{\partial x_i}$$

does not denote the synapse from the ith neuron to itself. It denotes a *feedback* parameter and is not subject to Dale's principle.

We use the mathematical definition of Dale's principle throughout this book. Notice, though, that it differs from the previous two definitions. Indeed, suppose neuron a has two synapses with each of neurons b and c, as depicted in Figure 1.4. Suppose that each synapse uses its own neuro-

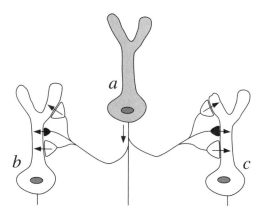

Figure 1.4. The mathematical definition of Dale's principle is weaker than the original one. The mathematical definition allows for individual synapses to have different actions, but the "averaged synapses" from a neuron to other neurons must have the same action. For example, if the excitatory synapses (white) are much stronger than the inhibitory synapses (black), then the averaged synapses are excitatory.

transmitter, so the original Dale's principle is violated. Suppose also that each neurotransmitter has a different effect on the postsynaptic neurons; that is, the second version of Dale's principle is also violated. Now suppose that the inhibitory effect is much weaker than the excitatory one. For example, the number of excitatory synapses is much larger than that of the inhibitory ones, so that the cumulative action is excitatory. Then the mathematical version of Dale's principle can still hold, because $s_{ba} > 0$ and $s_{ca} > 0$ in this case. We see that the mathematical definition of Dale's principle can be interpreted as requiring that the *averaged* actions of one neuron on the other neurons have the same sign, not the actions of each individual synapse. Finally, we note that Dale's principle for nonscalar x_i has yet to be defined.

1.3 Weakness of Synaptic Connections

The key assumption of the theory developed in this book is that the synaptic contacts between neurons are weak. Mathematically this assumption means that the dynamical system describing brain activity can be written in the "weakly connected form"

$$\dot{x}_i = f_i(x_i) + \varepsilon g_i(x_1, \ldots, x_n, \varepsilon) , \qquad \varepsilon \ll 1, \quad i = 1, \ldots, n .$$

Here x_i is the activity of the ith isolated neuron, f_i governs its dynamics, g_i describes connections converging from the entire system to the ith neuron, and the parameter ε describes the (dimensionless) strength of synaptic connections.

We assume that ε is small. This is a purely mathematical assumption made in the spirit of other mathematical assumptions such as "lines are straight" or "angles are right", even though there are no such things as straight lines or right angles in the universe. These are mathematical idealizations made to simplify analysis, and they have worked pretty well so far. The same applies to the assumption $\varepsilon \ll 1$, which we write sometimes in the form $\varepsilon \to 0$. We use it to make rigorous analysis of neural networks possible. In fact, the condition $\varepsilon \to 0$ is too strong. Most of our theorems can be expressed in the form "there is positive ε_0 such that for all $\varepsilon < \varepsilon_0$ the result is true"; i.e., they are valid for small, but not infinitesimal, ε.

1.3.1 How Small Is ε?

Notice that the question, "What is the value of ε in the real brain?" has a restricted sense. For suppose we are given certain ε; then we can always rescale functions $g_i = \delta \bar{g}_i$ so that the new rescaled weakly connected system

$$\dot{x}_i = f_i(x_i) + \varepsilon \delta \bar{g}_i(x_1, \ldots, x_n, \varepsilon) , \quad i = 1, \ldots, n.$$

has $\bar{\varepsilon} \doteq \varepsilon\delta$, which could be smaller or bigger than ε depending on $\delta > 0$.

We can ask the following question: How small need ε be? To determine its size, we require that the continuous functions g_i be all of order 1, i.e., $g_i(x_1, \ldots, x_n, \varepsilon) = \mathcal{O}(1)$. We can also require that $\sup g_i(x_1, \ldots, x_n, 0) = 1$ for $(x_1, \ldots, x_n)$ from some domain. Then no rescaling can change the order of ε. Unfortunately, even this constraint does not help much. For example, in Section 8.2.1 we find that $0.004 < \varepsilon < 0.008$. What does it mean? Is it small enough and for what? We see that the idealized mathematical condition $\varepsilon \ll 1$ is too abstract and not so easy to interpret in terms of real synaptic connections between neurons.

1.3.2 Size of PSP

There are many ways to estimate the strength of connections between neurons. One of them is based on the analysis of cross correlograms obtained from pairs of neurons. Performing such an analysis, Abeles (1988) concluded that interactions between adjacent neurons in the cortex are weak (his estimate is $\varepsilon = 0.06$), and interactions between distant neurons are even weaker.

Another reasonable way to characterize weakness of synaptic connections is to consider amplitudes of postsynaptic potentials (PSPs) measured in the soma (cell body) of neurons while the neuron membrane potential is far below the threshold value. Why below the threshold value? Because membrane properties of a postsynaptic neuron (the function f_i) may be assumed to be locally linear in this case. Then the size of the PSP reflects the order of weakness of connections. When the membrane potential is close to the threshold, the nonlinear properties of f_i come to play. The PSP can be augmented substantially in this case; it can even grow to a full-scale action potential. We will discuss this issue in Section 1.3.3.

The soma is a preferable site to measure PSPs since it is relatively large (in comparison with dendrites and axons) and hence easier to stick an electrode into. Besides, it is near the axon hillock – the place of initiation of the action potential. Therefore, soma PSP amplitudes best characterize the magnitude of the postsynaptic neuron response under normal conditions. We illustrate in Figure 1.5 typical PSPs measured at the vicinity of the synapse, at the middle of the dendritic tree, and at the soma. There is a great deal of attenuation of membrane potential along dendrites. This diminishes the soma PSP, which is usually less than 1 mV. Apparently, it is small in comparison with the amplitude of an action potential, which is usually almost 100 mV. The primary goal of the discussion below is to give the reader some intuition of how small the PSP really is.

Current information about electrophysiological processes taking place in neural tissues was obtained by studying giant axons of squids, neuromuscular junctions of frogs, synapses of goldfish, etc. It is very difficult and often impossible to study live brain cells of mammals. The major excuse

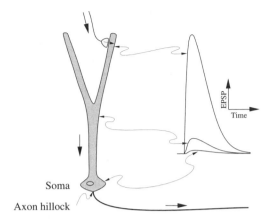

Figure 1.5. Excitatory postsynaptic potentials (EPSP) in three different locations. The EPSP size at the soma is much smaller than the one in the vicinity of the synapse. All EPSPs are simulations of nerve cable equations and presented here for illustrational purposes only.

is that the cell sizes are extremely small, making the experiments expensive. Thus, most of our knowledge of synaptic functions of, say, cortex neurons of humans is an extrapolation of facts obtained by studying more primitive animals. This is one of explanations that comes to mind of why mathematicians working in neurosciences discard the observation that connections between neurons are weak – they may not be weak in squids and frogs.

PSP in Hippocampal Neurons

There is a brain structure that is relatively well studied – the hippocampus. The reason for its popularity is that it plays an important role in the formation of new memories. The mechanism of the learning processes in the hippocampus is not completely understood, but it is believed to be connected to the long-term potentiation (LTP) (Dliss and Lynch 1988). LTP is investigated using rats, mice, and rabbits, which are closer to humans than fish and squids. Their neurons are relatively small and resemble those of humans. Furthermore, the hippocampus is a cortical structure; hence its dynamics could be typical in the sense that some of its features could be observed in other cortical structures, such as neocortex.

Studying the hippocampus, and LTP in particular, yields extensive quantitative data about the synaptic processes in the brain. These data are summarized by Traub and Miles (1991), who are a good source for various other quantitative facts about neuron dynamics. In fact, the hippocampus is one of the best-studied structures of the mammalian brain; That is, there are many quantitative facts about neuron dynamics that are available only for the hippocampus. Yet, we are still far away from understanding it.

We do not intend to discuss the synaptic organization of the hippocampus here, though we mention that its primary neurons are granule and pyramidal cells. One of the most important observations that we use in this book is that the excitatory PSP (EPSP) sizes in the hippocampal granule cells are as small as 0.1 ± 0.03 mV (McNaughton et al. 1981). This is extremely small in comparison with both the amplitude of action potentials and the mean EPSP size necessary to discharge a hippocampus cell, which is 24 ± 9 mV, indicating that the postsynaptic neuron membrane potential is far below the threshold value.

If we assume the linear summation of EPSPs, then at least 240 cells firing simultaneously should be required to make a given cell fire. The actual summation of EPSPs is nonlinear; that is, a pair of EPSPs each of size 0.1 mV produce a single EPSP of size less than 0.2 mV. Making correction for the nonlinear summation, McNaughton et al. (1981) assessed that the number of cells firing simultaneously must be at least 400, which leads to the estimate $\varepsilon = 0.0025$. (In Section 8.2.1 we use mathematical modeling and data above to obtain a slightly larger estimation: $0.004 < \varepsilon < 0.008$.)

Is 400 a large number? Although this number is not big, if we take into account that a typical granule cell of the hippocampus receives signals from approximately $10,000$ other cells, it clearly indicates that a firing of a single neuron produces small changes in the membrane potential of another neuron, provided that the other neurons are silent. It should be noted, though, that membrane changes could be dramatic if the other neurons are not quiet, as we discuss later.

There have been many studies of single PSPs in other hippocampal cells (see Traub and Miles (1991) pp. 49–51 for a summary). For example, firing of a hippocampal CA3 pyramidal cell elicits EPSP in other CA3 pyramidal cells ranging from 0.6 to 1.3 mV (Miles and Wong 1986). Firing of the same CA3 cell can evoke EPSP in CA1 pyramidal cells (via Schaffer collateral) of amplitude 0.13 mV (Sayer et al. 1990). In Figure 1.6 we summarize such an experiment. A short stimulation is applied to a CA3 pyramidal cell to evoke a unitary spike, which is depicted at the top right corner in Figure 1.6. The CA3 cell excites hippocampal CA1 pyramidal cell via Schaffer collateral (the arrow in Figure 1.6). The EPSP sizes vary, since the synaptic transmission is a random process. The distribution of 71 unitary EPSPs is depicted at the right-hand side of Figure 1.6, where we also depict four examples of individual EPSPs of the same CA1 neuron and the average of 1780 such EPSPs. Notice that all EPSPs are so small that they are hardly distinguishable from the background noise. In order to obtain reliable estimates, one needs to implement various statistical signal processing methods to separate the useful signal from the background noise. Inhibitory PSP (IPSP) amplitudes in hippocampus are also small – usually less than 1 mV.

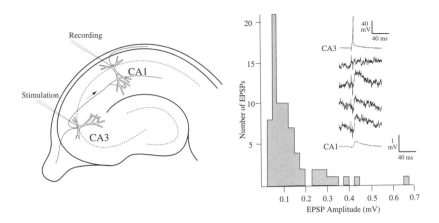

Figure 1.6. Electrode position, EPSP amplitude distribution, and four examples of individual and one example of averaged EPSP of CA1 pyramidal neurons in response to stimulation of a CA3 pyramidal neuron in the hippocampus (Sayer et al. 1990). Data reproduced with the authors' permission.

Mason et al. (1991) studied synaptic transmission and PSP amplitudes between pyramidal neurons of the rat visual cortex. They discovered that the majority of PSPs are less than 0.5 mV in amplitude, with the range 0.05 − 2.08 mV. As the scientists point out in their discussion section, there is an underestimate of the true range because PSPs smaller than 0.03 mV would have gone undetected. This remark may also apply to the hippocampus data reviewed above.

1.3.3   A Common Misinterpretation

There is a common misconception that the weakness of synaptic connections precludes one neuron from evoking the action potential of another one. As we discussed above, a single neuron cannot do this if it fires alone. But this situation is rare in the normal brain. Each firing of a neuron is superimposed upon the firings of other neurons. Suppose a certain fraction of the 10, 000 neurons having synaptic contacts with the granule cell generate action potentials such that the total EPSP is, say, 23.9 mV, which is still below the threshold value of 24 mV. Then a firing of a single neuron or small number of neurons can make the granule cell fire.

The consideration above serves as an illustration to the fact that no matter how small influences from presynaptic neurons may be, the post-synaptic neuron can generate an action potential if it is close enough to the threshold. When a neuron membrane potential is close to threshold value, the neuron activity is close to a bifurcation point. Then the example above illustrates the Fundamental Theorem of Weakly Connected Neural Network Theory (Chapter 5), which says that only those neurons participate nontrivially in the brain processes whose dynamics are at a bifurcation point.

There could be many mechanisms that keep the neuron membrane potential close to the threshold value. The tonic background activity of other neurons is the most feasible mechanism, suggested by Steriade (1978) and reported by Hirsch et al. (1983). They made intracellular recordings of membrane potentials of thalamo-cortical neurons in cats during the sleep-waking cycle. The potential decreased (hyperpolarization) by 4 ± 2.31 mV as the cat shifted from quiet wakefulness to slow sleep. There was a dramatic increase of the membrane potential (10.2 ± 1.3mV) as the cat shifted from the slow to the paradoxical sleep stage. During paradoxical sleep the membrane potential is closer to the threshold value, and the neuron fires more frequently, thereby contributing to the increase of membrane potential of the other neurons.

There could be many other mechanisms for keeping neuron dynamics at the bifurcation point corresponding to the generation of action potentials. The increased background activity is an external mechanism. There could be internal ones, but our knowledge of them is still too vague. In this book we do not study these mechanisms; we just assume that they exist and that neurons are near thresholds.

1.3.4 Spike Delays

We see that the size of a PSP reflects the strength of synaptic connections only when the membrane potential of the postsynaptic neuron is far below the threshold value (e.g., during the slow sleep stage or in some *in vitro* preparations). The delay of a spike could reflect the synaptic strength when the potential is near the threshold value. This issue was brought up in the context of weakly connected neural networks by Kopell (1986, p.174) and in the general context of weakly connected systems by Peponides and Kokotovic (1983).

To illustrate this, suppose $x \in \mathbb{R}$ is the membrane potential of a neuron at equilibrium. When the neuron generates an action potential, a bifurcation of the equilibrium occurs. One of the simplest bifurcations to occur is the saddle-node, which we study later. Such a route to generating action potential corresponds to Class 1 neural excitability (Section 2.8). At the bifurcation, the neuron is described by a local canonical model

$$\dot{x} = a + x^2 \, ,$$

where small $a \in \mathbb{R}$ is a bifurcation parameter measuring distance to the bifurcation point. Its value depends on many factors, including the background activity of other neurons. The value $a = 0$ is the bifurcation value.

When $a < 0$, the equilibrium $x = -\sqrt{a}$ is stable. When influences from other neurons perturb a slightly, the membrane potential can make small excursions (PSPs), and eventually it returns to the rest potential $-\sqrt{a}$. When $|a|$ is small but comparable with the strength of connections, the influences from other neurons can cross the bifurcation value $a = 0$, and x can grow further, leaving a small neighborhood of the origin. Such an increase is the beginning of the action potential. We discuss this issue in Chapter 2.

When $|a|$ is small, we can rescale $a = \varepsilon a_0$ and $x = \sqrt{\varepsilon} y$, where $\varepsilon \ll 1$. Then the equation above can be written in the form

$$\dot{y} = \sqrt{\varepsilon}(a_0 + y^2) .$$

Obviously, any changes in y are slow. In particular, $y(t)$ spends time of order $\mathcal{O}(1/\sqrt{\varepsilon})$ in a neighborhood of the origin. This indicates that there is a delay of initiation of the action potential, as we illustrated in Figure 1.7. The magnitude of the delay is at least $\mathcal{O}(1/\sqrt{\varepsilon})$, which is large. It would

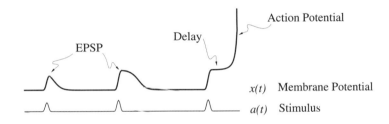

Figure 1.7. Solution of a local canonical model for Class 1 neural excitability, $\dot{x} = a + x^2$, exhibits spike initiation delay when $|a|$ is small ($a = -1/20$ and $t \in [0, 120]$ in the simulation).

be incorrect to ascribe the delay to the synaptic transmission delay, since it results from the *weak* (not slow) synaptic connections.

1.3.5   Changes of Synaptic Efficacy

There are many factors that can alter the efficacy of synaptic connections. Efficacy can increase because

- The synapse is close to the soma of a neuron. In fact, the amplitude of PSP is big in the vicinity of a synapse, and it decreases with the distance, see Figure 1.5. Therefore, the increase of PSP size observed in the soma can be substantial when the synapse is close to it.

- A neuron could make many synapses with another neuron. The summation of PSPs could be almost linear if the synapses are far away from each other, that is, in different branches of the dendritic tree.

- A neuron can have bursting activity; i.e., it can generate multiple spikes. We discuss this issue in Section 2.9.

- The LTP (long-term potentiation) can increase (up to three times) the amplitude of PSP.

- Certain chemicals can augment the nonlinear excitatory properties of neuron membranes.

- The membrane potential can be close to the threshold value. The non-linearity of the membrane (due to the existence of voltage-dependent channels) can substantially increase the size of PSP or even convert it to the full-scale action potential.

The efficacy of synaptic transmission can decrease because

- The synapse is far away from the soma.

- The synapse can be in the dendritic spine, which enhances its isolation from the rest of the neuron.

- There is a depletion of neurotransmitter. This can reduce or completely eliminate the PSP.

- There could be synaptic response failures, which occur when no neurotransmitter vesicles are released to the synaptic cleft. This happens because the neurotransmitter release is a random process.

- There is an effect of aging.

- Certain chemicals and hormones can hinder release of neurotransmitter and/or partially block receptors in presynaptic and postsynaptic parts, respectively.

Finally, the synaptic efficacy can change as a result of learning, as was postulated by Hebb (1949). This can increase or decrease the synaptic strength, but it is not clear whether the order of the connection weakness (the small parameter ε) can be changed substantially.

1.4 Empirical Models

Empirical models of the brain arise when one constructs a simple dynamical system reflecting one or more important neurophysiological observations. The invention of such models is an art rather than a science. The number of empirical models is huge, since there is no commonly accepted opinion of which properties of neurons are important and relevant to a particular phenomenon and which are not. Besides, the same property of a neuron can be expressed mathematically in more than one way. This amplifies the importance of the canonical model approach taken here that provides a rigorous way to derive empirical models and to substantially reduce their number.

In this section we discuss some empirical models. We use them for illustrational purposes only. Our choice of the empirical models is personal and is not based on their applicability or plausibility.

1.4.1 Activities of Neurons

Most of the empirical models of the brain are neural networks of the form

$$\dot{x}_i = f_i(x_1, \ldots, x_n), \quad x_i \in \mathbb{R}, \quad i = 1, \ldots, n, \tag{1.1}$$

where the scalar x_i denotes "activity" of the ith neuron. The precise definition of activity follows from our definition of "observation", which we give later.

The activity of a neuron could be (but is not restricted to) one of the following:

- Membrane potential at the axon hillock or soma.

- Averaged number of action potentials generated per unit time.

- Probability of generation of the action potential.

- Frequency of oscillations.

- Phase of oscillations.

- Amount of chemical neurotransmitter released by synaptic terminals.

- Averaged concentration of neurotransmitter in the synaptic gaps at dendritic endings.

The case of one-dimensional neuron activity is probably the simplest one. It is also feasible to consider multidimensional variables describing activity of a singe neuron. They could be

- Amplitude and phase of oscillations.

- Membrane potential and activity of ion channels at the axon hillock.

- Spatial distribution of membrane potential and ion channel activities.

- The above plus variables describing biochemical reactions at the neuron soma, synapses, etc.

We see that the more neurophysiological data are taken into account by the neuron activity variable, the closer the empirical model is to a comprehensive one.

1.4.2 Redundancy of the Brain

Evidence suggests that the brain is a redundant system (see Freeman 1975 and Wilson and Cowan 1973). It is believed that the functional unit of many brain systems, such as the olfactory bulb, cerebral cortex, or thalamus, is not a single neuron, but a local population of neurons. For example, in the olfactory bulb such local populations correspond to the glomerulus. In the cerebral cortex they are cortical columns. In them neurons are strongly interconnected, have approximately the same pattern of synaptic connections, and respond similarly to very nearly identical stimuli. This redundancy might have various purposes, for example, to increase reliability.

Because of redundancy, we can study networks of local populations of neurons, and in this case the variable x_i in (1.1) represents the averaged activity of the ith local population of neurons. It could denote, for example, the average number of action potentials generated by the neurons from the ith cortical column or glomerulus per unit time, or any other averaged characteristic mentioned above.

Rhythms are ubiquitous in the brain. When the averaged activity of a population of neurons $x_i(t)$ is a continuous function oscillating with a small amplitude, it does not necessarily mean that activity of each neuron from the local population oscillates with a small amplitude. Specifically, when all neurons from the local population fire with a certain probability, then the average activity x_i of the population reflects the probability, provided that the number of neurons in the population is large enough. For example, when neurons fire completely randomly, the averaged activity is almost a constant. When firings of neurons are synchronized, the averaged activity oscillates. Measuring the average activity x_i from the local field potential provides information about probabilities of firings of individual neurons from the local population; see Figure 1.8. When x_i is a constant, neurons

Figure 1.8. Averaged activity of a local population of neurons relates to the probability of firing of a single ("representative") neuron from the population. The continuous curve is the averaged activity, and spikes are firings of the "representative" neuron.

fire randomly; when $x_i(t)$ oscillates, neurons tend to fire when $x_i(t)$ is maximum and tend to be silent when $x_i(t)$ is minimum. Notice that the averaged activity provides only statistical information and does not tell anything about the particular firing of a single neuron.

We see that a one-dimensional variable x_i can be treated as an average activity of the ith local population of neurons or as the probability of firing of a "representative" neuron from the population or as any other characteristic considered above. When we discuss empirical models, we do not worry about precise definitions of neuron activity. This gives us some flexibility in interpreting results. Indeed, any elaborated and explicit definition of neuron activity implies a deliberate restriction of the set of phenomena that could be described by our mathematical modeling. Moreover, in the realm of empirical modeling the notion of neuron activity or of a local population of neurons cannot be strictly defined and is taken as an axiom (like the notion of the *set* in mathematics).

Finally, we abuse notation, as is widely accepted in the neural network literature: We refer to a population of neurons simply as a neuron, and the activity of the population is the activity of the neuron. Thus, a neuron denotes the "averaged" or "representative" neuron of a population unless stated otherwise. This allows us to replace messy pictures of populations of neurons (left-hand side of Figure 1.9) by neat pictures of the "averaged" neurons (right-hand side of Figure 1.9).

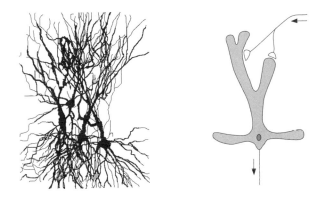

Figure 1.9. Messy picture of a local population of neurons replaced by the picture of an "averaged" neuron.

1.4.3 Additive Neural Network

One of the simplest continuous-time neural network models is an additive network of the form

$$\dot{x}_i = -x_i + S(\rho_i + \sum_{j=1}^{n} c_{ij}x_j), \quad x_i \in \mathbb{R}, \quad i = 1, \ldots, n, \qquad (1.2)$$

where each scalar x_i denotes the ith neuron activity; $\rho_i \in \mathbb{R}$ denotes the overall input from external receptors (sense organs) to the ith neuron; the coefficients c_{ij} , $i \neq j$, describe synaptic connections; c_{ii} describe feedback parameters, and the saturation (gain) function $S : \mathbb{R} \to \mathbb{R}$ has S-shape as depicted in Figure 1.10. A typical example of such a function is

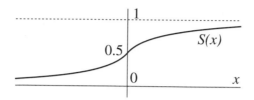

Figure 1.10. Sigma shaped function $S(x)$.

$$S(x) = \frac{1}{1 + e^{-x}}.$$

Function S is continuous, monotone increasing and satisfies $S(-\infty) = 0$ and $S(+\infty) = 1$. Therefore, all neuron activities take values in $(0, 1)$, which is consistent with the hypothesis that x_i could denote the probability that the ith neuron generates an action potential (or fires).

The neural network model (1.2) is often written in the form

$$\dot{y}_i = \rho_i - y_i + \sum_{j=1}^{n} c_{ij} S(y_j) \, ,$$

which can be obtained from (1.2) (Grossberg 1988) by setting

$$y_i = \rho_i + \sum_{j=1}^{n} c_{ij} x_j \, .$$

Notice that if each x_i denotes the average number of spikes per unit time, then y_i could denote the average amount of neurotransmitter in the synaptic clefts at the dendrite endings. This is the same as the total excitation converging from the entire network and external receptors to the ith neuron.

The idea that lies behind the additive model (1.2) is simple: The more excitation converges to a neuron, the more active it is. The empirical model (1.2) is one of many models that can illustrate this idea. For example, the linear model

$$\dot{x}_i = \rho_i + \sum_{j=1}^{n} c_{ij} x_j$$

also illustrates the same idea, though it is linear and hence has trivial dynamics.

1.4.4   Wilson-Cowan Model

Consider an interconnected pair of excitatory and inhibitory neurons as depicted in Figure 1.11. Keep in mind that each neuron in the oscillator

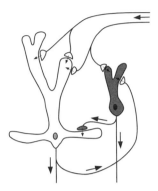

Figure 1.11. Neural Oscillator. Excitatory neuron (white) excites inhibitory neuron (black), which reciprocally inhibits the excitatory one.

can actually be a local population of neurons. They can be motoneurons and Renshaw interneurons in the spinal cord; Pyramidal and basket cells in the hippocampus; Mitral and granule cells in the olfactory bulb; Pyramidal cells and thalamic inter-neurons in cortico-thalamic system, etc. Such pairs of interacting excitatory and inhibitory populations of neurons can also be found in the cerebellum, olfactory cortex, and neocortex (Shepherd 1976). It is believed this is one of the basic mechanisms for the generation of oscillating activity in the brain. We call such a pair a *neural oscillator*.

It should be stressed that there is a difference between an oscillatory neuron and a neural oscillator. The former is a neuron or a population of neurons that exhibits periodic activity. It may consist entirely of either excitatory or inhibitory neurons. The latter consists of two distinguishable populations of neurons (excitatory and inhibitory) and may or may not exhibit a periodic activity. We use the adjective *neural* to stress the physiological distinction between neurons. Usage of the word *oscillator* is somehow misleading, since neural oscillators may have a broad dynamic repertoire ranging from equilibrium state (no oscillations) to chaotic dynamics.

Wilson and Cowan (1973) suggested a model for a network of such neural oscillators. Let x_i and y_i denote activity of the excitatory and inhibitory neurons, respectively. Then the Wilson-Cowan model has the form

$$\begin{cases} \mu_x \dot{x}_i = -x_i + (1 - \tau_x x_i) S(\rho_{xi} + \sum_{j=1}^n a_{ij} x_j - \sum_{j=1}^n b_{ij} y_j) \\ \mu_y \dot{y}_i = -y_i + (1 - \tau_y y_i) S(\rho_{yi} + \sum_{j=1}^n c_{ij} x_j - \sum_{j=1}^n d_{ij} y_j) \end{cases} , \quad (1.3)$$

where $\mu_x, \mu_y > 0$ are the membrane time constants, and τ_x and τ_y are refractory periods of excitatory and inhibitory neurons, respectively. The positive coefficients a_{ij}, b_{ij}, c_{ij}, and d_{ij} for $i \neq j$ are called synaptic. The positive coefficients b_{ii} and c_{ii} are synaptic too, since they denote interactions between excitatory and inhibitory neurons within the ith neural oscillator. In contrast, a_{ii} and d_{ii} are feedback parameters, and they can be positive or negative. Parameters ρ_{xi} and ρ_{yi} denote the external input from sensory organs and other regions of the brain to the ith excitatory and inhibitory neurons, respectively; S has sigma shape as in Figure 1.10.

We consider the Wilson-Cowan model in the special case $\tau_x = \tau_y = 0$ and $\mu_x = \mu_y = 1$, that is

$$\begin{cases} \dot{x}_i &= -x_i + S(\rho_{xi} + \sum_{j=1}^n a_{ij}x_j - \sum_{j=1}^n b_{ij}y_j) \\ \dot{y}_i &= -y_i + S(\rho_{yi} + \sum_{j=1}^n c_{ij}x_j - \sum_{j=1}^n d_{ij}y_j) \end{cases} . \qquad (1.4)$$

Obviously, this is like the additive model discussed in the previous section with the only difference being that there is a distinction between the neurons: They are separated into excitatory and inhibitory ones. Notice also the implementation of Dale's principle in (1.4).

We use the two models presented above in the next chapter, where we illustrate bifurcation phenomena in neural networks.

2
Bifurcations in Neuron Dynamics

Bifurcations play important roles in analysis of dynamical systems; they express qualitative changes in behavior. A typical example of a bifurcation in neuron dynamics is the transition from rest to periodic spiking activity. Generation of a single spike is a bifurcation phenomenon too: It can result when the rest potential is close to a bifurcation point, which is often referred to as its threshold value. We discuss these and similar issues in this chapter.

2.1 Basic Concepts of Dynamical System Theory

A bifurcation is a change of qualitative behavior of a dynamical system. To make this precise we must explain what is *qualitative behavior* of a dynamical system. First, we introduce some notations and terminology; then we illustrate geometrically the most important topics. We cannot provide a comprehensive background on dynamical system theory here. For this the reader should consult other books. For example, Guckenheimer and Holmes's book *Nonlinear Oscillations, Dynamical Systems, and Bifurcations of Vector Fields* (1983) is a good introduction to the qualitative theory of dynamical systems, and bifurcation theory is presented systematically and in a complete form by Kuznetsov (1995) in *Elements of Applied Bifurcation Theory*. The reader who is familiar with the basic concepts of dynamical system theory may skip this section and start from Section 2.2.

2.1.1 Dynamical Systems

Below we consider systems of ordinary differential equations (ODEs) of the form

$$\dot{X} = F(X) , \qquad X = (X_1, \ldots, X_m)^\top \in \mathbb{R}^m , \qquad (2.1)$$

where $\dot{X} = dX/dt$. We refer to (2.1) as a *dynamical system*. (An exact definition of a dynamical system, viz., a homomorphism of an Abelian group, e.g., $\mathbb{Z}$ or $\mathbb{R}$, to a group of all automorphisms of a space, e.g., $\mathbb{R}^m$, is not used in this book). We assume implicitly that the vector-function $F = (F_1, \ldots, F_m)^\top : \mathbb{R}^m \to \mathbb{R}^m$ is sufficiently smooth for our purposes below.

Flows and Orbits

A *flow* of the dynamical system (2.1) is a function $\Phi_t : \mathbb{R}^m \to \mathbb{R}^m$ parametrized by t such that $X(t) = \Phi_t(X_0)$ is a solution of the system starting from X_0; that is,

$$\left. \frac{d\Phi_t(X_0)}{dt} \right|_{t=\tau} = F(\Phi_\tau(X_0)) , \qquad \Phi_0(X_0) = X_0 .$$

The flow might not be defined for all t, since solutions can escape in a finite time: A typical example of such a dynamical system is $\dot{x} = 1 + x^2$, whose solutions are of the form $x(t) = \tan(t + c)$.

Fix $X \in \mathbb{R}^m$, and let $I \subset \mathbb{R}$ be the interval for which $\Phi_t(X)$, $t \in I$, is defined. Then $\Phi(X) : I \to \mathbb{R}^m$ defines a *solution curve*, or *trajectory*, or *orbit*, of X.

Equivalence of Dynamical Systems

Two dynamical systems

$$\dot{X} = F(X) , \ X \in \mathbb{R}^m , \qquad \text{and} \qquad \dot{Y} = G(Y) , \ Y \in \mathbb{R}^m ,$$

are said to be *topologically equivalent* if there is a homeomorphism (continuous mapping with continuous inverse) $h : \mathbb{R}^m \leftrightarrow \mathbb{R}^m$ taking each orbit $\Phi_t(X), X \in \mathbb{R}^m$, of one of the systems to an orbit $\Psi_t(Y), Y \in \mathbb{R}^m$ of the other one. The homeomorphism is not required to preserve parametrization by time; that is, for any X and t there is a t_1, which could differ from t, such that

$$h(\Phi_t(X)) = \Psi_{t_1}(h(X)) .$$

If it preserves time parametrization ($t = t_1$), then the equivalence is called *conjugacy*. When the homeomorphism is defined only locally, the dynamical systems are said to be locally equivalent. Two dynamical systems are said to have *qualitatively similar behavior* if they are topologically equivalent. A dynamical system is *structurally stable* if it is topologically equivalent to any ε-perturbation $\dot{X} = F(X) + \varepsilon P(X)$ where $\varepsilon \ll 1$ and P is smooth enough.

When h is not a homeomorphism, dynamical system $\dot{Y} = G(Y)$ is said to be a *model* of $\dot{X} = F(X)$; see Chapter 4. In this case Y usually belongs to a space smaller than $\mathbb{R}^m$.

Bifurcations

Consider a dynamical system

$$\dot{X} = F(X, \lambda) , \qquad X \in \mathbb{R}^m ,$$

where $\lambda \in \Lambda$ summarizes system parameters and is an element of some Banach space Λ. Often, Λ is simply a Euclidean space $\mathbb{R}^l$, for some $l > 0$. A parameter value λ_0 is said to be *regular*, or *nonbifurcational*, if there is an open neighborhood W of λ_0 such that any system $\dot{X} = F(X, \lambda)$ for $\lambda \in W$ is topologically equivalent to $\dot{X} = F(X, \lambda_0)$. Thus, qualitative behavior of such a dynamical system is similar for all λ near the nonbifurcation value λ_0.

Let $\Lambda_0 \subset \Lambda$ denote the set of all nonbifurcation values of λ. The set $\Lambda_b = \Lambda \setminus \Lambda_0$ is called the *bifurcation set*. Any $\lambda_b \in \Lambda_b$ is said to be a *bifurcation value*. A bifurcation set can have quite complicated geometry; for example, it can be fractal.

Thus, a dynamical system $\dot{X} = F(X, \lambda)$ is at a bifurcation point $\lambda = \lambda_b$ if any neighborhood of λ_b contains some λ_1 such that the qualitative behavior of the system is different for λ_b and λ_1. Since bifurcations reflect changes in a system's dynamical behavior, bifurcation sets are often accessible in laboratory experiments.

A dynamical system at a bifurcation is not structurally stable. The converse (a structurally unstable system is at a bifurcation) is not valid, since the definition of a bifurcation depends on the parametrization of F by λ. When the parametrization is selected unwisely, the dynamical system $\dot{X} = F(X, \lambda)$ can be structurally unstable at nonbifurcational values of λ. For example, the equation

$$\dot{x} = (1 + a^2)x^2 , \qquad x, a \subset \mathbb{R} ,$$

is structurally unstable for any value of $a \in \mathbb{R}$, but no value of a is a bifurcation point.

Equilibria of Dynamical Systems

As a first step toward determining the behavior of the system $\dot{X} = F(X)$, $X \in \mathbb{R}^m$, one would find its equilibria and study their stability. A point $X_0 \in \mathbb{R}^m$ is an *equilibrium* of the dynamical system if

$$F(X_0) = 0 .$$

The orbit $\Phi(X_0)$ of the equilibrium consists of one point, namely X_0. Equilibria of ODEs are sometimes called *equilibrium points*, or *rest points*.

Local behavior at the equilibrium X_0 depends crucially on the eigenvalues of the *Jacobian matrix*

$$L = D_X F = \left(\frac{\partial F_i}{\partial X_j} \right)_{i,j=1,\dots,m}$$

evaluated at X_0.

Hyperbolic Equilibria

When L has no eigenvalues with zero real part, it is said to be *hyperbolic*. When all eigenvalues of L have negative real part, L is said to be a *stable* hyperbolic matrix. In both cases it is nonsingular; that is, $\det L \neq 0$. A dynamical system is said to be *locally hyperbolic (stable)* at an equilibrium if the Jacobian matrix at the equilibrium is hyperbolic (stable). Local hyperbolicity does not imply global hyperbolicity.

Nonhyperbolic Equilibria

A matrix L is said to be *nonhyperbolic* if at least one eigenvalue has zero real part. It still can be nonsingular, which is the case for Andronov-Hopf bifurcations.

Hartman-Grobman Theorem

Local behavior of locally hyperbolic dynamical systems is almost linear. This follows from the theorem below.

Theorem 2.1 (Hartman-Grobman)   *A locally hyperbolic dynamical system*

$$\dot{X} = F(X)$$

is locally topologically conjugate to its linearization

$$\dot{X} = LX .$$

That is, there is a local homeomorphism h defined in a neighborhood of the equilibrium that maps orbits of the nonlinear flow to those of the linear flow (see Figure 2.1).

This theorem plays an essential role in determining local behavior of hyperbolic dynamical systems. Its proof can be found, for example, in Palis and de Melo (1982).

From the theorem it follows that locally hyperbolic dynamical systems are locally structurally stable; that is, any ε-perturbation $\dot{X} = F(X) + \varepsilon G(X)$ is locally topologically equivalent to $\dot{X} = F(X)$. The proof of this fact is well known: Since the Jacobian matrix L at the equilibrium is nonsingular, the implicit function theorem guarantees the existence of an equilibrium for the perturbed system if ε is sufficiently small. The Jacobian

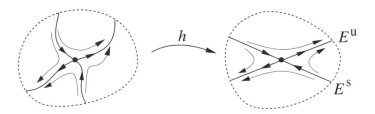

Figure 2.1. Illustration for the Hartman-Grobman theorem.

matrix $L + \mathcal{O}(\varepsilon)$ near the equilibrium is hyperbolic. Moreover, these matrices have ε-close eigenvalues and generate topologically equivalent linear flows. The result follows from the Hartman-Grobman theorem.

Since local dynamics at a hyperbolic equilibrium are structurally stable, the equilibrium cannot bifurcate. Thus, the only candidates for equilibria that can bifurcate are nonhyperbolic ones.

Center Manifold Theory

When an equilibrium is nonhyperbolic, the Jacobian matrix has eigenvalues with zero real parts. Such equilibria often correspond to bifurcations in the system, and local behavior near them can be quite complicated. Nevertheless, it is possible to simplify locally the dynamical system by reducing its dimension. The reduction is essentially a process of eliminating the directions along which the dynamics are trivial, and it is called the center manifold reduction. Our treatment of the reduction in this book is based on that of Iooss and Adelmeyer (1992, Section I.2). In particular, we use some of their notation.

Consider the dynamical system

$$\dot{X} = F(X), \quad X \in \mathbb{R}^m, \tag{2.2}$$

at a nonhyperbolic equilibrium $X = 0$. Let L be the Jacobian matrix at the equilibrium. Suppose that $v_1, \ldots, v_m \subset \mathbb{R}^m$ is the set of generalized eigenvectors of L such that the following hold:

- Eigenvectors $v_1, \ldots, v_{m_1} \subset \mathbb{R}^m$ correspond to the eigenvalues of L having negative real part. Let

$$E^{\mathrm{s}} = \mathrm{span}\,\{v_1, \ldots, v_{m_1}\}$$

be the eigensubspace spanned by these vectors. It is called the *stable* subspace, or the stable linear manifold, for the linear system $\dot{X} = LX$.

- Eigenvectors $v_{m_1+1}, \ldots, v_{m_2} \subset \mathbb{R}^m$ correspond to the eigenvalues of L having positive real part. Let

$$E^{\mathrm{u}} = \mathrm{span}\,\{v_{m_1+1}, \ldots, v_{m_2}\}$$

be the corresponding *unstable* subspace.

- Eigenvectors $v_{m_2+1}, \ldots, v_m \subset \mathbb{R}^m$ correspond to the eigenvalues of L having zero real part. Let

$$E^c = \text{span}\, \{v_{m_2+1}, \ldots, v_m\}$$

be the corresponding *center* subspace.

The phase space $\mathbb{R}^m$ is a direct sum of the subspaces; that is,

$$\mathbb{R}^m = E^s \oplus E^u \oplus E^c \,.$$

Below we set $E^h = E^s \oplus E^u$, where ^{h} stands for "hyperbolic". As we mentioned, the dynamics along E^h are simple. The activity is either exponentially converging to $X = 0$ (along E^s) or exponentially diverging from $X = 0$ (along E^u). The dynamics along E^c can be quite complicated, but it is possible to separate them from the other dynamics. The main tool in such a separation is the following theorem.

Theorem 2.2 (Center Manifold Theorem) *There exists a nonlinear mapping*

$$H : E^c \to E^h, \quad H(0) = 0, \quad DH(0) = 0 \,,$$

and a neighborhood U of $X = 0$ in $\mathbb{R}^m$ such that the manifold

$$M = \{x + H(x) \mid x \in E^c\} \,,$$

which is called the center *manifold, has the following properties:*

- (Invariance) *The center manifold M is locally invariant with respect to (2.2). More precisely, if an initial state $X(0) \in M \cap U$, then $X(t) \in M$ as long as $X(t) \in U$. That is, $X(t)$ can leave M only when it leaves the neighborhood U.*

- (Attractivity) *If the unstable subspace $E^u = \{0\}$, that is, if no eigenvalues of L have positive real part, then the center manifold is locally attractive. More precisely, all solutions staying in U tend exponentially to some solution of (2.2) on M.*

The center manifold M is parametrized by $x \in E^c$, and therefore it has the same dimension as the center subspace E^c. Moreover, it passes through the origin $X = 0$ and is tangent to E^c at the origin (see Figure 2.2). The center manifold is not unique, though any two such manifolds have the same initial terms in their Taylor series.

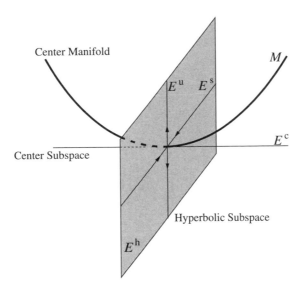

Figure 2.2. Center manifold M tangent to the center subspace E^c.

Center Manifold Reduction

To find the center manifold and the restriction of (2.2) to it, we introduce a few subsidiary objects. Let Π^h and Π^c be projectors from $\mathbb{R}^m$ to the subspaces E^h and E^c, respectively. We require that

$$\ker \Pi^h = E^c \quad \text{and} \quad \ker \Pi^c = E^h ,$$

that is, $\Pi^h E^c = 0$ and $\Pi^c E^h = 0$. The projectors can easily be found if we know the basis dual to $v_1, \ldots, v_m$. Recall that the set of row-vectors $w_1, \ldots, w_m$ is dual to the set of column-vectors $v_1, \ldots, v_m$ if

$$w_i v_j - \sum_{k=1}^{m} w_{ik} v_{jk} = \delta_i^j = \begin{cases} 1 & \text{if } i = j , \\ 0 & \text{otherwise.} \end{cases}$$

It is easy to check that $w_1, \ldots, w_m$ is the set of left eigenvectors of L. The projectors are given by

$$\Pi^h = \sum_{i=1}^{m_2} v_i w_i \quad \text{and} \quad \Pi^c = \sum_{i=m_2}^{m} v_i w_i ,$$

where the product of a column m-vector to a row m-vector is an $m \times m$ matrix. Notice that $\Pi^h + \Pi^c = I$, where I is the identity matrix. The projectors also commute with the Jacobian matrix L; that is,

$$\Pi^h L = L\Pi^h \quad \text{and} \quad \Pi^c L = L\Pi^c .$$

Let $X(t) \in \mathbb{R}^m$ be a solution of (2.2) such that $X(0) \in M \cap U$. Since by invariance the solution stays on M for some time t, we can represent it as

$$X(t) = x(t) + H(x(t)) \,,$$

where $x(t) = \Pi^c X(t)$ is the projection onto the center subspace E^c. Differentiating this with respect to t and using (2.2) gives

$$\dot{x} + DH(x)\dot{x} = F(x + H(x)) \,.$$

Now we can project both sides of the equation above to E^c and to E^h and obtain the two equations

$$\dot{x} = \Pi^c F(x + H(x)) \equiv f(x) \tag{2.3}$$

and

$$DH(x)f(x) = \Pi^h F(x + H(x)), \tag{2.4}$$

where we substitute for $\dot{x}$ in the second equation.

The dynamics of (2.2) on the center manifold M are governed by the ordinary differential equation (2.3), and the quasi-linear partial differential equation (2.4) is used to determine the unknown function H.

2.2   Local Bifurcations

Below we use the additive and the Wilson-Cowan models to illustrate some local bifurcations that are ubiquitous in mathematical neuroscience.

2.2.1   Saddle-Node

Consider the additive model (1.2) for only one neuron:

$$\dot{x} = -x + S(\rho + cx),$$

where $x \in \mathbb{R}$ is the activity of the neuron, $\rho \in \mathbb{R}$ is the external input to the neuron, and the feedback parameter $c \in \mathbb{R}$ characterizes the nonlinearity of the system.

We see that the neuron's dynamics have the following properties: When the input $\rho \to -\infty$ (inhibition), the neuron activity $x \to 0$ (hyperpolarization). When the input is strongly excitatory ($\rho \to +\infty$), the activity is depolarized ($x \to 1$). Thus, the neuron activity is trivial for extreme values of ρ. When ρ assumes intermediate values, the dynamics can exhibit *bistability*: The activity x can have three equilibrium values, two stable and one unstable (see Figure 2.3) provided that c is large enough.

When ρ increases and passes through a value ρ_1, a new pair of equilibria appears: a saddle and a node. Such a bifurcation, called the *saddle-node*

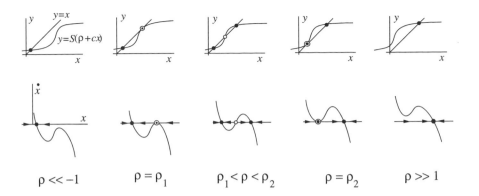

Figure 2.3. Additive neuron dynamics could have one, two, or three equilibria for various values of ρ.

bifurcation, is one of the simplest bifurcations to occur. When ρ increases further and passes a value ρ_2, a saddle and a node coalesce and annihilate each other; see Figures 2.3 and 2.4. Thus, the saddle-node bifurcation refers to the appearance or annihilation of a saddle and a node.

In order to understand the qualitative behavior of the neuron for $\rho_1 < \rho < \rho_2$, it suffices to understand it for ρ close to the bifurcation values ρ_1 or ρ_2. Behavior can be quantitatively different but qualitatively the same – having the same number of equilibria, stability type, etc. This observation serves as a guideline for our study of the brain: We study neural network models near bifurcation points in the hope that this will elucidate their behavior far away from the bifurcations.

General One-Dimensional Systems

In general, a one-dimensional dynamical system

$$\dot{x} = f(x, \lambda), \quad x \in \mathbb{R}, \quad \lambda \in \mathbb{R}^n, n > 0 ,$$

is at a *saddle-node* (sometimes called a *fold*) bifurcation point x_b for some value λ_b if the following conditions are satisfied.

- Point x_b is an equilibrium; that is,

$$f(x_b, \lambda_b) = 0 .$$

- x_b is a nonhyperbolic equilibrium; that is,

$$\frac{\partial f}{\partial x}(x_b, \lambda_b) = 0 .$$

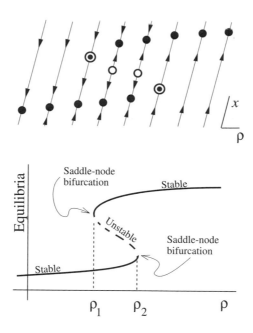

Figure 2.4. Bifurcations in the additive neuron model.

- The function f has a nonzero quadratic term at the bifurcation point; that is,

$$\frac{\partial^2 f}{\partial x^2}(x_{\mathrm{b}}, \lambda_{\mathrm{b}}) \neq 0 \ .$$

- The function f is nondegenerate with respect to the bifurcation parameter $\lambda = (\lambda_1, \ldots, \lambda_n) \in \mathbb{R}^n$; that is, the n-dimensional vector

$$\mathbf{a} = \frac{\partial f}{\partial \lambda}(x_{\mathrm{b}}, \lambda_{\mathrm{b}}) = \left(\frac{\partial f}{\partial \lambda_1}, \ldots, \frac{\partial f}{\partial \lambda_n} \right)^{\top} \neq 0 \in \mathbb{R}^n.$$

This is referred to as the *transversality condition*. When the bifurcation parameter λ is a scalar, this condition is $\partial f / \partial \lambda \neq 0$.

The conditions above can easily be generalized for an arbitrary multidimensional dynamical system; see Section 2.5. They guarantee that the nonhyperbolic equilibrium x_{b} either disappears or bifurcates to a pair of new equilibria: a saddle and a node; see Figure 2.5.

A typical example of a dynamical system at a saddle-node bifurcation is

$$\dot{x} = a + x^2 \tag{2.5}$$

considered at $x = 0$ and $a = 0$. Obviously, it has two equilibria $x = \pm\sqrt{|a|}$ for $a < 0$, one equilibrium $x = 0$ for $a = 0$, and no equilibria for $a > 0$. In

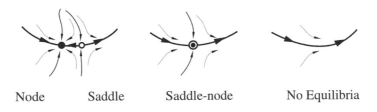

Node   Saddle   Saddle-node   No Equilibria

Figure 2.5. Saddle-node bifurcation in $\mathbb{R}^2$.

multidimensional cases the two equilibria can correspond to a saddle and a node, and the nonhyperbolic equilibrium is called a saddle-node point, which motivates the name for this bifurcation. At the saddle-node point the saddle and the node coalesce and then disappear.

From the four conditions above it follows that any system at a saddle-node bifurcation has a Taylor series of the form

$$\dot{x} = a + bx^2 + \text{higher order terms,}$$

where a is given by the dot product

$$a = \mathbf{a} \cdot \lambda$$

and

$$b = \frac{1}{2} \frac{\partial^2 f}{\partial x^2} \neq 0.$$

Although λ is a multidimensional parameter, only its projection on the vector $\mathbf{a}$ is relevant (to the leading order).

Now consider an ε-neighborhood ($\varepsilon \ll 1$) of the bifurcation value $a = 0$ for (2.5). Let us rescale variables $a = \varepsilon \alpha / b$, $x = \sqrt{\varepsilon} y / b$ and introduce a new slow time variable $\tau = \sqrt{\varepsilon} t$. If f is sufficiently smooth, then the rescaled dynamical system is

$$\frac{dy}{d\tau} = \alpha + y^2 + \mathcal{O}(\sqrt{\varepsilon}), \qquad (2.6)$$

where $\mathcal{O}(\sqrt{\varepsilon})$ is Landau's "Big Oh" function denoting terms bounded above by $R\sqrt{\varepsilon}$ for some constant R. That is, the ratio $\mathcal{O}(\sqrt{\varepsilon})/\sqrt{\varepsilon}$ remains bounded as $\varepsilon \to 0$.

We see that by an appropriate rescaling we can make the higher-order terms as small as we wish. Using bifurcation theory (Golubitsky and Schaeffer 1985) it is possible to prove that the qualitative behavior of (2.6) and of (2.5) is the same. Therefore, to understand the saddle-node bifurcation it suffices to understand the dynamics of (2.5). This observation is powerful and significant in applications. Indeed, it follows that many complicated systems are governed by simple dynamical systems of the form (2.5) when they are near a saddle-node bifurcation. We will deal with plausibility of such bifurcations in brain dynamics later.

Equation (2.5) is a local canonical model for the saddle-node bifurcation in the sense that any dynamical system at the bifurcation can be transformed to the form (2.5) by an appropriate local change of variables. In this book we study local dynamics of a WCNN in the form

$$\dot{x}_i = f_i(x_i, \lambda) + \varepsilon g_i(x_1, \ldots, x_n)$$

in the case when each equation $\dot{x}_i = f_i(x_i, \lambda)$ is at a saddle-node bifurcation. The WCNN is said to be at multiple saddle-node bifurcations in this case. When an adaptation condition (to be stated later) is satisfied, the WCNN has a local canonical model

$$\dot{x}_i = a_i + x_i^2 + \sum_{j=1}^{n} c_{ij} x_j ,$$

where we will show that the synaptic coefficients c_{ij} are essentially the partial derivatives $\partial g_i / \partial x_j$. Studying the canonical model sheds light on neurocomputational properties of all WCNNs at multiple saddle-node bifurcations.

The Additive Model: Bifurcation Sets

Let us return to the additive neuron model

$$\dot{x} = -x + S(\rho + cx)$$

and find those values of x, ρ, and c that correspond to saddle-node bifurcations. In our analysis we assume that $c \neq 0$ and ρ are bifurcation parameters.

First of all, x is an equilibrium if

$$x = S(\rho + cx) . \tag{2.7}$$

Nonhyperbolicity of the equilibrium requires that

$$-1 + S'(\rho + cx)c = 0 . \tag{2.8}$$

The presence of the second-order terms requires that

$$S''(\rho + cx)c^2 \neq 0 ,$$

which is satisfied everywhere but at the inflection point of S, which is at the origin. Therefore,

$$\rho + cx \neq 0 . \tag{2.9}$$

The transversality condition has the form

$$\mathbf{a} = \left(\begin{array}{c} S'(\rho + cx) \\ xS'(\rho + cx) \end{array} \right) \neq 0 ,$$

and it is satisfied for all finite values of x, c, ρ.

Thus, we have two equations, (2.7) and (2.8), and three parameters, x, c, ρ. If we consider $x \in (0, 1)$ as a running parameter, then we can find functions $c(x)$ and $\rho(x)$ satisfying (2.7) and (2.8). This gives us a curve in the $c\rho$-plane, namely the cusp-like curve in Figure 2.6.

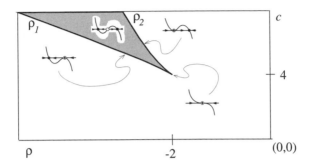

Figure 2.6. Bifurcation diagram for the additive neuron model. Shaded area is the region of bistability. The bold curve is the saddle-node bifurcation set. The point $(\rho, c) = (-2, 4)$ is the cusp bifurcation.

Due to the special nature of the S-shaped function S that we use, we can find the curve analytically without resort to computers. In fact, if $y = \rho + cx$, then

$$S(y) = \frac{1}{1 + e^{-y}} = \frac{1 + e^{-y} - e^{-y}}{1 + e^{-y}} = 1 - e^{-y}S(y)$$

and

$$S'(y) = \frac{e^{-y}}{(1 + e^{-y})^2} = e^{-y}S^2(y) = (1 - S(y))S(y) .$$

The inverse of $z = S(y)$ is given by the formula

$$y = S^{-1}(z) - \ln\left(\frac{z}{1 - z}\right) .$$

From (2.8) it follows that $1 - (1 - S(y))S(y)c$. Since $S(y) = x$ (equation (2.7)), we obtain $1 = (1 - x)xc$, and solving for c gives

$$c = \frac{1}{x(1 - x)} .$$

Then (2.7) can be rewritten as

$$\rho = S^{-1}(x) - cx = \ln\left(\frac{x}{1 - x}\right) - cx .$$

The curve $\rho = \rho(x)$, $c = c(x)$ given by the last two equations is depicted in Figure 2.6. It consists of two branches, ρ_1 and ρ_2 (we use the same notation as in Figure 2.4). Let us fix c and increase ρ. When we cross the bifurcation value ρ_1, the saddle-node bifurcation corresponds to the birth of a pair of equilibria. When we cross ρ_2, the saddle and the node equilibria coalesce and annihilate each other.

The two curves have a common point, $\rho = -2$, $c = 4$. It is easy to check that $x = 1/2$ for these values. At this point condition (2.9) is violated, which means that the quadratic term vanishes. Therefore, this point does not correspond to the saddle-node bifurcation, but to another bifurcation, called the cusp bifurcation, which we study next.

2.2.2   The Cusp

Let us consider a one-dimensional dynamical system of the form

$$\dot{x} = f(x, \lambda), \qquad x \in \mathbb{R}, \quad \lambda \in \mathbb{R}^n, \ n > 1 ,$$

at an equilibrium point x_{b} for some value λ_{b}; so $f(x_{\text{b}}, \lambda_{\text{b}}) = 0$. The equilibrium point corresponds to a *cusp* bifurcation (sometimes called cusp singularity or cusp catastrophe) if the following conditions are satisfied:

- The point x_{b} is a nonhyperbolic equilibrium.

- The function f does not have a quadratic term but has a cubic term. That is,

$$\frac{\partial^2 f}{\partial x^2} f(x_{\text{b}}, \lambda_{\text{b}}) = 0 \quad \text{but} \quad \sigma = \frac{1}{6} \frac{\partial^3 f}{\partial x^3}(x_{\text{b}}, \lambda_{\text{b}}) \neq 0 .$$

- The two n-dimensional vectors

$$\mathbf{a} = \frac{\partial f}{\partial \lambda}(x_{\text{b}}, \lambda_{\text{b}}) \ \text{ and } \ \mathbf{b} = \frac{\partial^2 f}{\partial \lambda \partial x}(x_{\text{b}}, \lambda_{\text{b}})$$

are transversal; i.e., they are linearly independent. This implies, in particular, that neither of them is the zero vector. This is the transversality condition.

A typical example of such a system is

$$\dot{x} = a + bx + \sigma x^3, \qquad \sigma \neq 0 ,$$

at the origin $a = b = x = 0$. This system is a local canonical model for the cusp bifurcation.

Notice that the original system has a multidimensional bifurcation parameter λ. The canonical form has only two: $a = \mathbf{a} \cdot \lambda$ and $b = \mathbf{b} \cdot \lambda$, which are projections of λ onto the vectors $\mathbf{a}$ and $\mathbf{b}$, respectively. The transversality condition ensures that these are independent bifurcation parameters.

The Additive Model

Consider again the additive neuron model

$$\dot{x} = f(x, \rho, c) = -x + S(\rho + cx)$$

in a neighborhood of the nonhyperbolic point $x = 1/2$ for $\rho = -2$ and $c = 4$. It is easy to check that $f = f'_x = f''_{xx} = 0$ and $f'''_{xxx} = c^3 S'''(0) = -8$ at this point. The parameter $\lambda = (\rho, c) = (-2, 4) \in \mathbb{R}^2$ in this case. The vectors **a** and **b** are given by

$$\mathbf{a} = \left(\begin{array}{c} \partial f/\partial r \\ \partial f/\partial c \end{array} \right) = \left(\begin{array}{c} S'(0) \\ xS'(0) \end{array} \right) = \left(\begin{array}{c} 1/4 \\ 1/8 \end{array} \right)$$

and

$$\mathbf{b} = \left(\begin{array}{c} \partial^2 f/\partial r \partial x \\ \partial^2 f/\partial c \partial x \end{array} \right) = \left(\begin{array}{c} cS''(0) \\ S'(0) + cxS''(0) \end{array} \right) = \left(\begin{array}{c} 0 \\ 1/4 \end{array} \right),$$

which are linearly independent. Therefore, the point $(x, \rho, c) = (1/2, -2, 4)$ is a cusp bifurcation point.

Let us study a small neighborhood of the cusp point. We can introduce local coordinates $x = 1/2 + \tilde{x}$, $\rho = -2 + \tilde{\rho}$, $c = 4 + \tilde{c}$ so that the origin $(\tilde{x}, \tilde{\rho}, \tilde{c}) = (0, 0, 0)$ coincides with the cusp point. In the new coordinates $\lambda = (\tilde{\rho}, \tilde{c})$, and the additive model has the form (for convenience of notation we remove $\tilde{}$)

$$\dot{x} = 0 + 0x + \mathbf{a} \cdot \lambda + 0x^2 + \mathbf{b} \cdot \lambda x - \frac{4}{3}x^3 + \text{h.o.t.},$$

where "h.o.t." stands for high-order terms in x and in the components of λ. If we write

$$a = \mathbf{a} \cdot \lambda = r/4 + c/8 ,$$
$$b = \mathbf{b} \cdot \lambda = c/4,$$

then the model can be concisely written as

$$\dot{x} = a + bx - \frac{4}{3}x^3 + \text{h.o.t.}$$

By an appropriate rescaling, the high-order terms can be made as small as desired, so that the model has the form

$$\dot{x} = a + bx - \frac{4}{3}x^3.$$

Any dynamical system at the cusp bifurcation can be reduced to this form (with possibly a different coefficient of the cubic term) by an appropriate change of variables. Therefore, it is a canonical model for the cusp bifurcation.

Let us check that there are saddle-node bifurcations in any neighborhood of the cusp point. The bifurcation sets of the canonical model can easily be found. Differentiating $a + bx - \frac{4}{3}x^3$ with respect to x gives $b - 4x^2$. Equating both of these expressions to zero and eliminating x gives the saddle-node bifurcation set

$$\{(a, b) \mid a = \pm \frac{1}{3} b\sqrt{b}, \ b \neq 0\},$$

which is depicted at the bottom of Figure 2.7. The set

$$\left\{(a, b, x) \mid a + bx - \frac{4}{3}x^3 = 0\right\}$$

is depicted in the same picture.

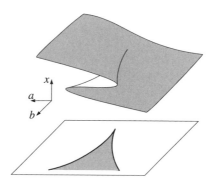

Figure 2.7. Cusp surface.

The bifurcation sets in Figures 2.7 and 2.6 are similar. This is not surprising: The bifurcation set for the canonical model is a rescaled version of a small neighborhood of the cusp bifurcation point of the additive model. We can use either of the bifurcation sets to obtain qualitative information about the additive model. The only difference is that the former set was obtained by a local analysis at the cusp point, while the latter requires global information about the neuron's behavior. If the function S had been more complicated, we would not have been able to find global bifurcation sets analytically.

It is always possible to observe a saddle-node bifurcation in any neighborhood of a cusp bifurcation point. For example, we can fix any $b > 0$ and vary a as a bifurcation parameter (in the original variables this corresponds to fixed c and variable ρ). A saddle-node bifurcation occurs every time a crosses the values $\pm b\sqrt{b}/3$. The bifurcation diagrams in this case are depicted at the top of Figure 2.8. Many other bifurcation diagrams can be obtained by assuming that the bifurcation parameter $\lambda = \lambda(s)$ is a segment or a curve in Λ. This projects to a segment or a curve in the ab-plane parametrized by s (that is, $a = a(s)$ and $b = b(s)$). A summary of some special cases is depicted in Figure 2.8 showing that there can be many interesting dynamical regimes in the vicinity of a cusp bifurcation point.

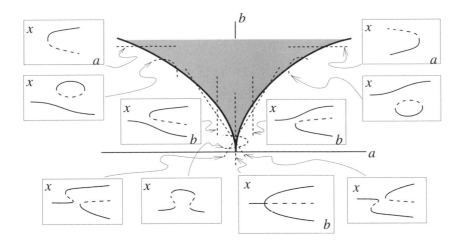

Figure 2.8. Summary of special cases for the cusp bifurcation. Dotted segments and curves crossing the bifurcation sets represent one-dimensional bifurcation parameters. Bifurcation diagrams are depicted in boxes: Continuous curves represent stable solutions, dashed curves represent unstable solutions.

In this book we study the local dynamics of the WCNN

$$\dot{x}_i = f_i(x_i, \lambda) + \varepsilon g_i(x_1, \ldots, x_n)$$

at a multiple cusp bifurcation, that is, when each equation $\dot{x}_i = f_i(x_i, \lambda)$ is at a cusp bifurcation. Such a WCNN has a local canonical model

$$\dot{x}_i = a_i + b_i x_i + \sigma_i x_i^3 + \sum_{j=1}^{n} c_{ij} x_j \,,$$

provided that some additional conditions are satisfied. We study neurocomputational properties of this canonical model in Chapter 11.

2.2.3 The Pitchfork

Consider the canonical model for a cusp bifurcation in the special case $a = 0$, that is,

$$\dot{x} = bx - \frac{4}{3}x^3,$$

where b is now the bifurcation parameter. In the canonical model variables this corresponds to moving along the line $2r + c - 0$, which passes through the cusp point.

For $b < 0$ there is only one equilibrium, $x = 0$, which is stable. At $b = 0$ the equilibrium bifurcates, giving birth to a new pair of equilibria, $\pm\sqrt{3b}/2$. When $b > 0$, the origin is unstable and the new equilibria are stable. The bifurcation diagram depicted at the bottom of Figure 2.8 looks like a pitchfork.

In general, a one-dimensional system

$$\dot{x} = f(x, \lambda), \quad x \in \mathbb{R}, \quad \lambda \in \mathbb{R}^n,$$

having an odd function f $(f(-x, \lambda) = -f(x, \lambda))$ is at a *pitchfork* bifurcation point $x_b = 0$ for some value $\lambda_b \in \mathbb{R}^n$ if the following conditions are satisfied:

- The point $x_b = 0$ is a nonhyperbolic equilibrium.

- The function f has a nonzero cubic term at the bifurcation point.

- The n-dimensional vector

$$\frac{\partial^2 f}{\partial \lambda \partial x}(x_b, \lambda_b) \in \mathbb{R}^n,$$

  is not the zero vector. This is referred to as the transversality condition.

The condition $f(-x, \lambda) = -f(x, \lambda)$ implies that $x = 0$ is an equilibrium for any λ. Moreover, it prevents the Taylor expansion of f from having terms of any even power. Five typical perturbations of a pitchfork bifurcation are depicted in the lower half of Figure 2.8.

2.2.4 Andronov-Hopf

Andronov-Hopf bifurcations are among the most important bifurcations observed in neuron dynamics, since they describe the onset (or disappearance) of periodic activity, which is ubiquitous in the brain.

As before, we first study the bifurcation for a simple (two-dimensional) case and discuss the general case later. Consider a dynamical system of the form

$$\begin{cases} \dot{x} &= f(x, y, \lambda) \\ \dot{y} &= g(x, y, \lambda) \end{cases}, \quad x, y \in \mathbb{R}, \quad \lambda \in \mathbb{R}^n, n > 0$$

near an equilibrium point $(x_b, y_b) \in \mathbb{R}^2$ for λ near some bifurcation value λ_b. The dynamical system is said to be at an *Andronov-Hopf bifurcation* if the following conditions are satisfied:

- The Jacobian matrix

$$L = \frac{D(f,g)}{D(x,y)} = \begin{pmatrix} \frac{\partial f}{\partial x} & \frac{\partial f}{\partial y} \\ \frac{\partial g}{\partial x} & \frac{\partial g}{\partial y} \end{pmatrix}$$

evaluated at the bifurcation point has a pair of purely imaginary eigenvalues. This is equivalent to the conditions

$$\operatorname{tr} L = \frac{\partial f}{\partial x} + \frac{\partial g}{\partial y} = 0 \quad \text{and} \quad \det L = \frac{\partial f}{\partial x}\frac{\partial g}{\partial y} - \frac{\partial f}{\partial y}\frac{\partial g}{\partial x} > 0 .$$

- Since $\det L > 0$, the Jacobian matrix is not singular. The implicit function theorem ensures that there is a unique family of equilibria $(x(\lambda), y(\lambda)) \in \mathbb{R}^2$ for λ close to λ_b. If $\alpha(\lambda) \pm i\omega(\lambda)$ denotes the eigenvalues of the Jacobian matrix $L(\lambda)$ evaluated at $(x(\lambda), y(\lambda)) \in \mathbb{R}^2$, then the n-dimensional vector

$$\frac{d\alpha}{d\lambda}(\lambda) = \left(\frac{\partial \alpha}{\partial \lambda_1}, \dots, \frac{\partial \alpha}{\partial \lambda_n} \right)$$

is nonzero, which is referred to as the transversality condition.

A typical example of a dynamical system at the Andronov-Hopf bifurcation is

$$\begin{cases} \dot{x} &= \alpha x - \omega y + (\sigma x - \gamma y)(x^2 + y^2) \\ \dot{y} &= \omega x + \alpha y + (\gamma x + \sigma y)(x^2 + y^2) \end{cases}$$

at $(x,y) = (0,0)$ and $\alpha = 0$. Here $\omega, \sigma \neq 0$ and γ are scalar (nonbifurcational) parameters. The system is the normal form for the bifurcation; see the definitions in Section 4.2. Often it is written concisely in complex coordinates $z = x + iy$ as

$$\dot{z} = (\alpha + i\omega)z + (\sigma + i\gamma)z|z|^2.$$

Rewriting this system using polar coordinates r and φ with $z = re^{i\varphi}$ (that is, $x = r\cos\varphi$ and $y = r\sin\varphi$) results in

$$\begin{cases} \dot{r} &= \alpha r + \sigma r^3 \\ \dot{\varphi} &= \omega + \gamma r^2 \end{cases}.$$

The frequency of oscillations is determined by ω and γr^2. The parameter γ determines how the frequency depends on the amplitude r of oscillations. The value of γ is often irrelevant and assumed to be zero in applications. (Dynamics of the system above for $\gamma = 0$ and $\gamma \neq 0$ are topologically equivalent.) When parameter $\sigma > 0$, the bifurcation is subcritical, and supercritical otherwise. Supercritical and subcritical bifurcations are discussed in detail in Section 2.6. Below we consider the special case $\sigma < 0$.

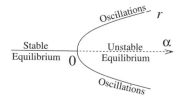

Figure 2.9. r-dynamics at Andronov-Hopf bifurcation undergoes a pitchfork bifurcation.

Notice the similarity of the first equation of the system above with the canonical form for the pitchfork bifurcation. In fact, the r-dynamics, which are independent from the φ-dynamics, are at a pitchfork bifurcation when $\alpha = 0$; see Figure 2.9. For $\alpha < 0$ the equilibrium $r = 0$ is asymptotically stable, and so is the equilibrium $(x, y) = (0, 0)$ of the original system. The rate of convergence to the origin is exponential. When $\alpha = 0$ there is a pitchfork bifurcation in the amplitude dynamics. Although the origin is still stable, the rate of approach is not exponential. When $\alpha > 0$, there are three equilibria, $r = 0$ and $r = \pm\sqrt{\alpha/|\sigma|}$. The equilibrium $r = 0$ corresponds to the origin $(x, y) = (0, 0)$, which is now unstable. The pair of stable equilibria $r = \pm\sqrt{\alpha/|\sigma|}$ corresponds to a limit cycle of radius $\sqrt{\alpha/|\sigma|}$ in the xy-plane, which is orbitally asymptotically stable with the exponential rate of approach. We depict typical phase portraits of a supercritical Andronov-Hopf bifurcation in Figure 2.10.

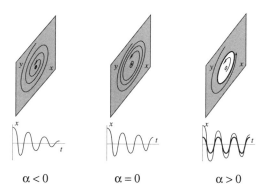

Figure 2.10. Typical phase portraits at supercritical Andronov-Hopf bifurcation for $\alpha = 0$.

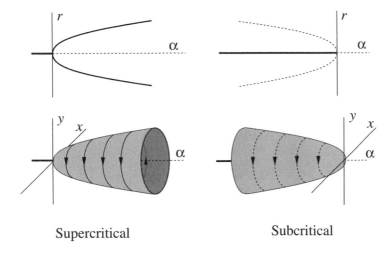

Supercritical   Subcritical

Figure 2.11. Andronov-Hopf bifurcation at $\alpha = 0$.

Bifurcation Sets for the Wilson-Cowan Model

Consider the Wilson-Cowan model of a neural oscillator

$$\begin{cases} \dot{x} & = & -x + S(\rho_x + ax - by) \\ \dot{y} & = & -y + S(\rho_y + cx - dy) \end{cases} , \quad x, y \in \mathbb{R}, \qquad (2.10)$$

for fixed parameters a, b, c, and d and bifurcation parameters ρ_x and ρ_y. With our choice of the function S (see Section 2.2.1) we can find values of ρ_x and ρ_y corresponding to the Andronov-Hopf bifurcation.

The point (x, y) is an equilibrium of (2.10) if

$$\begin{cases} x = S(\rho_x + ax - by) \\ y = S(\rho_y + cx - dy) \end{cases} , \qquad (2.11)$$

or equivalently,

$$\begin{cases} \rho_x = S^{-1}(x) - ax + by \\ \rho_y = S^{-1}(y) - cx + dy \end{cases} . \qquad (2.12)$$

The Jacobian matrix of the model (2.10) is

$$L = \begin{pmatrix} -1 + aS'(\rho_x + ax - by) & -bS'(\rho_x + ax - by) \\ cS'(\rho_y + cx - dy) & -1 - dS'(\rho_y + cx - dy) \end{pmatrix}.$$

Using the fact that $S' = S(1 - S)$ and (2.11), we can represent L in a more convenient form:

$$L = \begin{pmatrix} -1 + ax(1 - x) & -bx(1 - x) \\ cy(1 - y) & -1 - dy(1 - y) \end{pmatrix}.$$

Then the conditions for the Andronov-Hopf bifurcation are

$$\operatorname{tr} L = -2 + ax(1-x) - dy(1-y) = 0 \qquad \text{and} \qquad \det L > 0 \,.$$

Treating x as a curve parameter, we can eliminate y from the equation above and plot the curve given by (2.12) on the $\rho_x \rho_y$-plane as shown in Figure 2.12 (thin curve); The saddle-node bifurcation set (bold curve) is given by $\det L = 0$, and the saddle separatrix loop curve is shown as a dashed curve; see Figure 2.13 for an explanation. The first two curves represent local bifurcations and can be determined by local analysis at equilibria. The last curve represents global bifurcation and usually cannot be found by local considerations.

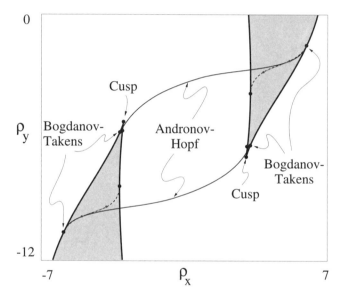

Figure 2.12. Bifurcation sets of Wilson-Cowan neuron oscillator model for $a = b = c = 10$ and $d = -2$. Bold curve is saddle-node bifurcation set; Thin curve is Andronov-Hopf bifurcation set; Dashed curve is saddle separatrix loop bifurcation set. Light segment is double limit cycle bifurcation set.

Figure 2.13 shows phase portraits of the Wilson-Cowan model for various values of ρ_x and ρ_y. Due to the obvious symmetry of the bifurcation sets we reproduce only half of Figure 2.12 in Figure 2.13. A detailed bifurcation analysis of the Wilson-Cowan model in its original form (1.3) was performed by Borisyuk and Kirillov (1992). The bifurcation sets they obtained are essentially the same as those depicted here. In both cases there is a region of parameter values in which the stable limit cycle and stable equilibrium coexist. Such coexistence of rest and periodic activity is observable experimentally (see the bibliography in Marder et al. (1996)). Borisyuk (1991),

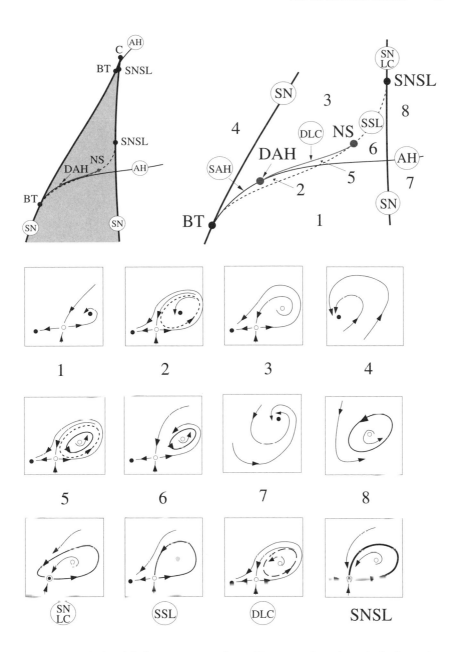

Figure 2.13. Refined bifurcation sets from Figure 2.12 and typical phase diagrams. (Saddle separatrix loop and double limit cycle curves are distorted for the sake of clarity). *Nomenclature*: AH – supercritical Andronov-Hopf, BT – Bogdanov-Takens, C – cusp, DAH – degenerate Andronov-Hopf, DLC – double limit cycle, NS – neutral saddle, SAH – subcritical Andronov-Hopf, SN – saddle-node, SNLC – saddle-node on limit cycle, SNSL saddle-node separatrix-loop, SSL – saddle separatrix loop.

Abbott and Marder (1995), and Marder et al. (1996) hypothesize that such bistable regimes might have relevance to many interesting phenomena such as selective attention and short-term memory.

Finally, the Wilson-Cowan model of a neural oscillator exhibits periodic activity when the parameters ρ_x and ρ_y are inside the area containing "Andronov-Hopf" in Figure 2.12 (see Figure 2.13 for details). For other values of the parameters the activity converges to an equilibrium. Therefore, being a neural oscillator does not necessarily imply being an oscillator. This is in accordance with the well-known principle (M. Hirsch 1976 *Differential Topology*, p. 22) that in mathematics a *red herring* does not have to be either *red* or a *herring*.

In this book we study the local dynamics of the WCNN

$$\begin{cases} \dot{x}_i = f_i(x_i, y_i, \lambda) + \varepsilon p_i(x_1, \ldots, x_n, y_1, \ldots, y_n) \\ \dot{y}_i = g_i(x_i, y_i, \lambda) + \varepsilon q_i(x_1, \ldots, x_n, y_1, \ldots, y_n) \end{cases}$$

when each neural oscillator is at an Andronov-Hopf bifurcation. Then the WCNN has a local canonical model,

$$\dot{z}_i = (\alpha_i + i\omega_i)z_i + (\sigma_i + i\gamma_i)z_i|z_i|^2 + \sum_{j=1}^{n} c_{ij}z_j ,$$

where the c_{ij} are complex-valued coefficients that depend on the matrices

$$S_{ij} = \frac{\partial(p_i, q_i)}{\partial(x_j, y_j)} = \begin{pmatrix} \frac{\partial p_i}{\partial x_j} & \frac{\partial p_i}{\partial y_j} \\ \frac{\partial q_i}{\partial x_j} & \frac{\partial q_i}{\partial y_j} \end{pmatrix}.$$

This dependence reveals an interesting relationship between synaptic organizations and dynamical properties of networks of neural oscillators, which we study in Chapter 13.

2.2.5 Bogdanov-Takens

The Andronov-Hopf and saddle-node bifurcation sets depicted in Figure 2.13 intersect each other. Since the Wilson-Cowan model can have more than one equilibrium, the intersections could correspond to simultaneous bifurcations of different equilibria.

When both bifurcations happen to be of the same equilibrium, then the Jacobian matrix at the equilibrium satisfies $\det L = 0$ and $\operatorname{tr} L = 0$. Both eigenvalues of L are then zero. The Jordan form of matrix L in generic case is

$$\begin{pmatrix} 0 & 1 \\ 0 & 0 \end{pmatrix}$$

and the bifurcation is called a *Bogdanov-Takens* bifurcation. Its detailed analysis can be found, for example, in Guckenheimer and Holmes (1983) or in Kuznetsov (1995).

There is a change of variables that transforms any system

$$\left(\begin{array}{c} \dot{x} \\ \dot{y} \end{array} \right) = \left(\begin{array}{cc} 0 & 1 \\ 0 & 0 \end{array} \right) \left(\begin{array}{c} x \\ y \end{array} \right) + \left(\begin{array}{c} f(x, y, \lambda) \\ g(x, y, \lambda) \end{array} \right)$$

at the bifurcation to the form

$$\begin{cases} \dot{u} = v + \mathcal{O}(|u, v|^3) \\ \dot{v} = a + bu + u^2 + \sigma uv + \mathcal{O}(|u, v|^3), \end{cases} \tag{2.13}$$

where a and b are bifurcation parameters and $\sigma = \pm 1$. One can treat (2.13) as a canonical model for the Bogdanov-Takens bifurcation. Bifurcation diagram and phase portraits for various a, b and σ are depicted in Figure 2.14. It is easy to see that the case $\sigma > 0$ can be reduced to $\sigma < 0$ by the

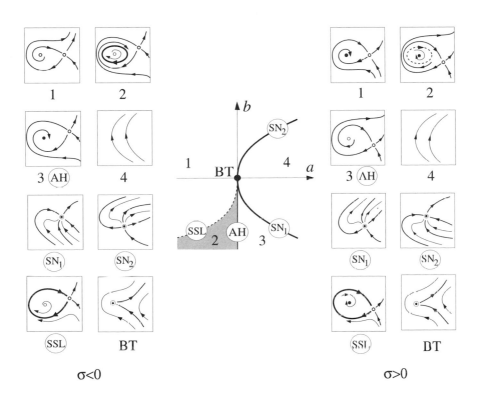

Figure 2.14. Bogdanov-Takens bifurcation diagram.

substitution $t \to -t$ and $v \to -v$.

We study weakly connected networks of neurons near multiple Bogdanov-Takens bifurcations in Section 5.5, though our analysis is rather sketchy. A detailed analysis of such problems goes beyond the scope of this book.

2.2.6 Quasi-Static Bifurcations

In our analysis of the Wilson-Cowan model (1.4) of neural oscillators we assumed that the membrane time constants μ_x and μ_y for the excitatory and inhibitory neurons were identical. Sometimes it is desirable to consider the case when the constants are not equal, in particular, when $\mu_x \ll \mu_y$. This means that the excitatory neuron activities x_i are much faster than those of the inhibitory neurons y_i. Any such system can be written in the form

$$\begin{cases} \mu x' &= f(x, y, \mu) \\ y' &= g(x, y, \mu) \end{cases} , \tag{2.14}$$

where $' = d/d\tau$, τ is a slow time, and $\mu = \mu_x/\mu_y \ll 1$ is a small parameter. Since the small parameter multiplies the derivative x', the system is *singularly perturbed*.

Singularly perturbed systems of the form (2.14) often describe dynamics of relaxation oscillators. In the context of neural networks we call such an oscillator a *relaxation neuron*. For example, if x denotes the activity of fast ion channels (e.g., Na^+) and y denotes the activity of slow ion channels (e.g., K^+), then (2.14) could describe a mechanism of generation of action potentials by a neuron, which is a relaxation process. The typical examples of such dynamical systems are the Hodgkin-Huxley equations (Hodgkin and Huxley 1954) and the Fitzhugh-Nagumo equations (Fitzhugh 1969).

Much useful information about singular perturbed dynamical systems (2.14) can be obtained by analyzing the "unperturbed" ($\mu = 0$) algebraic-differential system

$$\begin{cases} 0 &= f(x, y, 0) \\ y' &= g(x, y, 0) \end{cases} \tag{2.15}$$

and the fast system

$$\begin{cases} \dot{x} &= f(x, y, \mu) \\ \dot{y} &= \mu g(x, y, \mu) \end{cases} , \tag{2.16}$$

where the dot denotes d/dt, and $t = \tau/\mu$ is a fast time. We see that $\dot{y} \approx 0$ when $\mu \to 0$, which motivates the adjective "slow" for y and (relatively) "fast" for x.

Suppose for the sake of illustration that x and y are scalars. To analyze the behavior of the singularly perturbed dynamical system, let us consider the fast system (2.16) first. Since $\dot{y} = \mathcal{O}(\mu)$, we may assume y to be a quasi-static variable. Then the first equation describes fast dynamics for fixed y. The variable converges to an attractor, which depends on y as a parameter. The attractor is often (but not always, see Section 2.9) an equilibrium. It can be determined from the algebraic equation $f(x, y, 0) = 0$. Let $x = h(y)$ be a local solution of $f(h(y), y, 0) = 0$. Then the algebraic-differential system (2.15) can be reduced to the equation

$$y' = g(h(y), y, 0) ,$$

where $x = h(y)$ is a local solution to the algebraic equation $f(x, y, 0) = 0$. When $f_x = 0$, the solution $x = h(y)$ can disappear, and x can jump (possibly) to another local solution of $f(x, y, 0) = 0$. The point x at which the condition $f_x = 0$ is satisfied looks geometrically like a minimum or a maximum of the set $f(x, y, 0) = 0$ in the xy-plane; see Figure 2.15. To

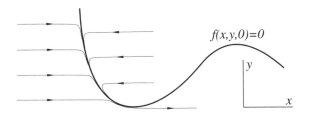

Figure 2.15. The dynamics of singularly perturbed systems consist of two parts: approaching the set $f(x, y, 0) = 0$ and sliding along the set until a leave point (a minimum or a maximum of the set $f(x, y, 0) = 0$) is reached.

study dynamics near such points one must consider the full system (2.14) or (2.16). Such points interest us since they can indicate occurrence of a quasi-static bifurcation, which we discuss below.

Suppose $(x, y) = (x_b, y_b)$ is an equilibrium point for (2.14) when $\mu = 0$; that is,

$$\begin{cases} f(x_b, y_b, 0) = 0 \\ g(x_b, y_b, 0) = 0 \end{cases}.$$

Consider the reduced ($\mu = 0$) fast system

$$\dot{x} = f(x, y, 0), \tag{2.17}$$

where $y \approx y_b$ is treated as parameter. We say that the singularly perturbed dynamical system (2.14) (or fast system (2.16)) is at a *quasi-static bifurcation point* (x_b, y_b) for $\mu \ll 1$ if the corresponding reduced fast system (2.17) is at a bifurcation point $x = x_b$ when $y = y_b$ is treated as a quasi-static bifurcation parameter.

In general, a bifurcation of a dynamical system is quasi-static when the bifurcation parameter is not a constant but a slow variable at an equilibrium. In this book we study quasi-static multiple saddle-node, cusp, and Andronov-Hopf bifurcations in singularly perturbed (relaxation) WCNNs.

Let us plot the sets

$$\begin{aligned} N_x &= \left\{ (x, y) \in \mathbb{R}^2 \mid f(x, y, 0) = 0 \right\}, \\ N_y &= \left\{ (x, y) \in \mathbb{R}^2 \mid g(x, y, 0) = 0 \right\}, \end{aligned}$$

which are called *nullclines* for the system. Intersections of the nullclines are equilibria of (2.14). An example of the nullclines of a typical relaxation

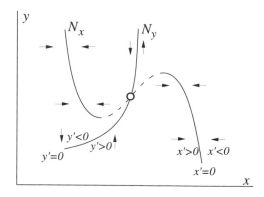

Figure 2.16. Nullclines of a typical relaxation oscillator.

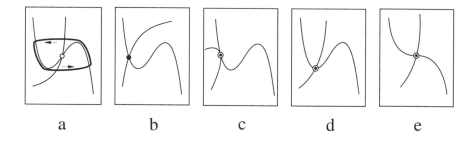

Figure 2.17. Nullclines determine the dynamics of relaxation systems. a. Periodic orbit corresponding to relaxation oscillations. b. Transversal intersection corresponding to hyperbolic stable equilibrium. c. Nontransversal intersection corresponding to a bifurcation that is not quasi-static. d. Quasi-static saddle-node bifurcation. e. Quasi-static cusp bifurcation.

oscillator is depicted in Figure 2.16, when there is only one intersection corresponding to the unique equilibrium, which is unstable. Such a relaxation system has a limit cycle attractor, depicted in Figure 2.17a.

A singularly perturbed system is at a quasi-static bifurcation point when $f_x = 0$. This means geometrically that the tangent to the nullcline N_x at the intersection is horizontal (see Figure 2.17d and e). One of the most ubiquitous bifurcations in applications is the quasi-static saddle-node bifurcation, which we discuss next.

Quasi-Static Saddle-Node Bifurcation

Quasi-static saddle-node bifurcations of (2.16) when x and y are scalars have the same linear part as the Bogdanov-Takens bifurcations, though, local dynamics at the equilibrium are essentially different. For example, there is always a unique equilibrium near a quasi-static saddle-node bifurcation, while there could be none or more than one equilibrium near a Bogdanov-Takens bifurcation point.

There is always an Andronov-Hopf bifurcation near a quasi-static saddle-node bifurcation. To see this, consider the dynamical system

$$\begin{cases} \dot{x} &= f(x,y,\lambda) \\ \dot{y} &= \mu g(x,y,\lambda) \end{cases},$$ (2.18)

which depends on two parameters, μ and λ. Suppose the nullclines of (2.18)

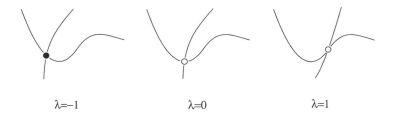

$$\lambda = -1 \qquad\qquad \lambda = 0 \qquad\qquad \lambda = 1$$

Figure 2.18. Intersections of nullclines of (2.18) for various λ.

intersect as in Figure 2.18 for three different values of λ, say $\lambda = -1, 0$, and 1, so that the equilibrium is stable for $\lambda = -1$, there is quasi-static saddle-node bifurcation for $\lambda = 0$, and the equilibrium is unstable for $\lambda = 1$. Now suppose the parameter μ is a positive constant. When μ is small but fixed and λ is moved from -1 to $+1$, the equilibrium loses stability at some λ_{b} between -1 and 1. Generically, a loss of stability of an equilibrium can be via either a saddle-node or an Andronov-Hopf bifurcation, since these are the only codimension 1 bifurcations of equilibria. If it were a saddle-node bifurcation, then there would be a change in the number of equilibria. According to our geometrical construction, there is no such a change, and therefore the equilibrium loses stability via an Andronov-Hopf bifurcation. Since we do not make any restrictions on the value of μ, the Andronov-Hopf bifurcation can be observed in an arbitrarily small neighborhood of the quasi-static saddle-node bifurcation.

The key element in our considerations is that μ is kept small but *fixed*, and λ is varied. Now let us see what happens when λ is kept fixed, say $\lambda = 0$, and μ is allowed to be arbitrarily small. The Jacobian matrix

$$L = \begin{pmatrix} 0 & f_y \\ \mu g_x & \mu g_y \end{pmatrix}$$

at the equilibrium satisfies $\det L = -\mu g_x f_y \to 0$ as $\mu \to 0$. This violates one of the fundamental assumptions of the Andronov-Hopf bifurcation, namely that $\det L \neq 0$, which guarantees that the complex-conjugate eigenvalues

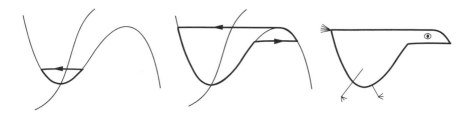

Figure 2.19. French duck (canard) in quasi-static saddle-node (singular Hopf) bifurcation.

of L have nonzero imaginary parts when they cross the imaginary axis. In order to stress this fact, some suggest naming such a bifurcation a *singular* Hopf bifurcation to contrast it with the regular Andronov-Hopf bifurcation having $\det L \neq 0$. In this book we use the name quasi-static saddle-node bifurcation instead of singular Hopf to distinguish it from the quasi-static cusp and quasi-static Andronov-Hopf bifurcations. The former is illustrated in Figure 2.17e, and the latter can be observed only when the fast variable x is a vector.

As one expects, the behavior of dynamical systems at regular Andronov-Hopf bifurcations and quasi-static saddle-node bifurcations is different. For example, the limit cycle in the latter has a triangular shape, as in Figure 2.19, and grows in amplitude much faster than $\sqrt{\lambda}$ (Baer and Erneux 1986, 1992). The quasi-static saddle-node bifurcation has many other interesting features not present in the Andronov-Hopf bifurcation. One of the most prominent peculiarities is the existence of *French ducks* (canards), see Eckhaus (1983) and Figure 2.19. One can observe both subcritical and supercritical Andronov-Hopf bifurcations and a double limit cycle bifurcation in a neighborhood of the quasi-static saddle-node bifurcation. Another prominent feature is that local analysis at the bifurcation provides some global information about the behavior of singularly perturbed systems. This might be useful in studies of Class 2 neural excitability and elliptic bursting, which we discuss later in this chapter.

2.3   Bifurcations of Mappings

Many dynamical systems can be described by or reduced to mappings of the form
$$X \mapsto P(X) \,,$$
which can also be written as
$$X^{k+1} = P(X^k) \,,$$

where X^k denotes the kth iterate of the variable X. Such systems usually arise as a Poincaré mapping or time-T mapping, which we discuss below.

2.3.1 Poincaré Mappings

Consider a dynamical system of the form

$$\dot{X} = F(X), \qquad X \in \mathbb{R}^m \tag{2.19}$$

having a periodic solution $X(t) \in \mathbb{R}^m$ with period $T > 0$. That is, $X(t) = X(t + T)$ for all $t \geq 0$. Such a periodic solution is called a *periodic orbit* or a *limit cycle*. We denote it by $\gamma \subset \mathbb{R}^m$. Let $\Sigma \subset \mathbb{R}^m$ be an $(m - 1)$-dimensional local cross section, namely a hypersurface or a manifold intersecting γ transversally as well as all orbits close to γ; see Figure 2.20. Without loss of generality we may assume that γ intersects Σ in a unique point p. Such a set Σ is called the *Poincaré section* for the dynamical system (2.19).

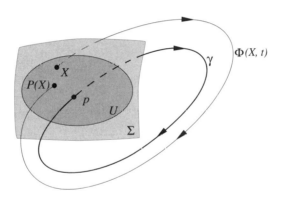

Figure 2.20. The Poincaré mapping $P : U \to \Sigma$.

Let $U \subset \Sigma$ be a neighborhood of p, so small that any orbit $\Phi_t(X)$ starting from a point $\Phi_0(X) = X \in U$ returns to Σ in a finite time $T(X) \approx T$. Since $p \in \gamma$ is a periodic point with period $T > 0$, such a neighborhood U exists, though it may be small. Then the *Poincaré*, or *first return*, mapping $P : U \to \Sigma$ is defined for all points X from U as

$$P(X) = \Phi_{T(X)}(X).$$

Poincaré mappings are defined locally: They usually cannot be defined globally when the dynamical system (2.19) has nontrivial dynamics.

Let W be a subset of U defined by

$$W = \{X \in U \mid P^k(X) \in U \text{ for all } k \geq 0\}.$$

It is not empty since $p \in W$. Moreover, it contains the local stable manifold of p. Then the Poincaré mapping P induces the iteration

$$X \mapsto P(X), \qquad X \in W, \tag{2.20}$$

which can be written as $X^{k+1} = P(X^k)$, $X^0 \in W$.

A point $X_0 \in \Sigma$ is a *fixed point* of the mapping (2.20) if

$$X = P(X).$$

Obviously, $p \in \gamma$ is a fixed point of P. Moreover, any fixed or periodic point Y (i.e., $P^k(Y) = Y$ for some $k \in \mathbb{Z}_+$) of P lies on some periodic solution of the original system (2.19). Thus, there is a correspondence between periodic solutions of (2.19) that are close to γ and fixed points of the Poincaré mapping P.

2.3.2   Time-T (Stroboscopic) Mappings

Consider a periodically forced dynamical system of the form

$$\dot{X} = F(X, Q(t)), \qquad X \in \mathbb{R}^m, \quad Q(t) \in \mathcal{Q}, \tag{2.21}$$

where $\mathcal{Q}$ is some space and the function $Q(t)$ is periodic with a period $T > 0$. Let $\Phi_t(X)$ be the flow defined by the system above. Let us define a function $P : \mathbb{R}^m \to \mathbb{R}^m$ by

$$P(X) = \Phi_T(X).$$

See Figure 2.21 for an illustration. This function is called *time-T* or *stroboscopic* mapping.

There is another, equivalent, definition of the time-T mapping. Consider system (2.21) in the autonomous form

$$\begin{cases} \dot{X} = F(X, Q(\theta)) \\ \dot{\theta} = \frac{T}{2\pi} \end{cases} \qquad X \in \mathbb{R}^m, \quad \theta \in \mathbb{S}^1,$$

where $\mathbb{S}^1$ is the unit circle. Then the time-T function is the Poincaré map of the m-dimensional cross section $\theta = \text{const}$.

Both Poincaré and time-T mappings provide a convenient way for studying flows: Instead of considering a local neighborhood of a limit cycle in $\mathbb{R}^m$, one needs to study a local neighborhood of a fixed point of the mapping P.

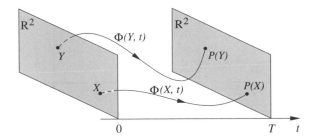

Figure 2.21. Time-T mapping $P : \mathbb{R}^2 \to \mathbb{R}^2$.

2.3.3 Periodically Forced Brain Systems

Since oscillations are ubiquitous in the brain, it is not surprising that there are many brain structures that are periodically forced. We discuss some of them here. It should be noted, though, that the forcing in the brain is not strictly periodic. It is, rather, stochastic with a distinguishable power spectrum. We study one such case in Section 5.4.5.

Olfactory Bulb → Piriform Cortex

The olfactory bulb receives its input from the olfactory receptors, processes the information using a network of neural oscillators (see Figure 3.2 from Section 3.1.1) and sends its oscillatory output to the piriform cortex. The output of the bulb consists of the outputs from all neural oscillators, whose phases are synchronized but whose amplitudes differ. In this case the space Q above is multidimensional.

It should be noted that the olfactory bulb itself receives an oscillatory input from olfactory receptors. There the oscillations are the sniffing cycles of an animal.

Thalamus → Cortex

It is believed that the thalamo-cortical interactions are responsible for such vital processes as sleep, arousal, or attention (Kilmer 1996, Steriade et al. 1990, 1993; and Steriade and Llinas 1988). Even though such interactions are not completely understood, one fact is indisputable: they have an oscillatory nature.

The thalamus together with the reticular thalamic nucleus consists of neural oscillators capable of sustaining periodic activity. The oscillators project their output directly to the cortical pyramidal neurons. Various oscillatory patterns of thalamus activity induce various behaviors in the cortex. For example, the thalamic activity that produces spindle waves (waxing-and-waning field potentials of 7 to 14 Hz) synchronizes cortical activity and puts an animal to sleep. Blockade of the high-amplitude spindles disrupts the EEG-synchronized sleep and either puts an animal to rapid eye movement sleep (REM sleep, during which dreams occur) or awakes the animal (Steriade 1988). We do not study these phenomena in this book. In Section 5.4.5 we discuss how a periodic or chaotic input from the thalamus can organize the cortical columns into dynamically linked ensembles.

Septum → Hippocampus

One of the most distinguishable features of the hippocampus is the oscillatory nature of its dynamics (for review see Traub and Miles 1991). EEG signals recorded from the hippocampus have a sinusoidal-like shape with the frequency in the range $4 - 12$ Hz. Such periodic activity is known as the *theta rhythm*. Even though the hippocampal neurons are capable of producing periodic activity by themselves (the hippocampus consists of neural oscillators), it is believed that the theta rhythm is induced by inputs from the medial septum and entorhinal cortex, which we are brain structures acting as pacemakers and projecting their output directly to the hippocampus. Under normal conditions the theta rhythms in the hippocampus and septum are synchronized and phase locked. There are numerous experiments involving lesions of connections between the hippocampus and septum (for references see Traub and Miles 1991). Being disconnected, the septum activity continues to be oscillatory with theta frequencies, while the hippocampus loses the theta rhythms. This observation suggests that hippocampal oscillations are driven by septum oscillations.

The interactions between septum, entorhinal cortex, and hippocampus are rather intricate and not completely understood. For example, it is not clear what kind of input there is from the septum to the hippocampus. Besides, there are reciprocal connections from hippocampus to the septum and entorhinal cortex. Moreover, unlike the septum, the entorhinal cortex is a spatially distributed cortical structure receiving and processing high-level information from brain systems other than the hippocampus. We do not study interactions between these brain structures in this book.

2.3.4 Periodically Forced Additive Neuron

Let us consider for the sake of illustration a periodically forced additive neuron model of the form

$$\dot{x} = -x + S(\rho + cx + \sin \frac{2\pi t}{T}) \,, \tag{2.22}$$

where $T > 0$ is the period of forcing. Using computer simulations it is easy to construct the time-T mapping P for various values of the parameters. In Figure 2.22 we fix $\rho = -3.15$ and $c = 6$ and use the period of forcing T as a bifurcation parameter to illustrate the striking dependence of behavior of the system on the frequency of the forcing. We do not want to

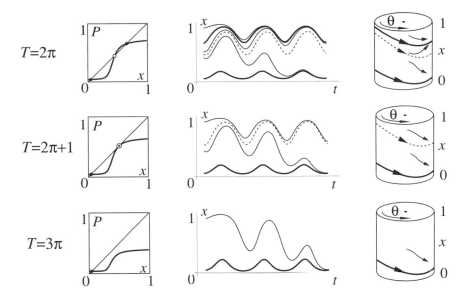

Figure 2.22. Periodically forced additive neuron model (2.22) for $\rho = -3.15$, $c = 6$ and three different values of period of the forcing T. From left to right: Time-T mapping P; Activity $x(t)$ for $t \in [0, 3T]$; Phase portrait in cylindrical coordinates $(x, \theta) \in [0, 1] \times \mathbb{S}^1$. Notice the saddle-node bifurcation for $T \approx 2\pi + 1$.

draw any conclusions about dependence of the hippocampus or thalamo-cortical system on their forcings here; we simply illustrate how dramatic the dynamical changes can be due to a simple modification of the forcing frequency.

To construct the function P, we solve equation (2.22) numerically for various initial conditions $x_0 \in [0, 1]$. Let $x(t)$ be such a solution. Then we take $P(x_0)$ to be $x(T)$, where T is the period of forcing. Graphs of P for various values of T are shown at the left-hand side of Figure 2.22. Every intersection of P and the graph $y = x$ corresponds to a fixed point of the mapping $x \mapsto P(x)$, and hence to a periodic or equilibrium solution of (2.22).

For $T = 2\pi$ there are three intersections, corresponding to two stable and one unstable limit cycles. Increasing T results in lowering the graph of P. For $T \approx 2\pi + 1$ the graph touches $y = x$. $P' = 1$ at this point, which means that the fixed point is nonhyperbolic: It corresponds to a saddle-node bifurcation for the mapping $x \mapsto P(x)$. Since the mapping has a nonhyperbolic fixed point, the dynamical system (2.22) has a nonhyperbolic limit cycle. The limit cycle is a result of coalescence of the stable and unstable limit cycles existing for $T < 2\pi + 1$. When $T > 2\pi + 1$, the nonhyperbolic limit cycle disappears and (2.22) has only one attractor.

In general, a mapping

$$X \mapsto P(X)$$

has a fixed point $X^\star$ if $X^\star = P(X^\star)$. The point is hyperbolic when no eigenvalues of the Jacobian matrix $L = DP$ evaluated at $X^\star$ lie on the unit circle (that is, they have modulus 1). In one-dimensional cases this means that $P' \neq \pm 1$.

2.3.5 Saddle-Node Bifurcation

Consider a one-dimensional mapping

$$x \mapsto f(x, \lambda)$$

near a point $x = x_b \in \mathbb{R}$ for some value $\lambda = \lambda_b \in \mathbb{R}^n$. The point corresponds to a *saddle-node* bifurcation if the following conditions are satisfied:

- x_b is a fixed point. That is,

$$x_b = f(x_b, \lambda_b) \ .$$

- x_b is a nonhyperbolic fixed point satisfying

$$\frac{\partial f}{\partial x}(x_b, \lambda_b) = 1 \ .$$

- The function f has a nonzero quadratic term at the bifurcation point. That is,

$$\frac{\partial^2 f}{\partial x^2}(x_b, \lambda_b) \neq 0 \ .$$

- The function f is nondegenerate with respect to the bifurcation parameter λ. That is, the transversality condition

$$\mathbf{a} = \frac{\partial f}{\partial \lambda} \neq 0$$

is satisfied at the point (x_b, λ_b).

A typical example of such a mapping is

$$x \mapsto x + a + x^2$$

at $(x, a) = (0, 0)$. In fact, any mapping at a saddle-node bifurcation can be transformed to this form by an appropriate change of variables.

As one expects, the behavior of the mapping above depends on the sign of a. When $a < 0$, there are two fixed points $\pm\sqrt{|a|}$. One of them is unstable and the other is stable (see Figure 2.23). In a multidimensional case the fixed points are a saddle and a node. When a increases and crosses 0, the

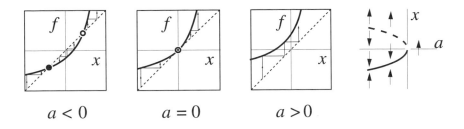

Figure 2.23. Mapping $f : x \mapsto x+a+x^2$ for various values of bifurcation parameter a. When $a = 0$ the saddle-node bifurcation occurs.

saddle and the node coalesce and disappear for $a > 0$. The bifurcation diagram coincides with the one for the saddle-node bifurcation of flows.

Recall that iterations arise as Poincaré or time-T mappings of the flows. When a mapping has a fixed point, the flow has a limit cycle solution. When the mapping undergoes the saddle-node bifurcation, there is a bifurcation in the flow: A stable and an unstable limit cycle coalesce and disappear. For an illustration see the right-hand side of Figure 2.22.

2.3.6 Flip Bifurcation

A one-dimensional mapping $x \mapsto f(x, \lambda)$ is at a *flip* bifurcation point $x = x_b \in \mathbb{R}$ for some value $\lambda = \lambda_b \in \mathbb{R}^n$ if

- x_b is a fixed point; that is,

$$x_b = f(x_b, \lambda_b) .$$

- x_b is a nonhyperbolic fixed point satisfying

$$\frac{\partial f}{\partial x}(x_b, \lambda_b) = -1 .$$

- The composition function $f \circ f$ has a nonzero cubic term at the bifurcation point. This is equivalent to the requirement that

$$\sigma = \frac{1}{6} \frac{\partial^3(f \circ f)}{\partial x^3} \neq 0 .$$

Careful calculation shows that

$$\sigma = -\frac{1}{3} \frac{\partial^3 f}{\partial x^3} - \frac{1}{2} \left(\frac{\partial^2 f}{\partial x^2} \right)^2 .$$

- The composition function $f \circ f$ is nondegenerate with respect to the bifurcation parameter λ. The transversality condition in this case is

$$\mathbf{a} = \frac{\partial^2 (f \circ f)}{\partial \lambda \partial x} \neq 0$$

at the point (x_b, λ_b). Again, it is easy to check that

$$\mathbf{a} = -\frac{\partial^2 f}{\partial x^2} \frac{\partial f}{\partial \lambda} - 2 \frac{\partial^2 f}{\partial \lambda \partial x} .$$

The simplest example of a mapping at a flip bifurcation is

$$x \mapsto -x + a + x^2$$

at $x = 0$ and $a = 0$, though it is not a canonical model. In fact, an effective way to study a mapping $x \mapsto f(x, \lambda)$ at a flip bifurcation is to consider the iterated mapping $x \mapsto f(f(x, \lambda), \lambda)$. Let us write the Taylor expansion of f at the bifurcation point as

$$f(x, \lambda) = -x + \alpha + \beta x + px^2 + qx^3 \; + \; \text{high-order terms},$$

where

$$\alpha = \frac{\partial f}{\partial \lambda} \cdot \lambda , \quad \beta = \frac{\partial^2 f}{\partial \lambda \partial x} \cdot \lambda , \quad p = \frac{1}{2} \frac{\partial^2 f}{\partial x^2} , \quad q = \frac{1}{6} \frac{\partial^3 f}{\partial x^3} .$$

It is easy to check that the Taylor expansion of $f \circ f = f(f(x, \lambda), \lambda)$ is

$$(f \circ f)(x, \lambda) = x - (2\alpha p + 2\beta)x - (2q + 2p^2)x^3 \; + \; \text{high-order terms}.$$

It is also easy to check that $-(2\alpha p + 2\beta) = \mathbf{a} \cdot \lambda$, which we denote below by a, and $\sigma = -(2q + 2p^2)$, where the vector $\mathbf{a}$ and nonzero constant σ are defined above. Thus, studying a mapping at a flip bifurcation reduces to studying its truncated composition

$$x \mapsto x + ax + \sigma x^3 ,$$

which is the canonical model for the flip bifurcation at $x = 0$ when $a = 0$. This mapping itself is not at a flip bifurcation. It is rather at a *pitchfork bifurcation*; meaning that there is a correspondence between these two types of bifurcations: The pitchfork is a composition of flips.

It is easy to see that $x = 0$ is always a fixed point of this mapping. It is stable when $a < 0$ and unstable for $a > 0$. The value $\sigma \neq 0$ defines an attribute of the bifurcation: It is *subcritical* for $\sigma > 0$ and *supercritical* otherwise. When the bifurcation is subcritical, there are two unstable fixed points $(\pm\sqrt{|a/\sigma|})$ for negative a that correspond to an unstable cycle of period 2 for the original mapping $x \mapsto f(x, \lambda)$. The points coalesce when

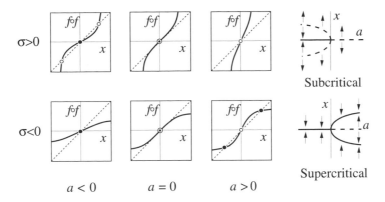

Figure 2.24. Flip bifurcation described by the mapping $x \mapsto x + ax + \sigma x^3$.

$a = 0$ and disappear when $a > 0$. When the bifurcation is supercritical, there is only one fixed point $x = 0$ for $a < 0$ and an additional pair of stable fixed points for $a > 0$. The pair corresponds to a stable cycle of period 2 in the original mapping. All these regimes are illustrated in Figure 2.24.

Recall that a fixed point of a mapping represents a periodic solution of the corresponding flow. When, say, the supercritical flip bifurcation occurs, a stable limit cycle loses stability and gives a birth to another stable limit cycle having approximately doubled period. As a result, the bifurcation is often referred to as being a *period-doubling* bifurcation. This bifurcation does not have an analogue in two-dimensional space. It can be observed in three- (or higher-) dimensional autonomous systems or in two- (or higher-) dimensional periodically forced systems. In Figure 2.25 we illustrate such bifurcations.

2.3.7   Secondary Andronov-Hopf Bifurcation

A fixed point of a one-dimensional mapping $x \mapsto f(x, \lambda)$ can become non-hyperbolic only when $f'_x = \pm 1$, since the unit circle in $\mathbb{R}^1$ is the pair of points ± 1. To study other cases we must go to higher dimensions.

Let us consider a two-dimensional mapping

$$X \mapsto F(X, \lambda), \quad X \in \mathbb{R}^2, \quad \lambda \in \mathbb{R}^n$$

at a fixed point $X = X_b$ for some $\lambda = \lambda_b$. The (2×2)-dimensional Jacobian matrix DF has two eigenvalues. The cases when DF has zero eigenvalues goes beyond the scope of this book. Below we consider the case when the eigenvalues are complex numbers (since DF is a real matrix, they are conjugate). The bifurcation that we discuss has many names. Some call it the *Neimark-Sacker* bifurcation, while others call it the *secondary Andronov-Hopf* bifurcation due to its similarity with that for flows discussed in Sec-

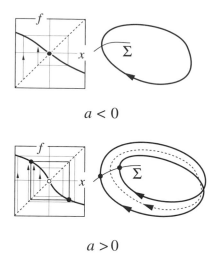

$a < 0$

$a > 0$

Figure 2.25. Poincaré mapping f of Σ for a flow at the flip bifurcation. A stable limit cycle (solid loop, $a < 0$) becomes unstable (dotted loop, $a > 0$), giving birth to a new stable limit cycle of approximately doubled period.

tion 2.2.4. It has the adjective "secondary" since it arises frequently when one studies a Poincaré return map of a periodic orbit, which itself can appear as a result of the Andronov-Hopf bifurcation of an equilibrium.

We say that the mapping above is at a *secondary Andronov-Hopf bifurcation* when the following conditions are satisfied:

- The Jacobian matrix DF at the bifurcation fixed point $(X, \lambda) = (X_b, \lambda_b)$ has a pair of complex-conjugate eigenvalues μ and $\bar{\mu}$ on the unit circle; that is, $|\mu| = |\bar{\mu}| = 1$.

- The eigenvalues depend continuously on the parameter λ, so we denote them by $\mu(\lambda)$ and $\bar{\mu}(\lambda)$. The transversality condition is that they cross the unit circle with a nonzero speed when λ changes. This is guaranteed by the requirement that the n-dimensional vector $d|\mu|/d\lambda$ be nonzero at $\lambda = \lambda_b$.

A typical example of a mapping at the secondary Andronov-Hopf bifurcation is

$$\begin{cases} x \mapsto \alpha x - \omega y + (\sigma x - \gamma y)(x^2 + y^2) \\ y \mapsto \omega x + \alpha y + (\gamma x + \sigma y)(x^2 + y^2) \end{cases}, \quad \omega, \sigma \neq 0, \qquad (2.23)$$

at $(x, y) = (0, 0)$, where α and ω satisfy $\alpha^2 + \omega^2 = 1$. It is easy to check that the Jacobian matrix of this system at the origin has two eigenvalues, $\alpha \pm i\omega \in \mathbb{C}$.

The mapping (2.23) is often written in the complex form

$$z \mapsto \mu z + dz|z|^2 , \tag{2.24}$$

where $z = x + iy$, $\mu = \alpha + i\omega$, and $d = \sigma + i\gamma$. This is a truncated normal form for the bifurcation when there are no *strong resonances*: $\mu^3 = 1$ or $\mu^4 = 1$. If $\mu^3 = 1$, then the normal form is

$$z \mapsto \mu z + w\bar{z}^2 + dz|z|^2 + \mathcal{O}(z^4, \lambda)$$

for some $w, d \in \mathbb{C}$. When $\mu^4 = 1$, the normal form is

$$z \mapsto \mu z + w\bar{z}^3 + dz|z|^2 + \mathcal{O}(z^4, \lambda) .$$

If $\mu^k = 1$ for $k = 5, 6, \ldots$, then the resonant monomial $\bar{z}^{k-1}$ can be hidden in $\mathcal{O}(z^5, z^{k+1}, \lambda)$ and the truncated normal form can be written as (2.24). Detailed analysis of secondary Andronov-Hopf bifurcations can be found in Kuznetsov (1995).

2.4 Codimension of Bifurcations

The notion of codimension of bifurcations is related to the topological notion of transversality, which we do not consider in this book. Instead, we provide an algorithm how to determine codimensions of bifurcations that we discuss.

Suppose a dynamical system

$$\dot{x} - f(x, \lambda)$$

has an equilibrium x for a certain value of λ; that is,

$$f(x, \lambda) = 0 \qquad \text{(equilibrium point)}. \tag{2.25}$$

If the equilibrium is a bifurcation point, then $f(x, \lambda)$ satisfies a number of additional conditions besides (2.25); for example,

$$\frac{\partial f}{\partial x} = 0 \qquad \text{(nonhyperbolicity)}$$

or

$$\frac{\partial f}{\partial \lambda} \neq 0 \qquad \text{(transversality)}$$

The number of strict conditions given by equations (condition (2.25) is not counted) is called the *codimension* of the bifurcation. It coincides with the number of transversality conditions, which in turn coincides with the number of bifurcation parameters in the canonical models we have discussed.

Let us count the codimensions of the bifurcations we considered earlier. The saddle-node bifurcation has codimension 1 ($f'_x = 0$ for flows or $f'_x = 1$ for mappings). So does the Andronov-Hopf bifurcation ($\mathrm{tr} D_x f = 0$) and the flip bifurcation ($f'_x = -1$). The secondary Andronov-Hopf bifurcation has codimension 1 unless the eigenvalues are roots of unity, in which case the codimension is 2. The cusp bifurcation has codimension 2, since there are two conditions to be satisfied: $f'_x = 0$ and $f''_{xx} = 0$. So does the Bogdanov-Takens bifurcation ($\mathrm{tr} D_x f = 0$ and $\det D_x f = 0$).

There is a temptation to assign codimension 1 to the pitchfork bifurcation because of the condition $f'_x = 0$. This would be true if we considered dynamical systems with only the odd right-hand side $f(-x, \lambda) = -f(x, \lambda)$ (such systems are called $\mathbb{Z}_2$-equivariant). This is equivalent to an infinite chain of conditions

$$f(0) = 0, \quad \frac{\partial^2 f}{\partial x^2} = 0, \quad \ldots, \quad \frac{\partial^{2n} f}{\partial x^{2n}} = 0, \ldots$$

if f is smooth enough. Therefore, the pitchfork has infinite codimension.

The notion of codimension plays an important role in classifying bifurcations of dynamical systems. The higher the codimension, the more difficult the bifurcation is to analyze. In addition, bifurcations of high codimension are less likely to be encountered in the real world, since many parameters are needed to control the system to remain near the bifurcation point. Nevertheless, there are exceptions. For example, a dynamical system may describe a phenomenon having a symmetry, say in hydrodynamics, or Hamiltonian systems.

In this book we study weakly connected networks of "approximately" similar neurons. If a neuron's dynamics are near a bifurcation point, which corresponds to the neuron membrane potential being near threshold value, it may or may not follow that another neuron is near the same (or different) bifurcation. Nonetheless, if there is an electrophysiological mechanism or process that keeps a neuron's dynamics near a bifurcation point, then it is reasonable to assume that the same mechanism might be present in many other neurons, keeping them near the same bifurcation. In the weakly connected case this implies that it is feasible to consider the entire network to be at a multiple bifurcation of codimension kn, where k is the codimension of bifurcation of each neuron and n is the number of participating neurons. Since n can be a large number (up to 10^{11}), the codimension of the multiple bifurcation would be enormous even for $k = 1$. From a mathematical point of view such bifurcations may be very unlikely. Such bifurcations are not so exceptional from the neurophysiological point of view since all neurons are "similar". Besides, the neuron similarity may be considered as evidence of some sort of a symmetry in the network. Then, the multiple bifurcations in WCNNs need not be unlikely, even from a mathematical point of view.

2.5   Multidimensional Systems

We have provided definitions of various bifurcations in dynamical systems for scalar or two-dimensional x. We used this low-dimensional system for the sake of clarity. All definitions can be generalized to m-dimensional systems of the form

$$\dot{X} = F(X, \lambda), \qquad \lambda \in \mathbb{R}^l, \quad X \in \mathbb{R}^m ,$$

where m and l are positive integers. Here we discuss only saddle-node bifurcations; the other bifurcations can be treated similarly.

Suppose the system above has an equilibrium $X_b \in \mathbb{R}^m$ for some parameter value $\lambda = \lambda_b \in \mathbb{R}^l$; that is,

$$F(X_b, \lambda_b) = 0 .$$

The equilibrium is hyperbolic if the Jacobian matrix $L = D_X F$ evaluated at the equilibrium has no eigenvalues with zero real part. It is nonhyperbolic if at least one eigenvalue has zero real part. Let $\mu_1, \ldots, \mu_m \in \mathbb{C}$ denote the eigenvalues of L, let vectors $v_1, \ldots, v_m \in \mathbb{C}^m$ denote corresponding (right) eigenvectors of L, and let $w_1, \ldots, w_m \in \mathbb{C}^m$ denote the dual to the eigenvectors basis; that is, $w_i v_j = \delta_i^j = 0$ when $i \neq j$, and $\delta_i^i = 1$ for all $i = 1, \ldots, m$. It is easy to check that $w_1, \ldots, w_m \in \mathbb{C}^m$ is the set of left eigenvectors of L.

The dynamical system is at a saddle-node bifurcation point when the following conditions are satisfied.

- The equilibrium X_b is nonhyperbolic with precisely one eigenvalue with zero real part. Since L is real, the eigenvalue, say μ_1, is real and hence is zero. This is guaranteed if

$$D_X F(X_b, \lambda_b) v_1 = 0$$

and $\operatorname{Re} w_i D_X F(X_b, \lambda_b) v_i \neq 0$ for $i = 2, \ldots, m$.

- The function $F(X, \lambda_b)$ has nonvanishing quadratic terms along the eigenvector v_1; that is,

$$w_1 (D_X^2 F(X_b, \lambda_b) (v_1, v_1)) \neq 0 ,$$

where $D_X^2 F(X_b, \lambda_b)(\cdot, \cdot) : \mathbb{C}^m \times \mathbb{C}^m \to \mathbb{C}^m$ is a bilinear vector-form.

- The n-dimensional vector

$$\mathbf{a} = w_1 D_\lambda F(X_b, \lambda_b)$$

is nonzero, which is called the transversality condition.

Obviously, the definition above is too cumbersome in comparison with the one given in Section 2.2.1. Below, we discuss how to avoid such complex definitions.

One can use the center manifold reduction procedure (see Section 2.1.1) to analyze a multidimensional system at a saddle-node bifurcation point. This approach utilizes the fact that there is a local invariant (center) manifold M at the saddle-node point that is tangent to the eigenvector v_1; see Figure 2.26. The reduction is valid only for X near X_{b} unless global validity

Figure 2.26. There is a center manifold M tangent to the center subspace spanned by the eigenvector v_1 near a saddle-node bifurcation.

can be established. After reduction, one obtains scalar equation

$$\dot{x} = f(x, \lambda) \, ,$$

where $x \in \mathbb{R}$ is a variable along M and the equation above is the restriction of $\dot{X} = F(X, \lambda)$ to M. It contains all the necessary information about the local behavior of the original m-dimensional system. In particular, it is at the saddle-node bifurcation point.

While performing the center manifold reduction, one does not use the fact that the equation is near a certain type of bifurcation. The reduction can be accomplished for any system at a nonhyperbolic equilibrium. In this book we take advantage of this observation: We first perform the center manifold reduction of a nonhyperbolic weakly connected neural network without worrying about the type of bifurcation. Then we consider the reduced system (which is also a weakly connected neural network) and determine what kind of a bifurcation takes place for it.

2.6   Soft and Hard Loss of Stability

Bifurcations play important roles in the analysis of dynamical systems since their qualitative behavior changes when the parameters pass the bifurcation values. For an observer such qualitative changes might seem to be dramatic or to be not noticeable at all. For example, suppose that a dynamical system of the form

$$\dot{x} = f(x, a, b), \qquad a, b \in \mathbb{R} \, ,$$

has a stable equilibrium $x = 0$ for $a < 0$ that bifurcates (disappears or loses its stability) when a crosses the bifurcation value $a = 0$. Then there can be two different results of the bifurcation:

- (Soft) For small positive a there is a new stable equilibrium (or equilibria, or stable limit cycle, or other attractor) in a small neighborhood of the old one.

- (Hard) For $a > 0$ there is no attractor in a neighborhood of the origin.

In the first case, $x(t)$ may approach the new attractor and hence continue to be small. In the second case, $x(t)$ leaves the neighborhood of the origin and grows. If the dynamical system describes a physical process for which $x(t)$ needs to be small, then the first bifurcation type is harmless, while the second one has catastrophic character and can be devastating. Therefore, it is useful for us to distinguish the soft and hard loss of stability. The notion of soft and hard cases of loss of stability is closely related to the definition of stability of a dynamical system under persistent disturbances (see, e.g., Hoppensteadt 1993).

To have concrete examples in mind, we may think of the following canonical models

$$\dot{x} = a + x^2 , \qquad\qquad\qquad \text{(Saddle-node)}$$
$$\dot{x} = ax + \sigma x^3 , \qquad\qquad\qquad \text{(Pitchfork)}$$
$$\dot{x} = a + bx + \sigma x^3 , \qquad\qquad \text{(Cusp)}$$
$$\dot{z} = (a + i\omega)z + (\sigma + i\gamma)z|z|^2 , \qquad \text{(Andronov-Hopf)}$$
$$x \mapsto x + a + x^2 , \qquad\qquad\qquad \text{(Saddle-node)}$$
$$x \mapsto x + ax + \sigma x^3 , \qquad\qquad\quad \text{(Flip)}$$
$$z \mapsto z + (a + ib)z + (\sigma + i\gamma)z|z|^2 , \quad \text{(Secondary Andronov-Hopf)}$$

where $\sigma \in \mathbb{R}$ is a nonzero parameter.

Consider, for example, the pitchfork bifurcation diagrams for $\sigma = -1$ and $\sigma = +1$, as depicted in Figure 2.27. In the first case, the stable equilibrium

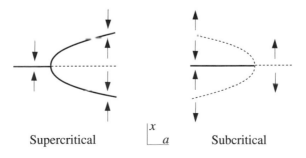

Figure 2.27. Soft and hard loss of stability via the pitchfork bifurcation.

$x = 0$ loses stability at $a = 0$, giving birth to a pair of new equilibria $\pm\sqrt{a}$, which are close to the origin. Such a bifurcation is called *supercritical*, and it corresponds to a soft loss of stability. In the second case, the stable equilibrium $x = 0$ is surrounded by unstable equilibria $\pm\sqrt{-a}$. They shrink down to the origin as a increases toward the bifurcation value $a = 0$. For $a > 0$ the origin $x = 0$ is not stable, and $x(t)$ must leave its neighborhood. Such a bifurcation is called *subcritical*, which corresponds to the hard loss of stability.

The pitchfork, cusp, flip, and Andronov-Hopf bifurcations are supercritical for negative σ and subcritical otherwise. The loss of stability via the saddle-node bifurcation is always hard, unless the bifurcation is close to the supercritical cusp bifurcation as depicted in Figure 2.8.

Supercritical bifurcations may be acceptable for engineering applications, since they provide soft loss of stability, but subcritical bifurcations can lead to failure. Nevertheless, subcritical bifurcations play essential roles in brain dynamics since they enable a neuron to achieve swift and large response to certain changes of parameters of the system, such as the external input.

The variable x could describe activity of a single neuron or a local population of neurons, as we discussed in the introduction. This influences the importance of soft and hard bifurcations in such neural networks. Suppose x describes the activity of a single neuron. When an equilibrium, which corresponds to the rest potential, loses stability via soft bifurcation, the neuron does not fire. It does not produce any macroscopic changes noticeable to other neurons. Biologists say that changes in activity of such a neuron are *graded*. When neurons communicate via action potentials, the network activity is not sensitive to graded (nonspiking) activities of neurons, no matter how strong synaptic connections between neurons are. Therefore, the soft bifurcation affects the dynamics of only a single neuron, but it does not affect the dynamics of the whole network. In contrast, when neurons communicate via gap junctions (electrical synapses) or have dendro-dendritic synapses, graded activity plays an essential role. The dynamics of other neurons can be affected by soft bifurcations even when the connections between neurons are weak.

Now suppose the variable x describes the averaged activity of a local population of neurons. Then x being at an equilibrium does not necessarily mean that each neuron from the local population is quiet. It rather means that the local neurons fire randomly with constant probabilities. Soft loss of stability of such an equilibrium can affect behavior of the whole network regardless of whether interactions between the local populations are spiking or graded. For example, activities of other populations can be near bifurcations too, and they can be sensitive to even small perturbations, such as those from x.

When a bifurcation is hard, it does not make any difference whether it is on the level of single neurons or local populations of neurons, or whether the interactions between neurons are graded or spiking. Hard bifurcations play crucial roles in all cases. We summarize bifurcations that are important to our study below.

INTERACTIONS BETWEEN NEURONS	BIFURCATIONS IN DYNAMICS OF	
	NEURONS	LOCAL POPULATIONS
Graded (nonspiking)	soft/hard	soft/hard
Spiking	hard	soft/hard

2.7 Global Bifurcations

Bifurcations indicate qualitative changes in the dynamic behavior of neurons. Are those changes local or global? Let us discuss this issue.

It is important to distinguish local and global bifurcations. Loss of stability or disappearance of an equilibrium is a local bifurcation. Indeed, when such a bifurcation takes place, the phase portrait of a dynamical system changes locally in a neighborhood of the nonhyperbolic point; see Figure 2.28. Qualitative changes of the vector field are localized in a small

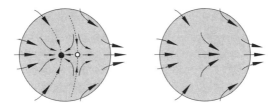

Figure 2.28. Saddle-node bifurcation. Qualitative changes of the vector field are localized in a small neighborhood.

neighborhood. Outside the neighborhood of the equilibrium, the vector field is qualitatively the same provided that there are no other bifurcations occurring simultaneously. Thus, for local bifurcations of a system $\dot{x} = f(x, \lambda)$, $x \in \mathbb{R}^m$, there is a small open ball in $\mathbb{R}^m$ containing the qualitative changes. In contrast, when a bifurcation is global, there is no such ball.

So far, we have considered only local bifurcations of neuron dynamics described by a continuous-time dynamical system $\dot{x} = f(x, \lambda)$ near an equilibrium. The case of a discrete-time dynamical system $x \mapsto f(x, \lambda)$ provides an important example where a global bifurcation (loss of stability or disappearance of a limit cycle) can be reduced to a local bifurcation of a fixed point of the Poincaré or time-T mapping.

Can a local bifurcation affect the global dynamic behavior of a neuron? The answer is YES. Below we discuss important examples that are relevant to the mechanism of generation of action potentials by neurons. The examples serve as a motivation for our definition of nonhyperbolic neural networks, which we discuss in the next chapter.

2.7.1 Saddle-Node on Limit Cycle

Consider a dynamical system $\dot{x} = f(x, \lambda)$, $x \in \mathbb{R}^2$, having a saddle-node bifurcation as depicted in Figure 2.26. If the initial condition $x(0)$ starts from the right-hand side of the saddle (open circle), then $x(t)$ increases and leaves a small neighborhood containing the saddle and the node. To study $x(t)$ far away from the neighborhood requires global knowledge of the function $f(x, \lambda)$. Depending on the function, $x(t)$ can go to infinity or be attracted to another stable equilibrium, limit cycle, chaotic attractor, etc.

It is also possible that $x(t)$ can return to the neighborhood. The simplest case for this occurs when the left and the right branches of the invariant manifold M make a loop; see Figure 2.29. When $x(0)$ is near the right

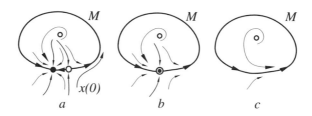

Figure 2.29. Saddle-node bifurcation on a limit cycle.

branch of M, it leaves a neighborhood of the saddle, makes a rotation along M, and approaches the node. Such an excursion in neuroscience applications is called an *action potential* or a *spike*.

Obviously, the node is an attractor for all x inside the limit cycle and could be an attractor for all points outside the cycle. The activity $x(t)$ of a dynamical system having phase portrait as in Figure 2.29a or b always converges to the node, at least for $x(0)$ close enough to the limit cycle. The invariant manifold M is itself an orbit, which starts from the saddle (open circle in Figure 2.29a) and ends in the node (shaded circle). Such an orbit is said to be *heteroclinic*.

When a saddle-node bifurcation takes place, the saddle and the node coalesce to form a nonhyperbolic equilibrium, the saddle-node; see Figure 2.29b. Now, M is a *homoclinic* orbit since it starts and ends in the same equilibrium. The nonhyperbolic equilibrium is sensitive to small perturbations of the parameters of the system. It could split back to the saddle and the node, or it could disappear, as in Figure 2.29c, in which case M becomes a limit cycle. All local solutions of the system are attracted to the limit cycle, and hence the activity $x(t)$ becomes oscillatory. We see that the saddle-node bifurcation can separate two qualitatively different regimes, rest and oscillatory activity. Unlike the Andronov-Hopf bifurcation, here oscillations are not a local, but a global phenomenon. It is a remarkable fact that they occur via a local bifurcation (some suggest to call such a bifurcation *semi-local*).

On the other hand, if we are given a dynamical system having a limit cycle attractor, and the parameters of the system are changed so that a nonhyperbolic (saddle-node) equilibrium appears on the limit cycle, then the equilibrium splits into a saddle and a node; see Figure 2.30. Both equi-

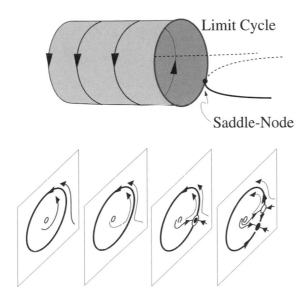

Figure 2.30. Saddle-node on limit cycle bifurcation.

libria are on an invariant circle that used to be a limit cycle attractor. This motivates the name for the bifurcation: the saddle-node on the limit cycle. Other names are *homoclinic orbit to saddle node equilibrium bifurcation* or *saddle-node (fold) on invariant circle* bifurcation.

VCON

The saddle-node bifurcation on a limit cycle cannot be observed in one dimensional systems $\dot{x} = f(x, \lambda)$, $x \in \mathbb{R}$, since there are no limit cycles in $\mathbb{R}$. Nevertheless, it is observable when x is a variable on a one-dimensional manifold such as the unit circle $\mathbb{S}^1 = (\mathbb{R} \bmod 2\pi) = [-\pi, \pi]/\{-\pi \equiv \pi\}$. The manifold $\mathbb{S}^1$ can be treated as a segment $[-\pi, \pi]$ having points $-\pi$ and π glued together. Thus, when x increases and passes through π, it appears at $-\pi$. One can treat the *phase variable* $x \in \mathbb{S}^1$ as an angle that parametrizes the unit circle $\mathbb{S}^1 = \{e^{ix}\} \subset \mathbb{C}$.

An example of a dynamical system on $\mathbb{S}^1$ is the voltage controlled oscillator neuron (VCON)

$$\dot{x} = \omega - \cos x \, ,$$

where $\omega > 0$ is a bifurcation parameter. When $\omega < 1$ there are two equilibria, $\pm \cos^{-1} \omega$; see Figure 2.31. The saddle-node bifurcation occurs when

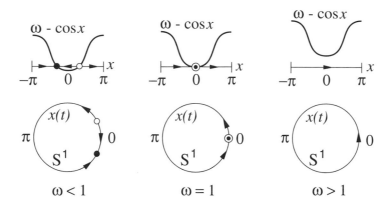

Figure 2.31. The VCON for various values of bifurcation parameter $\omega > 0$.

$\omega = 1$. The two equilibria $\pm \cos^{-1} \omega$ coalesce into one $(x = 0)$ when ω hits 1 and disappear for $\omega > 1$. For $\omega > 1$ the right-hand side is always positive, and $x(t)$ increases without bound going through $\pi \equiv -\pi$ again and again. Such variable x describes the phase of rotation on $\mathbb{S}^1$. Any function $P(x(t))$ oscillates with a certain frequency dependent on ω. In Chapter 8 we prove that the VCON is not only a typical system exhibiting the saddle-node bifurcation on a limit cycle, but also a canonical model. That is, any other system at the bifurcation can be transformed to the VCON by an appropriate change of variables.

Wilson-Cowan Oscillator

Saddle-node bifurcations on limit cycles are ubiquitous in two-dimensional systems. We use the Wilson-Cowan neural oscillator model

$$\begin{cases} \dot{x} &= -x + S(\rho_x + ax - by) \\ \dot{y} &= -y + S(\rho_y + cx - dy) \end{cases}$$

to illustrate this. For this we plot the nullclines $\dot{x} = 0$ and $\dot{y} = 0$ on the xy-plane. Each intersection of the nullclines corresponds to an equilibrium of the model. For certain values of parameters the nullclines intersect as

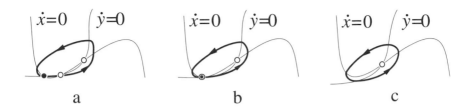

$$\dot{x}=0 \quad \dot{y}=0 \qquad \dot{x}=0 \quad \dot{y}=0 \qquad \dot{x}=0 \quad \dot{y}=0$$

a   b   c

Figure 2.32. Wilson-Cowan model exhibiting saddle-node bifurcation on the limit cycle. Parameter values are $a = 10$, $b = 10$, $c = 10$, $d = -2$, and $\rho_x = -3.12$. Simulations are performed for three values of ρ_y close to $\rho_y \approx -6.14$.

depicted in Figure 2.32, producing a phase portrait similar to the one in Figure 2.29. Even though the saddle-node bifurcation can be determined analytically, one has to use numerical simulations to show that the bifurcation occurs on the limit cycle (bold loop).

Relaxation Wilson-Cowan Oscillator

A saddle-node bifurcation on a limit cycle can also be observed in relaxation systems having nullclines intersected as in Figure 2.33. To illustrate this

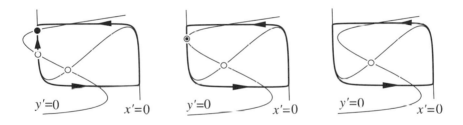

$y'=0$ $x'=0$ $y'=0$ $x'=0$ $y'=0$ $x'=0$

Figure   2.33.   Wilson-Cowan   relaxation   oscillator   exhibiting   saddle-node   bifurcation   on   the   limit   cycle.   Parameter   values   are $a = 10$, $b = 7$, $c = 1$, $d = -6$, $\mu = 0.1$, and $\rho_x - -0.5$. Simulations are performed for three values of ρ_y close to $\rho_y \approx -3.42$.

we use the relaxation Wilson-Cowan oscillator

$$\begin{cases} \mu x' = -x + S(\rho_x + ax - by) \\ \quad y' = -y + S(\rho_y + cx - dy) \end{cases} \tag{2.26}$$

where $\mu \ll 1$. Here x is a fast and y a slow variable. Such a system is capable of producing relaxation oscillations for a broad range of parameters. When the feedback constant d is negative and sufficiently large in absolute value, the y-nullcline $y' = 0$ has S-shape and can intersect the x-nullcline near the relaxation limit cycle (bold loop in Figure 2.33). Again, the phase portrait of such a system is qualitatively similar to the one depicted in Figures 2.29, 2.31, and 2.32.

We continue discussion of the bifurcation in Section 2.8 which is devoted to mechanisms of generation of action potentials by neurons.

2.7.2   Saddle Separatrix Loop

Next, we discuss another example of a global bifurcation that occurs in two-dimensional systems. This bifurcation is remarkable in the following sense: Even though it requires an equilibrium, the equilibrium is a hyperbolic one. In the two-dimensional case it must be a saddle having a homoclinic orbit.

Consider a dynamical system

$$\dot{x} = f(x, \lambda), \ x \in \mathbb{R}^2, \ \lambda \in \Lambda \,,$$

having a saddle $x = x_0 \in \mathbb{R}^2$ for some $\lambda = \lambda_0 \in \Lambda$. Let $L = D_x f$ be the Jacobian matrix at $(x, \lambda) = (x_0, \lambda_0)$. It has two real eigenvalues $\mu_1 < 0 < \mu_2$ and corresponding eigenvectors $v_1, v_2 \in \mathbb{R}^2$. There are stable $E^s \subset \mathbb{R}^2$ and unstable $E^u \subset \mathbb{R}^2$ one-dimensional eigensubspaces spanned by eigenvectors v_1 and v_2, respectively. There are also one-dimensional stable $M^s \subset \mathbb{R}^2$ and unstable $M^u \subset \mathbb{R}^2$ manifolds, which are tangent to E^s and E^u, respectively; see Figure 2.34a.

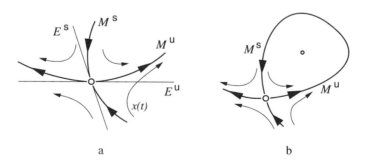

Figure 2.34. a. A saddle has stable and unstable eigensubspaces E^s and E^u and manifolds M^s and M^u tangent to the subspaces. b. The saddle separatrix loop.

The manifolds are also called *separatrices* due to the fact that they separate the phase space regions having different qualitative features. For example, if the initial state $x(0)$ is close to the saddle and is to the right of M^s, then $x(t)$ moves to the right along the unstable manifold M^u, as in Figure 2.34a. If it were at the left-hand side, it would move to the left along the left branch of M^u. Since the unstable manifold branches can go to different regions of the phase space, the behavior of the solution $x(t)$ depends crucially on whether $x(0)$ is at the right or at the left side of M^s.

The *saddle separatrix loop* bifurcation occurs when a branch of the stable manifold M^s conjoins with a branch of the unstable manifold M^u; see Figure 2.34b. Such a loop is a *homoclinic orbit to the saddle*, which motivates another name for the bifurcation – *homoclinic bifurcation*. It was studied by Andronov and Leontovich (1939) and sometimes carries their name – *Andronov-Leontovich* bifurcation. We do not use these names in this book, since the former can be confused with the saddle-node on a limit cycle bifurcation, which is also a homoclinic bifurcation, and the latter name is rarely used.

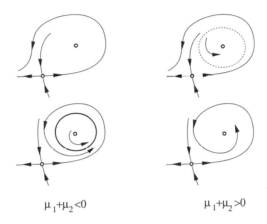

$$\mu_1 + \mu_2 < 0 \qquad\qquad \mu_1 + \mu_2 > 0$$

Figure 2.35. Perturbation of the saddle separatrix loop. Parameters $\mu_1 < 0 < \mu_2$ are eigenvalues of the Jacobian matrix at the saddle.

Small perturbations destroy the homoclinic orbit, making the branches of stable and unstable manifolds miss each other. Typical perturbations of the orbit are depicted in Figure 2.35. They depend on the sum of the eigenvalues, which is called the *saddle quantity*, in particular, whether it is positive or negative. If $\mu_1 + \mu_2 < 0$ the homoclinic orbit is stable from the inside, and there is a birth of a unique stable limit cycle (bold curve). If $\mu_1 + \mu_2 > 0$, the homoclinic orbit is unstable from the inside, and there is a death of a unique unstable limit cycle. The case $\mu_1 + \mu_2 = 0$ is called the *neutral saddle*. There is always a double limit cycle bifurcation (see below) in a small neighborhood of a neutral saddle separatrix loop bifurcation. We encountered such a saddle in Figure 2.13 when we studied bifurcations of Wilson-Cowan model.

Detailed information about the saddle separatrix loop bifurcation can be found in Kuznetsov (1995, p.183). We just mention here the fact that the period of the limit cycle attractor increases when the limit cycle approaches the saddle. This happens because the solution spends more and more time in the vicinity of the saddle equilibrium. When the limit cycle attractor becomes the homoclinic orbit to the saddle, the period becomes infinite.

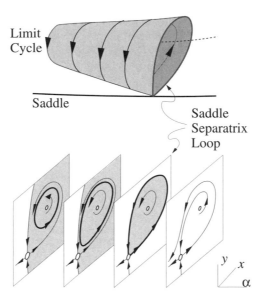

Figure 2.36. Saddle separatrix loop bifurcation for negative saddle quantity. α is a bifurcation parameter.

The stable and unstable manifolds of a saddle may conjoint as in Figure 2.37 producing the big saddle separatrix loop. This bifurcation is qualita-

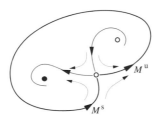

Figure 2.37. Big saddle separatrix loop.

tively similar to the one depicted in Figure 2.34b, and we do not consider it in this book. Notice, though, that perturbations of this bifurcation may produce a limit cycle that encompasses three equilibria, which leads to a new type of bursting, see Section 2.9.5.

Relaxation Wilson-Cowan Oscillator

We have already encountered the saddle separatrix loop bifurcation in the
Wilson-Cowan model; see Figure 2.13 on page 47. It can also be illustrated
using the relaxation Wilson-Cowan oscillator (2.26) or any other relaxation
system having nullclines intersecting as in Figure 2.33. For this we change
parameters so that the y-nullcline moves up (or x-nullcline moves down)
until the node tears the limit cycle; see Figure 2.38a. Moving the nullcline

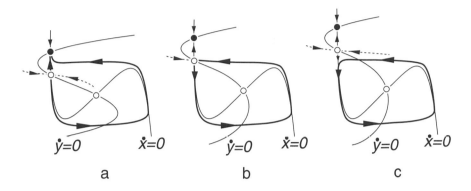

Figure 2.38. Saddle separatrix loop bifurcation in Wilson-Cowan relaxation oscillator.

further makes the stable manifold of the saddle and the piece of the limit
cycle coalesce; see Figure 2.38b. At this moment a bifurcation occurs, since
the limit cycle (bold curve) is the saddle separatrix loop. Lifting the y-
nullcline even further is a perturbation of the saddle separatrix loop. In the
case of the relaxation Wilson-Cowan oscillator, such a perturbation revives
the limit cycle; see Figure 2.38c. Another geometrical illustration of this
bifurcation is depicted in Figure 2.45.

2.7.3 Double Limit Cycle

The double limit cycle bifurcation occurs when a stable and an unstable
limit cycle coalesce and disappear, as depicted in Figure 2.39. This bifurcation is global for flows, but it is reducible to the saddle-node (fold)
bifurcation of the Poincaré first return map, which is local, and so another
name for the double limit cycle bifurcation is the *fold bifurcation of limit
cycles*; see Figure 2.40. Similarly, a flip bifurcation of limit cycles is the one
having a flip bifurcation of the corresponding Poincaré first return map; see
Figure 2.25.

An example of a dynamical system having double limit cycle bifurcation
is

$$\dot{z} = (\alpha + \mathrm{i}\omega)z + 2z|z|^2 - z|z|^4 , \quad z \in \mathbb{C} , \qquad (2.27)$$

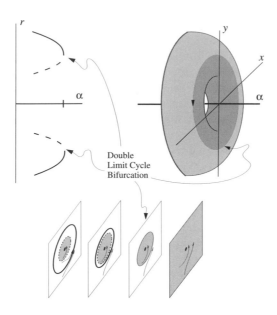

Figure 2.39. Double limit cycle bifurcation. Stable and unstable limit cycles coalesce and annihilate each other.

at $\alpha = -1$. It is easier to study this system when it is written in the polar coordinates $z = re^{i\varphi}$

$$\begin{cases} \dot{r} & = \alpha r + 2r^3 - r^5 \\ \dot{\varphi} & = \omega \end{cases} .$$

The phase variable $\varphi(t)$ rotates with the frequency ω and r determines the amplitude of oscillations. The limit cycles of (2.27) are in one-to-one correspondence with the positive equilibria of the dynamical system

$$\dot{r} = r(\alpha + 2r^2 - r^4) , \qquad (2.28)$$

which is easy to analyze.

Relaxation Wilson-Cowan Oscillator

The double limit cycle bifurcation can be observed in many relaxation systems, including the Wilson-Cowan model (2.26). If we treat ρ_x as a bifurcation parameter, then the Wilson-Cowan oscillator (2.26) behaves similarly to (2.27). In Figure 2.41 we have depicted typical phase portraits. When the external input ρ_x is below a bifurcation value, the unique equilibrium is a global attractor of the system. When ρ_x increases through a bifurcation value, stable and unstable limit cycles appear through the double limit cycle bifurcation. The relaxation Wilson-Cowan oscillator (2.26) can have two kinds of dynamics – rest and oscillation. The equilibrium seems to be

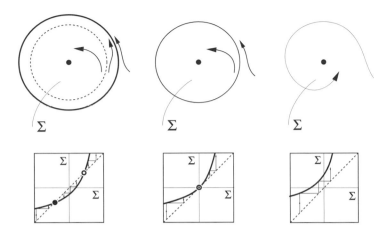

Figure 2.40. *Above*: Double limit cycle bifurcation. Stable and unstable limit cycles disappear. Σ is a Poincaré cross section. *Below*: Saddle-node (fold) bifurcation of the Poincaré first return map.

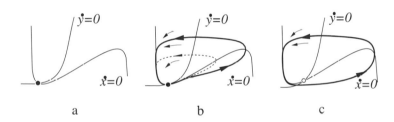

Figure 2.41. Double limit cycle bifurcation in relaxation Wilson-Cowan oscillator (2.26). Parameters: $a = 10$, $b = 10$, $c = 10$, $d = -2$, $\mu = 0.1$, $\rho_y = -5$ and three values of ρ_x close to $\rho_x = -2.96$.

on a limit cycle in Figure 2.41b, but magnification shows that it is inside the unstable limit cycle, albeit very close to it. Further increasing the input ρ_x shrinks the unstable limit cycle to a point, and the rest activity becomes unstable though a subcritical Andronov-Hopf bifurcation.

Remark. A noteworthy fact is that the double limit cycle and the subcritical Andronov-Hopf bifurcations in Figure 2.41 occur for two distinct but very close ($\pm 10^{-9}$) values of the parameter ρ_x. Therefore, the passage from the stable rest state to the relaxation oscillations while ρ_x increases should seem instantaneous for an observer. This happens because the relaxation system is near the quasi-static saddle-node bifurcation that we discussed in Section 2.2.6.

2.8   Spiking

There are two phenomena associated with the generation of action potentials – neural excitability and transition from rest to periodic spiking activity. The former is a single response to external perturbations, the latter is a qualitative change in dynamic behavior. The type of neural excitability depends intimately on the type of bifurcation from quiescent to oscillatory activity. In order to understand the spiking mechanism, we must understand the bifurcation.

Generation of action potentials is a complicated process involving many ions and channels. An interested reader should consult physiological references, such as Shepherd's *Neurobiology* (1983). A recent book of Johnston and Wu, *Foundations of Cellular Neurophysiology* (1995), is an excellent introduction to mathematical neurophysiology of membranes.

The basic idea behind the spiking mechanism is that a neuron membrane has various voltage-dependent channels operating on various time scales. Among them the two most important are the fast Na^+ and slow K^+ channels. An increase in the membrane potential opens Na^+ channels, resulting in Na^+ inflow and further increase in the membrane potential. Such positive feedback leads to sudden and abrupt growth of the potential. This opens slow K^+ channels, which leads to increased K^+ current and eventually reduces the membrane potential. These simplified positive and negative feedback mechanisms are depicted in Figure 2.42. The resemblance

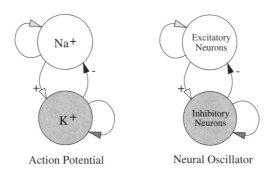

Figure 2.42. Qualitative similarity of action potential mechanism and neural oscillator.

to the neural oscillator suggests that we may use the Wilson-Cowan model to illustrate many bifurcation phenomena observable in dynamical systems modeling neuron excitability. It should be stressed that the mechanism of spiking we described above is simplified. It does not reflect activities of other ion channels (such as Cl^-) and currents that contribute to generation of action potentials. Some of these activities can alter the shape of the spike; others (such as Ca^{2+}) can change significantly the pattern of firing, for example transforming it to a bursting pattern of action potentials.

There is a wide spectrum of mathematical models of generation and propagation of action potentials in excitable membranes. The most prominent are Hodgkin-Huxley (Hodgkin and Huxley 1954), FitzHugh-Nagumo (FitzHugh 1969), Connor (Connor et al. 1977), and Morris-Lecar (Morris and Lecar 1981). We do not present them here. Instead, we discuss mechanisms of generation of spikes using a geometrical approach based on bifurcation theory. The advantage of this approach, which we adapted from Rinzel and Ermentrout (1989), is that it allows us to see the essence of the problem without being distracted by tedious electrophysiological and neurochemical details.

A simple way to study nonlinear excitable properties of neuron membranes is to penetrate the membrane with a micro electrode, inject a current, and measure the membrane potential. If we perturb a membrane potential (e.g., by applying a strong, brief impulse of current), it could return immediately to the equilibrium. This happens when the perturbation does not lead the activity beyond certain boundaries. Such perturbations are said to be *subthreshold*. The perturbations that evoke a spike are said to be *superthreshold*, and the neuron activity is said to be excitable in this case. Such perturbations do exist, and neurons are excitable because the equilibrium corresponding to the rest potential is nearly nonhyperbolic. That is, the neuron activity is near a bifurcation.

To analyze the bifurcation we perform the following experiment. We apply a weak current of increasing magnitude. When the current is too weak, nothing happens, and the membrane potential is at an equilibrium. When the current is strong enough, the membrane potential suddenly increases (Na^+ current), reaches its maximum, and then decreases (K^+ current), producing a spike in the membrane potential. If the application of the current persists, the neuron generates spikes repetitively with a certain frequency. The following classification was suggested by Hodgkin (1948); see Figure 2.43.

- CLASS 1 NEURAL EXCITABILITY. Action potentials can be generated with arbitrarily low frequency, depending on the strength of the applied current.

- CLASS 2 NEURAL EXCITABILITY. Action potentials are generated in a certain frequency band that is relatively insensitive to changes in the strength of the applied current.

Hodgkin also identified Class 3 neural excitability as occurring when action potentials are generated with difficulty or not at all. Class 3 neurons have a rest potential far below the threshold value. If the rest potential is somehow increased, these neurons could fire too, and thereby be classified as Class 1 or 2 neurons. Some researchers (Rinzel and Ermentrout 1989,

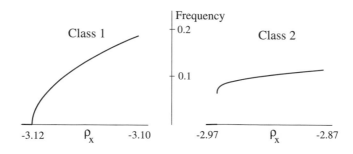

Figure 2.43. Dependence of frequency of oscillations on the strength of applied current ρ_x in the Wilson-Cowan model for $a = b = c = 10$, $d = -2$, $\mu = 1$, $\rho_y = -6.14$ (Class 1); and $\mu = 0.1$, $\rho_y = -5$ (Class 2).

Ermentrout 1996) call the two classes of neural excitability Type I and Type II excitability. We do not use such a notation in this book, since the word "Type" is loaded with various meanings, such as Type I and II phase resetting curves, Golgi Type I and II neurons, Type I and II synapses, and Type I, II and III burstings.

Class 1 neurons fire with a frequency that varies smoothly over a range of about $5 - 150$ Hz. The frequency band of the Class 2 neurons is usually $75 - 150$ Hz, but it can vary from neuron to neuron. The exact numbers are not important to us here. The qualitative distinction between Class 1 and 2 neurons is that the emerging oscillations have zero frequency in the former and nonzero frequency in the latter. This is due to different bifurcation mechanisms.

Let us consider the strength of applied current in Hodgkin's experiments as being a bifurcation parameter. When the current increases, the rest potential increases until a bifurcation occurs, resulting in loss of stability or disappearance of the rest potential, and the neuron activity becomes oscillatory. The bifurcation resulting in transition from a quiescent to an oscillatory state determines the class of neural excitability. Since there are only two codimension 1 bifurcations of stable equilibria, it is not surprising that

- Class 1 neural excitability is observed when a rest potential disappears by means of a saddle-node bifurcation.

- Class 2 neural excitability is observed when a rest potential loses stability via the Andronov-Hopf bifurcation.

The saddle-node bifurcation may or may not be on a limit cycle. The Andronov-Hopf bifurcation may be either subcritical or supercritical. The bifurcational mechanism describing spiking activity must explain not only the appearance but also the disappearance of periodic spiking when the applied current is removed. This imposes some additional restrictions on the bifurcations. We discuss this issue in a moment.

The mechanism of transition from rest to oscillatory activity in neurons is not restricted to the bifurcations we mentioned above, though those bifurcations are the most typical in neuron models. It is feasible for a neuron to undergo other bifurcations. Due to the fact that the number of bifurcations of equilibria is infinite, we concentrate our efforts on bifurcations of lower codimension, since they are most likely to appear in nature.

2.8.1   Class 1 Neural Excitability

Class 1 neural excitability arises when the rest potential is near a saddle-node bifurcation. Besides the Wilson-Cowan model, such excitability can also be observed, e.g., in the Morris-Lecar and Connor models (see Ermentrout 1996).

Consider the saddle-node bifurcation *on a limit cycle* first. Transition from a quiescent to an oscillatory state when the current increases is qualitatively similar to the behavior depicted in Figure 2.29. Looking at the figure from right to left suggests the mechanism of disappearance of the periodic spiking activity. The saddle-node bifurcation on a limit cycle explains both appearance and disappearance of oscillatory activity, and no further assumptions are required.

These oscillations have small frequencies at the bifurcation because after the nonhyperbolic equilibrium disappears (Figure 2.29c), the vector field is still small at the place the equilibrium was, and the activity variable spends most of its time there. Let us illustrate this using the VCON model

$$\dot{\varphi} = \omega - \cos\varphi \,, \qquad \varphi \in \mathbb{S}^1 \,,$$

which is a canonical model, meaning that any other system near the saddle-node bifurcation on a limit cycle can be transformed to the VCON. The bifurcation occurs at $\omega = 1$, and when $\omega = 1 + \delta$, $\delta \ll 1$, phase variable $\varphi(t)$ increases with the speed δ at the origin and with speed more than 1 when $|\varphi| > \pi/2$. A solution of the VCON for $\delta = 0.01$ is depicted in Figure 2.44. It is easy to see that spike duration is small in comparison with the

Figure 2.44. Function $1 + \delta - \cos\varphi$ and a solution of VCON $\dot{\varphi} = 1 + \delta - \cos\varphi$ for $\delta \ll 1$.

time $\varphi(t)$ spends in a neighborhood of the origin. Decreasing δ increases the period of oscillations and therefore decreases the frequency. We prove in Chapter 8 that the frequency behaves like $\sqrt{\delta}$ when $\delta \to 0$.

Now consider a general saddle-node bifurcation that *is not on a limit cycle*. Since the neuron activity is oscillatory after the equilibrium disappears, there must be a limit cycle attractor corresponding to periodic spiking. Such mechanism of transition from quiescent to periodic activity is depicted in Figure 2.45a, b, and c. Increasing the depolarizing current

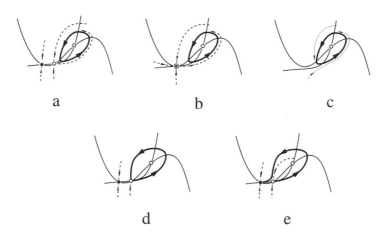

Figure 2.45. A possible mechanism for transition from rest to periodic activity and back in the Wilson-Cowan model. The stable rest point (a) becomes nonhyperbolic (b) and disappears (c) through the saddle-node bifurcation. The stable limit cycle (c) can disappear via a series of bifurcations. First a saddle and a node [as in (a)] appear via a saddle-node bifurcation [as in (b)], then the stable limit cycle is glued to the saddle separatrix loop (d) and becomes a heteroclinic orbit [bold curve in (e)].

destroys the stable equilibrium (Figure 2.45a) via saddle-node bifurcation (Figure 2.45b) and activity becomes periodic (Figure 2.45c). This explains the transition from rest to periodic activity. If we decrease the applied current, the activity remains periodic due to the coexistence of equilibrium and limit cycle attractor. An additional bifurcation is required that destroys the limit cycle attractor – the saddle separatrix loop bifurcation, as in Figure 2.45d – so that the activity becomes quiescent (Figure 2.45e).

When the limit cycle attractor is about to disappear via a saddle separatrix loop bifurcation, the frequency of oscillation is small, since the activity spends most of the time in a neighborhood of the saddle. Therefore, the scenario we discuss explains the zero frequency of disappearing oscillations. To be of Class 1 neural excitability, the frequency of emerging oscillations must be small too. This happens when the saddle-node bifurcation and the saddle separatrix loop bifurcation take place for nearby values of the bifurcation parameter, which is an additional assumption.

The two bifurcations take place almost simultaneously when the dynamical system is close to the saddle-node separatrix-loop bifurcation (Schecter 1987) at which the homoclinic orbit returns to the saddle-node point along the *noncentral* manifold, as we depict at the center of Figure 2.46 (see also Figure 2.13).

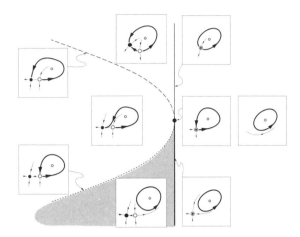

Figure 2.46. Bifurcation diagram for saddle-node separatrix-loop bifurcation. Bold line, thin line, and dotted line are the saddle-node, saddle-node on limit cycle, and the saddle separatrix loop bifurcation curves, respectively. The dashed line is not a bifurcation curve. There is a coexistence of rest and periodic activity in the shaded area.

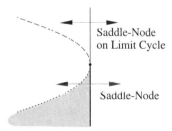

Figure 2.47. Saddle-node separatrix-loop bifurcation (Figure 2.46) is the organizing center for Class 1 neural excitability.

Incidentally, this bifurcation can be treated as an organizing center for Class 1 neural excitability, since its unfolding exhibits both types of saddle-node bifurcations we discussed above; see Figure 2.47. A canonical model for such dynamics has yet to be found. Finally, notice that both saddle-node bifurcations we discussed above can be observed far away from the organizing center.

We may identify Class 1 neural excitability with neuron dynamics near the organizing center when the distinction between two types of saddle-node bifurcation is not essential (Indeed, the corresponding phase portraits in Figure 2.48 are topologically equivalent). The difference is essential when we consider various types of bursting below.

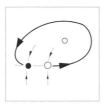 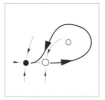

Figure 2.48. Phase portraits corresponding to two subtypes of Class 1 neural excitability are topologically equivalent.

2.8.2   Class 2 Neural Excitability

Class 2 neural excitability is observed when the rest potential is near an Andronov-Hopf bifurcation. Such a bifurcation has been scrutinized numerically and analytically in the Hodgkin-Huxley model (see, e.g., Hassard 1978, Troy 1978, Rinzel and Miller 1980, Hassard et al. 1981, Holden et al. 1991, Bedrov et al. 1992).

Consider the subcritical Andronov-Hopf bifurcation first. Transition from rest to oscillatory activity in this case is qualitatively similar to that depicted in Figure 2.41. Recall that the Andronov-Hopf bifurcation occurs when there is a pair of purely imaginary eigenvalues $\pm i\Omega$. The parameter Ω determines the frequency of local oscillations at the equilibrium. In the case of a subcritical Andronov-Hopf bifurcation, the local oscillations grow in amplitude quickly, but their frequencies remain relatively close to Ω.

Figure 2.49. Supercritical Andronov-Hopf bifurcation in the Wilson-Cowan model.

Next, consider a supercritical Andronov-Hopf bifurcation (Figure 2.49). The periodic activity in this case has small amplitude, which increases with the applied current. Since the bifurcation is supercritical, the loss of stability is soft. To be observable experimentally and to play any essential role in information processing by the neuron, the amplitude of oscillations (spiking) must grow quickly, as in relaxation systems near quasi-static saddle-node bifurcations. We hypothesize that the quasi-static saddle-node bifurcation is the organizing center for Class 2 neural excitability, since its unfolding exhibits both subcritical and supercritical Andronov-Hopf bifurcations. Even though this issue still poses some problems, we identify in what follows Class 2 neural excitability and neuron dynamics at the quasi-static saddle-node bifurcation.

2.8.3 Amplitude of an Action Potential

Class 1 and 2 neural excitability have other distinctions besides frequency of emerging oscillations. In particular, they differ in amplitudes of action potentials:

- Class 1 neural excitability exhibits action potentials of fixed amplitude.

- Class 2 neural excitability can exhibit action potentials of arbitrarily small amplitude.

Indeed, in the case of a saddle-node bifurcation, the amplitude of the action potential depends on the size of the limit cycle attractor (Figure 2.50a), which is not small (unless additional assumptions are made). In

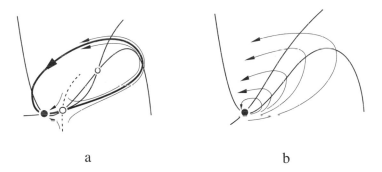

a   b

Figure 2.50. a. Class 1 neurons have action potentials of fixed amplitude. b. Class 2 neurons can have action potentials of arbitrary small amplitude.

the case of Andronov-Hopf bifurcation, the membrane potential can be perturbed in such a way that it produces a small amplitude spike; see

Figure 2.50b. We summarize principal differences between the excitability classes below.

NEURAL EXCITABILITY	CLASS 1	CLASS 2
Typical Bifurcation	Saddle-node on limit cycle	Subcritical Andronov-Hopf
Organizing Center	Saddle-node separatrix-loop	Quasi-static saddle-node
Frequency of emerging oscillations	Small	Nonzero
Amplitude of emerging oscillations	Fixed	Arbitrarily small
Coexistence of rest and periodic activity	Yes/No	Yes/No
Typical models exhibiting excitability	Morris-Lecar Connor	Morris-Lecar, FitzHugh-Nagumo, Hodgkin-Huxley

The dynamics of coupled neurons depends crucially on their excitability class. Hansel et al. 1995 observed numerically that Class 2 excitability can easily lead to synchronization of coupled oscillators. In contrast, synchrony is very difficult to achieve when the oscillators have Class 1 neural excitability. Ermentrout (1996) showed that this is a general property of Class 1 excitability that is relatively independent of the equations describing the neuron dynamics.

2.9 Bursting

Having understood the basic principles of how spiking activity can emerge, we are in a position to discuss bursting mechanisms. A neuron is said to be *bursting* when its activity changes periodically between rest and repetitive firing, as depicted in Figure 2.51. Since there is more than one mechanism for generation of action potentials, it is not surprising that there are many types of bursting. Some of them are classified by Wang and Rinzel (1995) and Bertram et al. (1995).

The theory of bursting is based on the observation that a neuron's dynamics have many time scales and can be described by a singularly perturbed dynamical system of the form

$$\begin{cases} \mu x' &= f(x,y) \\ y' &= g(x,y) \end{cases} \quad x \in \mathbb{R}^n, \ y \in \mathbb{R}^m, \ \mu \ll 1 . \qquad (2.29)$$

The fast subsystem $\dot{x} = f(x,y)$ describes the generation of spikes. It models the membrane potential, the Na$^+$ and K$^+$ conductances, and other vari-

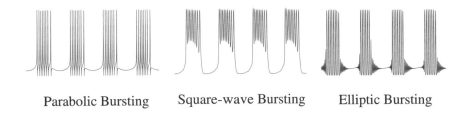

| Parabolic Bursting | Square-wave Bursting | Elliptic Bursting |

Figure 2.51. Major types of bursting.

ables having time constants comparable with the duration of a spike. The fast subsystem depends on bifurcation parameter y, which is assumed to be fixed. The slow subsystem $y' = g(x, y)$ describes the evolution of the bifurcation parameters. It models relatively slow processes, such as Ca^{2+} ion concentration and conductance. This approach, to consider fast and slow systems separately, is known as *dissection* of neural bursting (Rinzel and Lee 1987). We first discuss the dynamics of the slow variable y, assuming that there is a spiking mechanism for the fast variable x. After that, we discuss the peculiarities of bursting for the different spiking mechanisms.

System (2.29) is a singularly perturbed dynamical system, and therefore should be studied using standard techniques from the singular perturbation theory (Hoppensteadt 1993). A complication arises due to the fact that the fast variable x may converge to a periodic orbit instead of an equilibrium. This problem can be circumvented using averaging theory, which we discuss later.

2.9.1 Slow Subsystem

Suppose the fast system has periodic activity (i.e., it generates spikes) for the bifurcation parameter y in a certain region and is quiet for other values of y. That is, there is *no* coexistence of periodic and resting behavior. The simplest kind of bursting that occurs in this case is that in which the slow variable $y(t)$ oscillates, as we depict in Figure 2.52a. That is, $y(t)$ visits periodically the shaded area in which $x(t)$ fires repetitively and the white area in which $x(t)$ is at an equilibrium. This may happen if $g(x, y)$ does not depend on x and $y' = g(y)$ has a limit cycle attractor (bold loop in the figure), which can be treated as a slow subcellular oscillator.

In reality, $g(x, y)$ depends on x, but one can often use averaging to eliminate the fast dynamics. Indeed, suppose the fast system has a unique attractor that depends on y as a parameter. Then the slow system has the form

$$\dot{y} = \mu g(x(y, t), y), \quad \mu \ll 1, \tag{2.30}$$

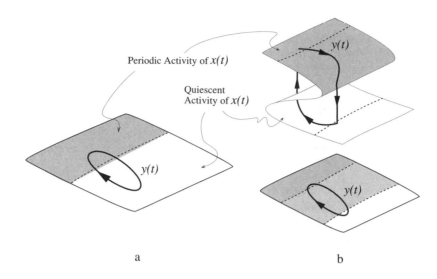

Figure 2.52. Bursting occurs when slow variable $y(t)$ periodically visits the region for which fast variable $x(t)$ oscillates (shaded area). a. There is no coexistence of oscillatory and quiescent activity of $x(t)$. b. Burstings occurs via hysteresis when there is such a coexistence.

and is amenable to the averaging theory. Justification of averaging for singularly perturbed systems[1] can be found in Mishchenko et al. (1994, Section I.7). This results in

$$\dot{y} = \mu \bar{g}(y) + \mathcal{O}(\mu^2) \,,$$

where $\bar{g}$ is the average of g. Such an approach was used e.g. by Rinzel and Lee (1986), Pernarowski et al. (1992), Smolen et al. (1993) and Baer et al. (1995). To exhibit bursting, the averaged system must have a limit cycle attractor. It should be noted that this does not imply existence of the underlying slow wave of subcellular activity, namely, a limit cycle attractor for the slow subsystem $y' = g(x, y)$ in (2.29) when x is at rest potential. We will see such examples in Sections 2.9.4 and 8.2.3.

Quite often, the fast variable x has many attractors for a fixed value y, in which case the averaging in (2.30) must be performed for each attractor separately. A typical example where this happens is when there is a coexistence of quiescent and periodic activity in the fast subsystem. We encounter such coexistence when we study spiking mechanisms. Bursting in this case may occur via hysteresis, as we illustrate in Figure 2.52b. Such behavior is simple and observable in many bursting models. The slow vari-

---

[1]There is a pitfall when the fast system passes slowly through an Andronov-Hopf bifurcation, see discussion and an example in Section 2.9.4

able y in this case may be one-dimensional, such as concentration of Ca^{2+} ions, which simplifies the analysis.

It is also reasonable to consider the case when y is an excitable variable itself, capable of generating slow spikes. We say that the bursting is *excitable* in this case. A typical example of excitable burst occurs when the dynamical system $y' = \mu g(x, y)$ is near a saddle-node bifurcation on a limit cycle for all values of x. Then the bursting occurs if $y(t)$ visits the shaded area during the slow spike. Since $y(t)$ eventually returns to nearly nonhyperbolic equilibrium, the state of repetitive firing of the fast variable x is followed by a long quiescent state until y fires again. We study this and other issues in Chapter 8.

Let us return to the spiking mechanism for the fast variable x. Among the many possible types of bursting, we would like to discuss three.

- PARABOLIC BURSTING arises in Class 1 neural excitability when the saddle-node bifurcation *is* on a limit cycle. The frequency of the emerging firing pattern is parabolic.

- SQUARE-WAVE BURSTING arises in Class 1 neural excitability when the saddle-node bifurcation is *not* on a limit cycle.

- ELLIPTIC BURSTING arises in Class 2 neural excitability. The shape of the emerging low amplitude firing is ellipsoid.

These names reflect only the bursting characteristics, not the type of equations that generate them. An alternative classification of bursting, one based solely on the interspike intervals, was suggested by Carpenter (1979). At the end of this section we discuss difficulties encountered in attempting to classify burstings.

2.9.2   Parabolic Bursting

Suppose the bifurcation parameter y assumes a value, say y_0, on the border of the area where $x(t)$, governed by

$$\dot{x} = f(x, y) ,$$

is resting (see Figure 2.52a) and the area for which $x(t)$ fires (the shaded region in Figure 2.52a). The dynamical system above is then at a bifurcation that determines the class of neural excitability of the fast variable. When it is Class 1 excitability corresponding to the saddle-node bifurcation on a limit cycle, the emerging spikes have zero frequency for $y = y_0$. The frequency increases while $y(t)$ moves away from y_0 into the shaded area. The graph of emerging frequencies shows a parabola-like profile similar to the one in Figure 2.53. This motivates the name for such a bursting. Incidentally, the curves are not parabolas at all. They are glued pieces of the function $1/\sqrt{|y - y_0|}$ for the interspike period and $\sqrt{|y - y_0|}$ for

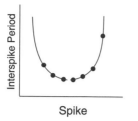
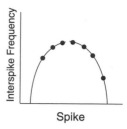

Figure 2.53. Interspike intervals and frequency in parabolic bursting from Figure 2.51.

the interspike frequency. This follows from Prop. 8.4. Guckenheimer et al. (1997) study in detail the rate at which the interspike periods slow near termination of firing. Their analysis is applicable to parabolic and square-wave bursting, which we discuss later.

Notice that in order to produce parabolic bursting it suffices to require that the size of the limit cycle in Figure 2.52a be small. As long as the limit cycle is partially in the white area, partially in the shaded area, bursting takes place. This leads to the following setting for parabolic bursting: Suppose the fast system depends weakly on the slow variable y; that is, suppose the burst model can be written in the form

$$\begin{cases} \mu x' = f(x) + \varepsilon p(x, y, \varepsilon) \\ y' = g(x, y, \varepsilon) \end{cases} , \quad x \in \mathbb{R}^n, \ y \in \mathbb{R}^m, \ \mu, \varepsilon \ll 1 ,$$

where the fast system $\dot{x} = f(x)$ is at a saddle-node bifurcation point $x = 0$ on a limit cycle, and the slow system $y' = g(0, y, \varepsilon)$ has a hyperbolic limit cycle attractor. Then there is a change of variables (Ermentrout and Kopell 1986; see also Section 8.1.4) putting any such system into the canonical form

$$\begin{cases} \dot{\varphi} = 1 - \cos \varphi + (1 + \cos \varphi) p(\theta) \\ \dot{\theta} = \omega \end{cases} \tag{2.31}$$

where the state variable $(\varphi, \theta) \in \mathbb{S}^1 \times \mathbb{S}^1$ is on a torus. Here φ describes the fast activity x, and θ describes the slow activity y. The frequency of slow oscillations is $\omega = \mathcal{O}(\mu/\sqrt{\varepsilon})$, and $p : \mathbb{S}^1 \to \mathbb{R}$ is a continuous function describing the influence of θ on φ. If p is negative for all θ, the fast variable φ is quiescent and there is no bursting. If $p(\theta) > 0$ for all θ, the fast variable fires all the time and there is no quiescent state at all. Bursting occurs when p changes signs. A typical example of such p is $p(\theta) = \cos \theta$. This choice leads to the *atoll model*

$$\begin{cases} \dot{\varphi} = 1 - \cos \varphi + (1 + \cos \varphi) \cos \theta \\ \dot{\theta} = \omega \end{cases} , \quad \varphi, \theta \in \mathbb{S}^1 . \tag{2.32}$$

Typical activity of this system is depicted in Figure 2.54. The name of the

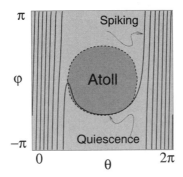

Figure 2.54. Activity of the atoll model (2.32) for $\omega = 0.05$. The dashed circle is the set of equilibria of the fast system.

model stems from the fact that the set of equilibria of the fast subsystem is a circle (dashed circle in Figure 2.54), and the vector of activity $(\varphi(t), \theta(t))$ "avoids" it. When the function $p(\theta)$ in the canonical model changes sign more than two times, there are many atolls on the torus, as, for example, in Figure 2.55b.

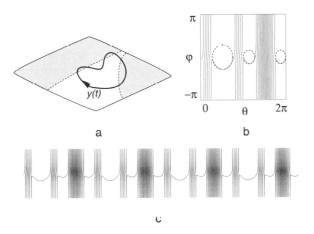

Figure 2.55. a. The bifurcation parameter y visits shaded area corresponding to spiking activity of x many times during a cycle. b. Canonical model for parabolic bursting having many atolls. c. Corresponding bursting activity. Simulations were performed for $p = 0.24 \cos \theta \sin \theta$ and $\omega = 0.01$.

As was pointed out by Soto-Trevino et al. (1996), the requirement that the slow system $y' = g(x, y, \varepsilon)$ has a limit cycle attractor for $x = 0$ is not necessary. One can derive the canonical model (2.31) under the milder assumption that the slow averaged system $y' = \bar{g}(y)$ has a periodic orbit, where $\bar{g}(y)$ is the average of $g(x(t,y), y, 0)$. We study this issue in Section 8.2.3.

It is easy to see that the slow variable must be at least two-dimensional to exhibit parabolic bursting.

2.9.3   Square-Wave Bursting

In contrast to parabolic bursting, there is a coexistence of rest and periodic activity in square-wave bursting, whose principal mechanism is depicted in Figures 2.45 and 2.56. The rest point disappears via a saddle-node bifur-

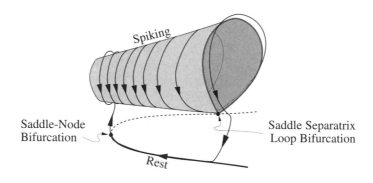

Figure 2.56. Square-wave bursting occurs via saddle-node and saddle separatrix loop bifurcations.

cation, and x is attracted to a stable limit cycle. The cycle disappears via a saddle separatrix loop bifurcation, and x is attracted to the rest point. Therefore, the frequency of disappearing oscillations is zero (with a logarithmic asymptote), which points to a similarity between square-wave and parabolic bursting.

The coexistence of rest and periodic activity permits square-wave bursting for systems having a one-dimensional slow variable y. Repetitive transition from rest to periodic activity and back occurs via hysteresis in this case; see Figure 2.52b. A canonical model for such bursting has yet to be found.

2.9.4   Elliptic Bursting

When y crosses a bifurcation value, the fast variable x becomes oscillatory (via Andronov-Hopf bifurcation) with a nonzero frequency but possibly with small amplitude. The periodic orbit, corresponding to repetitive firing, can disappear via another Andronov-Hopf bifurcation or double limit cycle bifurcation, thereby contributing to various subtypes of elliptic bursting. A typical elliptic burster is depicted in Figure 2.57. Since there is a coexistence

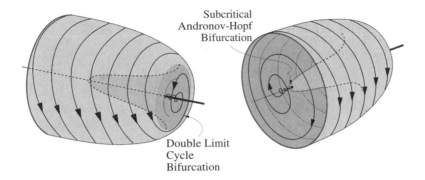

Figure 2.57. Elliptic burster.

of rest and periodic activities, the slow variable y may be one-dimensional.

The shape of such bursting resembles an ellipse, which motivates the name. Appearance of small amplitude oscillations at the onset and/or termination of bursting does not necessarily imply the supercritical Andronov-Hopf bifurcation. Such low amplitude oscillations can also be observed in the case of subcritical Andronov-Hopf bifurcation, as we illustrate in Figure 2.58. This phenomenon occurs when a parameter passes a bifurcation value

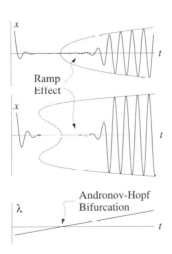

Figure 2.58. Slow passage through Andronov-Hopf and double limit cycle bifurcations in $\dot{z} = (\lambda + 5\mathrm{i})z \pm 2z|z|^2 - z|z|^4$, $z = x + \mathrm{i}y$.

slowly, and it is known as the *ramp effect*, *memory effect*, or effect of *slow passage* through a bifurcation (Nejshtadt 1985, Baer et al. 1989, Arnold et al. 1994).

The slow passage effect is an attribute of elliptic bursting that is not observable in parabolic or square-wave burstings. It conceals the actual occurrence of bifurcation and complicates substantially numerical or experimental analysis of elliptic bursters. For example, the slow passage effect allows elliptic bursting in a system that has a one-dimensional slow variable y and a two-dimensional fast variable x that is at a *supercritical* Andronov-Hopf bifurcation (and, hence, does not exhibit a coexistence of attractors), see illustration in Figure 2.59.

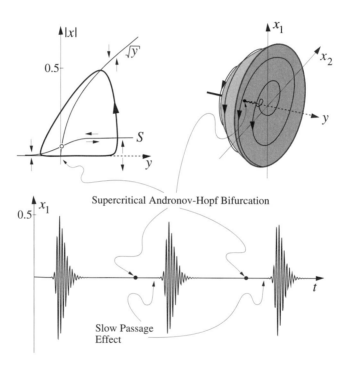

Figure 2.59. Elliptic bursting in $\dot{x} = (y+i)x - x|x|^2$, $\dot{y} = \mu(2aS(y/a-a) - |x|)$, where $x = x_1 + ix_2 \in \mathbb{C}$, $y \in \mathbb{R}$, $\mu = 0.05$, $a = \sqrt{\mu/20}$ and $S(\rho) = 1/(1+e^{-\rho})$.

Elliptic bursting is observable in many systems, including the celebrated Hodgkin-Huxley model, but a canonical model has yet to be found. This poses an interesting problem when there is a subcritical Andronov-Hopf or quasi-static saddle-node bifurcation, since the canonical models must combine local properties near the bifurcation with global properties at the large amplitude stable limit cycle. Moreover, the models should exhibit a double limit cycle bifurcation in order to reflect adequately the transition from periodic firing to a quiescent state. A local canonical model for a weakly connected network of elliptic bursters near multiple quasi-static Andronov-Hopf bifurcation is presented in Section 6.6.

2.9.5 Other Types of Bursting

Many types and subtypes of bursting are summarized by Wang and Rinzel (1995) and Bertram et al. (1995). The latter called square-wave, parabolic and elliptic bursters as type I, II and III bursters, respectively. Their list is necessarily incomplete, since there is no complete classification of spiking mechanisms, due to the infinite variety of bifurcations that a neuron can undergo. Besides, there could be mixed types of bursting if y visits different areas corresponding to different bifurcations in the fast dynamics. For example, the two shaded areas in Figure 2.55a may correspond to different classes of excitability. Then the bursting would consist of a mixture of two types. Moreover, the limit cycle $y(t)$ depicted in the figure may enter one shaded area corresponding to, say, Class 1 excitability and exit another area corresponding to Class 2 excitability. Then each burst starts as parabolic and ends up as elliptic, as we illustrate in Figure 2.60 using the Wilson-Cowan model.

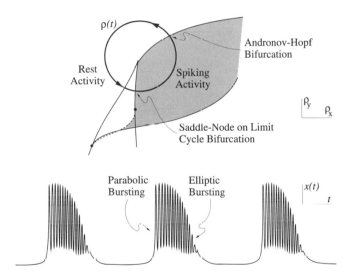

Figure 2.60. Hybrid type of bursting in the Wilson-Cowan model ($a = b = c - 10$, $d = -2$) when $\rho = (\rho_x, \rho_y)$ has a limit cycle attractor entering the shaded area via a saddle-node on limit cycle bifurcation and exiting via supercritical Andronov-Hopf bifurcation.

We see that the problem of bursting classification is challenging. Let us estimate the number of various burstings when the fast dynamics undergoes codimension 1 bifurcations. There are two codimension 1 bifurcations of equilibria – saddle-node and Andronov-Hopf. The saddle-node bifurcation can be on or off a limit cycle, and the Andronov-Hopf bifurcation can be subcritical or supercritical. Therefore, there are four different scenarios on how the rest activity can disappear:

- Saddle-node bifurcation *on* limit cycle.
- Saddle-node bifurcation *off* limit cycle.
- Subcritical Andronov-Hopf.
- Supercritical Andronov-Hopf.

Among many codimension 1 bifurcations of a limit cycle attractor there are four that may lead to disappearance of periodic activity:

- Saddle-node on limit cycle.
- Supercritical Andronov-Hopf.
- Double limit cycle.
- Saddle separatrix loop.

Therefore, there are at least 16 various types of burstings; see Figure 2.61. Some of them have names, others do not. Since there could be various kinds

Bifurcation of Limit Cycle / Bifurcation of Equilibrium	Saddle-Node on Limit Cycle	Supercritical Andronov-Hopf	Double Limit Cycle	Saddle Separatrix Loop
Saddle-Node on Limit Cycle	*Parabolic Bursting*			
Saddle-Node off Limit Cycle				*Square-Wave Bursting*
Subcritical Andronov-Hopf		*Elliptic Bursting*	*Elliptic Bursting*	
Supercritical Andronov-Hopf		*Elliptic Bursting*	*Elliptic Bursting*	

Figure 2.61. Various types of burstings.

of a saddle separatrix loop bifurcation (see Figures 2.34b and 2.37), some of the bursting types in Figure 2.61 may have subtypes. If we take into account that there are at least four types of dynamics for the slow variable:

- Limit cycle attractor for any x.
- Limit cycle attractor for the averaged system.
- Hysteresis.
- Excitability.

then the number of various types of burstings increases. Obviously, it would be difficult to give a useful name for each of them.

Finally, notice that a burster may have complicated dynamics, e.g. it can be chaotic. Quite often, however, it is periodic with a limit cycle attractor as in Figure 2.62. In this case, we can treat a burster as an oscillator with a

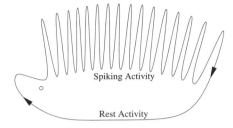

Figure 2.62. Limit cycle attractor corresponding to bursting oscillations.

periodic orbit in x, y-space having two regimes corresponding to quiescent and spiking activity. Such an approach is especially useful for bursters that cannot be described by the singularly perturbed system (2.29) for small μ. This has certain advantage in our study of WCNNs, since all the results that we present in Chapter 9 for weakly connected oscillators can also be applied to weakly connected networks of such bursters.

3

Neural Networks

3.1 Nonhyperbolic Neural Networks

Consider a neuron with its membrane potential near a threshold value. When an external input drives the potential to the threshold, the neuron's activity experiences a bifurcation: The equilibrium corresponding to the rest potential loses stability or disappears, and the neuron fires. This bifurcation is local, but it results in a nonlocal event – the generation of an action potential, or spike. These are observable global phenomena. To model them requires knowledge about global features of neuron dynamics, which usually is not available. However, to predict the onset of such phenomena, one needs only local information about the behavior of the neuron near the rest potential. Thus, one can obtain some global information about behavior of a system by performing local analysis. Our nonhyperbolic neural network approach uses this observation.

For illustration, consider Class 1 neural excitability: When a neuron is near a bifurcation point of saddle-node type, its activity is described locally by the canonical model

$$\dot{x} = a + x^2 , \tag{3.1}$$

where x is the neuron activity and a is some external input. When the input is inhibitory ($a < 0$) (see Figure 3.1), the equilibrium $x = -\sqrt{|a|}$ corresponding to the rest potential is asymptotically stable. If the initial state of the system is close to the equilibrium, the dynamics converge to it. Such a neuron is hyperpolarized and silent. When the input is excitatory, the rest potential and the threshold potential coalesce (when $a = 0$) and

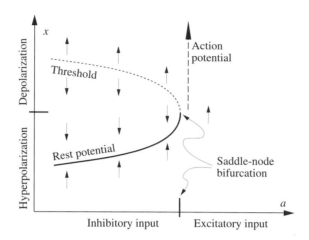

Figure 3.1. Nonhyperbolic neuron. Small changes of input (parameter a) may result in a saddle-node bifurcation. Sudden increasing of activity $x(t)$ corresponds here to generation of the action potential.

disappear (when $a > 0$). The activity $x(t)$ increases, and the neuron is depolarized and generates an action potential or spike.

As we mentioned, the action potential is a global phenomenon that is caused by a local event – a bifurcation at the equilibrium. This is possible because the equilibrium is nearly nonhyperbolic. After the activity $x(t)$ leaves a small neighborhood of the nonhyperbolic equilibrium, the neuron's dynamics cannot be described by the canonical model (3.1). To describe $x(t)$ far away from the nonhyperbolic equilibrium, we take into account global information about its behavior. In contrast, we need only local analysis in order to predict the generation of action potentials by neurons.

One can say that a neuron modeled by a nearly nonhyperbolic dynamical system (3.1) can perform a trivial pattern-recognition task: It can discriminate between excitatory ($a > 0$) and inhibitory ($a < 0$) inputs, no matter how small the amplitude of the input $|a|$ is. The neuron's reaction is simple – it either generates an action potential or remains inactive.

Now consider a network of such nonhyperbolic neurons. Suppose the input is inhibitory and strong. Then all neurons are hyperpolarized, and activity of the network converges to some global equilibrium. The network is silent. Next suppose the input is shifted slowly towards excitation until the global equilibrium loses its stability. The dynamics of the network might produce some macroscopic changes. For example, some of the neurons can generate action potentials while the others remain silent. The active neurons can send their signals to other parts of the network and trigger various behavioral responses, such as fight or flight. This active response of the neural network depends on how the equilibrium loses its stability, which in turn depends on the nature of the input and the connections between the neurons. We cannot say anything about the subsequent activity of the neurons that generate the action potentials: They can continue to fire or return to a rest state. Nevertheless, local analysis near the nonhyperbolic equilibrium can determine which neurons are to fire. This knowledge alone can be helpful for understanding many qualitative features of the network dynamics.

We see that the global behavior of the network depends crucially on local processes that take place when an equilibrium becomes nonhyperbolic. We call such networks *nonhyperbolic neural networks*. They give us examples of how even local information about the brain's dynamic behavior near a bifurcation point can be useful for understanding its neurocomputational properties.

3.1.1   The Olfactory Bulb

The Olfactory Bulb (OB) is a well-studied brain system. The neurons of OB receive signals from olfactory receptors, process information, and send it on to other parts of the brain, in particular to the piriform cortex. Anatomically, the OB consists of neural oscillators (see Figure 3.2). The processing

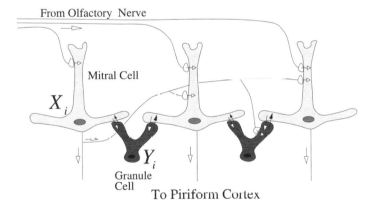

Figure 3.2. Schematic representation of the Olfactory Bulb (OB) anatomy. Activity of the mitral cells is denoted by X_i and of the granule cells by Y_i.

of information by the OB has been studied experimentally (see, for example, Shepherd 1976, 1983 and Skarda and Freeman 1987) and using mathematical models (see, for example, Baird 1986; Erdi et al. 1993; Izhikevich and Malinetskii 1992, 1993; Kerszberg and Masson 1995; and Li and Hopfield 1989).

It is believed that in the absence of signals from olfactory receptors the OB's activity is chaotic and of low amplitude. W. Freeman called this

background (chaotic) activity the dormant state. The signals from receptors make the chaotic activity coherent and periodic. Each inhaled odor
excites its own pattern of spatial oscillations: Local populations of neurons
in various parts of the OB oscillate with similar phases but with various
amplitudes. One may refer to such activity as amplitude modulation (AM)
of the local field potential of the OB, in contrast to frequency modulation
(FM), which we encounter in Section 5.4.5.

From a mathematical point of view, each odor is represented by a limit
cycle in the OB's phase space. Studying these limit cycles requires global information about the OB's dynamics, but studying how the chaotic attractor
corresponding to a dormant state changes requires only information about
local behavior near the attractor.

We hypothesize that it is possible to predict to which limit cycle the
OB's activity will be attracted simply by studying local bifurcations of the
attractor. If this is the case, the future of the OB dynamics is determined
by local events. This is the spirit of our approach using nonhyperbolic NNs.

Unfortunately, it is difficult to study chaotic attractors. To simplify analysis we may assume that when input from receptors is absent, the attractor
corresponding to the dormant state is an equilibrium point. In this case one
can think of low-amplitude chaos as being an irregular perturbation of an
equilibrium. The limitation of this approach is that we neglect possible
roles of deterministic chaos in information processing of the OB.

3.1.2 Generalizations

We study neural networks near equilibria (silent neurons) or near limit
cycles and tori (pacemaker neurons). Nevertheless, the notion of nonhyperbolic neural networks could be generalized to include the networks having
more complicated dynamics, such as chaotic.

The network's global activity depends on local events near a nonhyperbolic equilibrium because the equilibrium is a point of convergence of separatrices – the sets that separate attraction domains of various attractors.
Thus, separatrices bound regions having different qualitative dynamics.
Local dynamics near separatrices determines to which attractor the global
activity converges. Separatrices can have complicated geometry, and so can
the set of points of their accumulation; see Figure 3.3 for an illustration.

Since the set of accumulation points of separatrices is invariant, the vector of system activity can stay near it for a long period of time. This might
serve various purposes. For example, the system can process temporal information encoded in an input signal. The sets can play a certain role in
high level mental processes such as cognition and consciousness.

Complicated unstable invariant sets often give evidence of complicated
dynamics. Even though a system has trivial attractors, such as stable equilibria, convergence to the attractors can be prolonged and erratic. Indeed,
if the initial state of the system is in the invariant set, then its activity

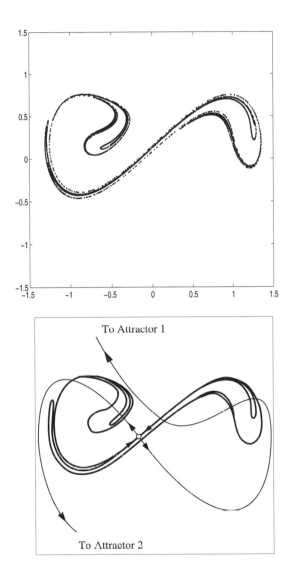

Figure 3.3. *Above*: $10,000$ iterations of the Poincaré mapping for the Duffing equation $\dot{u} = v$, $\dot{v} = u - u^3 - \delta v + \gamma \cos \omega t$ for $\delta = 0.25, \gamma = 0.3$, and $\omega = 1$ converge to a chaotic attractor that has fractal geometry, see Guckenheimer and Holmes (1983). *Below*: The attractor is a folded separatrix (stable manifold) for the Duffing equation when $t \to -t$.

never converges to the attractors. It wanders throughout the set and might even look chaotic. If the initial state is close to the invariant set or if there are external inputs that perturb the set, then the activity stays close to the set for a certain period of time and then leaves it, converging to an attractor. Which attractor is chosen depends on the initial state and the perturbations. An experimenter observes a transient erratic behavior followed by a regular motion. This transient chaos could play essential roles in information processing, for example, in the OB. Nevertheless, we disregard this mostly because there is a deficit of ideas of how it might be used in neurosystems and because there are no suitable mathematical techniques to analyze it. Information processing by transient chaos is a promising direction for future research in both computational neurosciences and dynamical systems.

3.2 Neural Network Types

Some believe that our memories correspond to (metastable) attractors in our brain's phase space, which is huge, since the human brain has more than 10^{11} neurons. Convergence to an appropriate attractor is called recognition. There are many neural network (NN) models that can mimic recognition by association processes.

In general, an NN is a network of interconnected simple neuron-like elements that performs computational tasks, such as memorization and recognition by association, on a given input (key) pattern. The pattern may be temporal or spatial or both. In the NN types discussed here, the input pattern depends upon time slowly; i.e., input is quasi-static.

Most of the NNs dealing with static input patterns can be divided into the following groups according to the way in which the pattern is presented:

MA type (*Multiple Attractor NN*) The input pattern is given as an initial state of the network, and the network converges to one of possibly many choices (Hopfield 1982, Grossberg 1988).

GAS type (*Globally Asymptotically Stable NN*) The key pattern is given as a parameter that controls the location and shape of a unique attractor (Hirsch 1989).

NH type (*Nonhyperbolic NN*) The input pattern is given as bifurcation parameter that perturbs a nonhyperbolic equilibrium.

In the first two cases the NN must converge to an equilibrium, although there have been many attempts to understand the role of limit cycles and chaotic attractors (Babloyantz and Lourenco 1994; Baird 1986; Basar 1990; Destexhe and Babloyantz 1991; Eckhorn et al. 1988, Gray 1994, Hoppensteadt 1989, 1986, Ishii et al. 1996, Izhikevich and Malinetskii 1992,1993;

Kazanovich and Borisyuk 1994; Skarda and Freeman 1987; Stollenwerk and Pasemann 1996; Tsuda 1992; see more references in review by Elbert et al. 1994). In the case of nonhyperbolic neural networks, only a local analysis near an equilibrium need be performed, and hence the type of attractor to which the activity converges is of secondary interest.

3.2.1   Multiple Attractor NN

Each MA-type network attractor corresponds to a prior memorized image. Each input pattern, considered as an initial state, lies in a domain of attraction of one of the attractors (see Figure 3.4a). A "good" MA-type NN

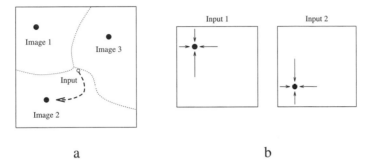

a   b

Figure 3.4. a. Multiple attractor (MA) neural network. b. Globally asymptotically stable (GAS) neural network.

is one that satisfies the following:

- It can memorize many images.

- It does not have spurious (or false) memory, i.e., attractors that do not correspond to any of the previously memorized images.

- All its attractors have "large" attraction domains.

A typical example of an MA-type NN is the Hopfield network (Hopfield 1982), but it is not a "good" network in this sense since it allows spurious memory. Some, though, argue that spurious memory reflects the creative role of neural networks in information processing (Caianiello et al. 1992; Ezhov and Vvedensky 1996).

3.2.2   Globally Asymptotically Stable NN

The GAS-type NN has only one attractor, whose behavior depends upon the input pattern as a parameter (see Figure 3.4b). Various inputs can place the attractor at various locations in the network's phase space. Learning in such an NN consists of adjusting the connections between the neurons so that the network can realize the mapping

$$\left\{ \begin{array}{c} \text{Set of patterns} \\ \text{to be memorized} \end{array} \right\} \;\rightarrow\; \left\{ \begin{array}{c} \text{Set of locations} \\ \text{of the attractors} \end{array} \right\}.$$

A "good" NN of this type should memorize such a mapping, so that if the input pattern to be recognized is near (in some metric) to a previously memorized pattern, then the resulting attractor is in the desired location.

3.2.3   Nonhyperbolic NN

The NH-type NN has a nonhyperbolic equilibrium that loses stability or disappears. Depending upon the external input that perturbs the equilibrium, dynamics can converge to various attractors or to other nonhyperbolic equilibria (see Figure 3.5). We discussed this NN type earlier.

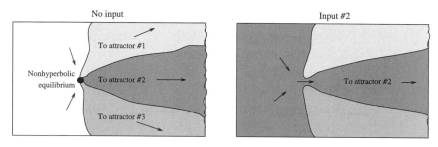

Figure 3.5. A basic principles of nonhyperbolic neural networks. The input is given as a parameter that perturbs the nonhyperbolic equilibrium. Local bifurcations of the equilibrium affect global dynamics of the network.

The MA and GAS NN types are widely studied and used. The NH-type is relatively neglected, and it interests us most here, since it motivates our local analysis of WCNNs. We presented the former NN types for the sake of completeness, but we will show in subsequent chapters that the canonical models that we derive for WCNNs can work as either MA or GAS types. The canonical models can behave like both MA- and GAS-type NNs simultaneously when an input stimulus is ambiguous.

4
Introduction to Canonical Models

Theoretical studies of the human brain require mathematical models. Unfortunately, mathematical analysis of brain models is of limited value, since the results can depend on particulars of an underlying model; that is, various models of the same brain structure could produce different results. This could discourage biologists from using mathematics and/or mathematicians. A reasonable way to circumvent this problem is to derive results that are largely independent of the model and that can be observed in a broad class of models. For example, if one modifies a model by adding more parameters and variables, similar results should hold.

It is essential to study families of models instead of one model. To carry out this task we develop an approach that reduces a family of models to a canonical model that has in some sense a universal character. Briefly, a model is canonical if there is a continuous change of variables that transforms any other model from the class into this one. The change of variables need not be invertible, meaning that we might lose information in return for generality. Such an approach (employing a simplifying change of variables) is not new to applications in the life sciences; for example, this can be found in Thom's use of normal forms and universal unfolding theories (Thom 1975).

In this chapter we provide precise definitions of canonical models and consider some methods for deriving them. In particular, we use normal form and invariant manifold theories. Using the former, one can simplify an equation without reducing its dimension, while using the latter, one can reduce its dimension without simplifying the equation.

4.1 Introduction

The process of deriving canonical models is rather an art than a science. A general algorithm for doing so is not known, although we would like to mention a "mystical" observation that has proved to be helpful: Mathematics works pretty well for problems having small parameters. Small parameters arise naturally when we study critical regimes, such as bifurcations in neuron or brain dynamics, extreme values of connections or external inputs, or lesions and neurological disorders.

4.1.1 Families of Dynamical Systems

Studying families of dynamical systems is not new in mathematical neuroscience. For example, the celebrated Cohen-Grossberg (1983) theorem analyzes the families of dynamical system of the form

$$\dot{x}_i = a_i(x) \left[b_i(x_i) - \sum_{j=1}^{n} c_{ij} d_j(x_j) \right] \tag{4.1}$$

that they found in studies of neural networks. Under some mild restrictions on the functions a_i, b_i, and d_i, system (4.1) admits a global Liapunov function of the form

$$U(x) = - \sum_{i=1}^{n} \int^{x_i} b_i(y_i) d_i'(y_i) \, dy_i - \frac{1}{2} \sum_{i,j=1}^{n} c_{ij} d_i(x_i) d_j(x_j)$$

if the synaptic matrix $C = (c_{ij})$ is symmetric. The existence of such a global Liapunov function could imply convergence of dynamics to equilibria (Cohen and Grossberg 1983). In this case, the whole family of dynamical systems of the form (4.1) with symmetric C has the same property, and this has some neurocomputational applications (Grossberg 1988).

In this book we adopt another approach to studying properties of families of dynamical systems. But first let us discuss how families of dynamical systems can arise.

We assume that brain dynamics can be described by a system of differential equations of the form

$$\dot{x} = F(x), \qquad x \in \mathcal{X}, \tag{4.2}$$

where $\mathcal{X}$ is the space of appropriate variables and F is some function. If we knew the correct space and function, then we could take (4.2) to be the mathematical model of the brain. Unfortunately, we do not have much information about (4.2).

The standard approach to this problem is to invent a simpler dynamical system

$$\dot{x} = F_1(x), \qquad x \in \mathcal{X}_1,$$

that mimics or illustrates some features of the brain, which we refer to as *empirical models*; see Section 1.1. There are two major drawbacks of such an approach: These models are far away from the reality, and the results obtained for them could depend on the model.

To cope with these drawbacks we incorporate more and more data into the model. For example, we could take into account all ions, channels, and pumps in a neuron's membrane. This leads to another (larger) dynamical system

$$\dot{x} = F_2(x) \,, \qquad x \in \mathcal{X}_2$$

that is more complex that the first one, but still does not reflect all peculiarities of the brain. Even if we construct a *comprehensive* model that takes into consideration all known data, it would fall behind as new facts are discovered every day. Besides, there could be some properties that are thought to be true now but will be shown to be invalid in the future.

The process of refinement is unending. It produces a family of dynamical systems $\mathcal{F} = \{F_k \mid k = 1, 2, \dots\}$. However, it might not be true that each $\dot{x} = F_k(x)$ reproduces all the results exhibited by $\dot{x} = F_{k-1}(x)$, especially when some "indisputable" facts are shown to be wrong. Moreover, the family $\mathcal{F}$ might not be countable. Nevertheless, there could be some results that are reproducible by all of the dynamical systems in $\mathcal{F}$. Such results do not depend on the mathematical model describing the brain, but they are rather a property of the brain, or more precisely, $\mathcal{F}$. How can they be studied? We return to this issue in a moment.

Another way to obtain a family $\mathcal{F}$ of models is to postulate it: Let $\mathcal{F}$ denote all dynamical systems describing completely the human brain. Obviously, such an $\mathcal{F}$ is given implicitly. Of course, there are philosophical issues left out, such as why $\mathcal{F}$ is not empty and what "describing completely" means.

Instead of studying each member of $\mathcal{F}$, we adopt the following approach: Suppose there is a dynamical system

$$\dot{y} = G(y) \,, \qquad y \in \mathcal{Y} \,,$$

such that any member of $\mathcal{F}$ can be converted to this system by a continuous change of variables. We call such a dynamical system a *canonical model* for the family $\mathcal{F}$. Of course, one does not have to actually determine the change of variables, especially when $\mathcal{F}$ is defined implicitly. It suffices to prove that such a change of variables exists. Then, studying the canonical model gives us some information about the dynamic behavior of *every* member of $\mathcal{F}$ simultaneously.

We define canonical models next. To do so we introduce the idea of an observation and a model. Our definitions are universal and can be useful outside the field of mathematical neuroscience.

4.1.2 Observations

Suppose a dynamical system of the form

$$\dot{x} = F(x) , \qquad x \in \mathcal{X} ,$$

describes a phenomenon. First, in order to study the phenomenon we must observe it. An *observation* of $x(t)$ is a function h defined on $\mathcal{X}$. The variable $y(t) \equiv h(x(t))$ is an *observable*. These notions play important role in control theory.

Let us provide some examples. The system above could describe the dynamics of molecules of a gas in a container. Then one of the observations could be the mass of the gas. Such an observation is given by a function $m : \mathcal{X} \to \mathbb{R}_+$ that is the sum of the masses of all the molecules. Another observation could be the temperature of the gas given by a function $T : \mathcal{X} \to \mathbb{R}$ that is essentially the average speed of the molecules.

Brain activity characteristics such as EEG (electroencephalogram), LFP (local field potentials), even neuron membrane potentials, are all observables too. For example, the EEG and LFP are functions that average activities of neurons in a vicinity of the electrode. If $\mathcal{X}$ denotes the phase space of the brain, then any function defined on $\mathcal{X}$ is an observation, though it might be difficult to interpret it in ordinary language.

4.1.3 Models

A major requirement for a model is that it be consistent with observations. We state this mathematically as, A dynamical system

$$\dot{x} = F(x) , \qquad x \in \mathcal{X} , \tag{4.3}$$

has

$$\dot{y} = G(y) , \qquad y \in \mathcal{Y} , \tag{4.4}$$

as a *model* if there is a continuous function (observation) $h : \mathcal{X} \to \mathcal{Y}$ such that if $x(t)$ is a solution of (4.3), then $y(t) = h(x(t))$ is a solution of (4.4) (see Figure 4.1). In this case, (4.4) is a factor of (4.3), and (4.3) is a covering of (4.4). Many properties of factors of dynamical systems are studied in ergodic theory (Ornstein 1974).

When h is a homeomorphism (continuous with continuous inverse), the dynamical systems (4.3) and (4.4) are said to be *conjugate*. Thus, the definition of a model is a generalization of the definition of conjugacy of dynamical systems, where we do not require h to be invertible.

We frequently relax the definition of a model and require only that the observation h map solutions of (4.3) to solutions of (4.4), preserving the sense but not necessarily the parametrization by time. That is, (4.4) is a model of (4.3) when the following holds: If $y(0) = h(x(0))$, then for every t there is t_1 such that $y(t_1) = h(x(t))$. We may use this definition, for

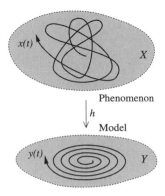

Figure 4.1. Dynamical system at bottom is a model of the dynamical system portrayed at the top if there is a continuous function h such that $y(t) = h(x(t))$.

instance, when we study periodic solutions of (4.3) and its model (4.4) and we are not interested in their periods. When h is a homeomorphism, dynamical systems (4.3) and (4.4) are *topologically equivalent*.

Not every observation can be a solution of a dynamical system. For example, when (4.3) has a periodic solution and $h : \mathcal{X} \to \mathbb{R}$, then the observation $h(x(t))$ cannot be a solution of a one-dimensional system unless h maps the limit cycle to a point. It is usually a difficult problem in applications to find a continuous function $h : \mathcal{X} \to \mathcal{Y}$ that maps solutions of (4.3) to those of (4.4).

It is natural to require that the function h map $\mathcal{X}$ onto $\mathcal{Y}$. We call such models *surjective*. These models do not have artifact, that is, behavior that is not inherited from the original system (4.3). Indeed, suppose there is an initial condition $y(0) \in \mathcal{Y}$ for which the activity of (4.4) is unusual. If the function h is surjective, we know that there is an initial condition $x(0) \in \mathcal{X}$ in the original system that produces the same or even more unusual behavior. If h is not surjective, then it is possible that $y(0) \notin h(\mathcal{X})$; that is, the behavior of the nonsurjective model is irrelevant to the original system (4.3). Since the set $h(\mathcal{X}) \subseteq \mathcal{Y}$ is invariant, any model can be converted to the surjective one by restricting $\mathcal{Y} = h(\mathcal{X})$.

It is usually assumed that models are simpler than the phenomenon they describe. This corresponds to the case when the space $\mathcal{Y}$ is "smaller" than $\mathcal{X}$ and the function h is many-to-one. In this circumstance we lose information about the phenomenon. Indeed, for any initial state $y(0) \in \mathcal{Y}$ there are many initial states $x_\alpha(0) \in \mathcal{X}$ such that $y(t) = h(x_{\alpha_1}(t)) = h(x_{\alpha_2}(t))$ but $x_{\alpha_1}(t)$ and $x_{\alpha_2}(t)$ can produce different qualitative behavior. Thus a model will not necessary reflect all peculiarities of a phenomenon. Rather, it concentrates on extracting some particularly useful dynamical features.

We see that each model is associated with a continuous function h : $\mathcal{X} \to \mathcal{Y}$, which we call an observation. When h is many-to-one, the space $\mathcal{X}$ is a disjoint union of equivalence classes corresponding to each preimage $h^{-1}(y), y \in \mathcal{Y}$. Instead of saying that the model loses information about the original phenomenon, we say that our model is insensitive to dynamics within an equivalence class $h^{-1}(y)$ and that it captures properties of (4.3) that are "transversal" to the partitioning

$$\mathcal{X} = \cup_{y \in \mathcal{Y}} h^{-1}(y) \ .$$

See Figure 4.2.

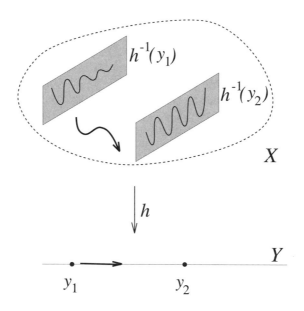

Figure 4.2. The model is simpler than the original phenomenon. It does not reflect behavior along equivalence classes $h^{-1}(y)$. It captures only behavior that is transversal to $h^{-1}(y)$ (bold curve).

There can be many models of the same phenomenon, depending on which properties are considered to be important. Moreover, a model of a model is also a model, etc. Models have some obvious properties:

- (Reflexivity) Any system $\dot{x} = F(x)$ is a model of itself. Indeed, take $\mathcal{Y} = \mathcal{X}$ and $h = \mathrm{id}$.

- (Transitivity) If P is a model of Q, and Q is a model of F, then P is a model of F.

- (Antisymmetry) If G is a model of F, and F is a model of G, then F and G are *weakly isomorphic*.

Weakly isomorphic models have similar properties, and we do not distinguish between them. We see that the set of models is a *partially ordered set*. In this book we do not study the set-theoretical properties of models.

It is easy to see that any dynamical system has a trivial model: $\dot{y} = 0$ for $y \in \mathcal{Y}$, where $\mathcal{Y}$ consists of a single point.

Let W be an open subset of $\mathcal{X}$. Then (4.4) is a *local model* of (4.3) if the observation h is defined (locally) on W. Obviously, a model of a local model is a local model. A typical example of a local model of a dynamical system is its linearization in a neighborhood of a hyperbolic equilibrium; see the Hartman-Grobman theorem (Theorem 2.1).

4.1.4 Canonical Models

Now suppose there is a family $\mathcal{F}$ of dynamical systems

$$\dot{x} = F_\alpha(x), \qquad x \in \mathcal{X}_\alpha, \qquad \alpha \in \mathcal{A},$$

describing the brain. If (4.4) is a model for every such system, then we call it canonical. Thus the dynamical system (4.4) is a *canonical model* for the family $\mathcal{F}$ of dynamical systems if for every member $F_\alpha \in \mathcal{F}$ there is a continuous observation $h_\alpha : \mathcal{X}_\alpha \to \mathcal{Y}$ such that solutions of $\dot{x} = F_\alpha(x)$ are mapped to the solutions of (4.4); see Figure 4.3. Studying the canonical

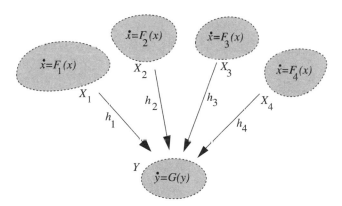

Figure 4.3. Dynamical system $\dot{y} = G(y)$, $y \in \mathcal{Y}$, is a canonical model for the family $\mathcal{F} = \{F_1, F_2, F_3, F_4\}$ of dynamical systems $\dot{x} = F_i(x)$, $x \in \mathcal{X}_i$.

model provides us with information about the behavior of every member of the family $\mathcal{F}$. For example, if the canonical model does not have an equilibrium, then neither has any member of $\mathcal{F}$, no matter how large the family is.

According to the definition, *every* dynamical system $\dot{x} = F(x)$ is a canonical model for some family, e.g., for $\mathcal{F} = \{F\}$ or for $\mathcal{F}$ consisting of all systems that are topologically equivalent to F. Thus, knowing that a model is canonical for a family $\mathcal{F}$ does not make it valuable unless the family is large. Such families arise naturally in mathematical neuroscience due to the fact that we do not have detailed information about brain dynamics and we are forced to consider general systems with few restrictions.

4.1.5   Example: Linear Systems

Suppose we are given a system of linear ordinary differential equations of the form

$$\dot{x} = Ax, \qquad x \in \mathbb{R}^m, \tag{4.5}$$

where A is a nonzero matrix. Let $\lambda \in \mathbb{R}$ be a real eigenvalue of A of multiplicity one and $v \in \mathbb{R}^m$ its corresponding unit eigenvector. The cases when A has no eigenvalues of multiplicity one or it has only complex eigenvalues can be considered similarly. Let us check that the ordinary differential equation

$$\dot{y} = \lambda y, \qquad y \in \mathbb{R}, \tag{4.6}$$

is a model of (4.5). Indeed, let $E \subset \mathbb{R}^m$ be the eigensubspace spanned by v and let $h : \mathbb{R}^m \to E$ be the canonical projector on E. It is easy to see that $hA = \lambda h$. Let us define $y(t) = hx(t)$; then

$$\dot{y} = h\dot{x} = hAx = \lambda hx = \lambda y .$$

That is, (4.6) is a model for (4.5).

When $\lambda > 0$, let us write $z = y^{1/\lambda}$. Then

$$\dot{z} = \frac{1}{\lambda} y^{\frac{1}{\lambda}-1} \dot{y} = \frac{1}{\lambda} y^{\frac{1}{\lambda}-1} \lambda y = y^{\frac{1}{\lambda}} = z .$$

So the dynamical system

$$\dot{z} = z$$

is a model for (4.6) and hence for (4.5). This example demonstrates the property that the dynamics of the linear system (4.5) grow exponentially (at least along direction of v), and it neglects dynamics along the other directions. If $\lambda < 0$, then the model is $\dot{z} = -z$, reflecting exponential decay.

The dynamical system $\dot{z} = z$ depends on neither the eigenvalue λ nor the eigenvector v of A. It does not even depend on the dimension of (4.5). The only property that we used to derive this model is that (4.5) is linear and has a positive eigenvalue of multiplicity one. Therefore, this is a model for the whole family of linear systems having the property stated above. We refer to such models as being canonical.

4.1.6 Example: The VCON

Consider a dynamical system

$$\dot{x} = F(x) , \qquad x \in \mathbb{S}^1 , \qquad (4.7)$$

where $\mathbb{S}^1$ is the unit circle. One can treat the phase variable $x \in \mathbb{S}^1$ as an angle that parametrizes $\mathbb{S}^1 = \{e^{ix}\} \subset \mathbb{C}$. Suppose that the function F satisfies

$$F(x) > 0 \quad \text{for all } x \neq 0 \qquad \text{and} \qquad F(0) = 0 .$$

This implies that $F'(0) = 0$ and $F''(0) > 0$. Such a system has a saddle-node equilibrium $x = 0$ on a limit cycle $\mathbb{S}^1$.

Let us show that the voltage controlled oscillator neuron (VCON) model

$$\dot{\theta} = 1 - \cos\theta , \qquad \theta \in \mathbb{S}^1 ,$$

is a canonical model for a family of dynamical systems of the form (4.7) having a saddle-node equilibrium on a limit cycle. First, note that the VCON model belongs to the family, since it also has the saddle-node equilibrium at $\theta = 0$, and $1 - \cos\theta > 0$ for all $\theta \neq 0$.

Let us construct an observation $h : \mathbb{S}^1 \to \mathbb{S}^1$ that maps solutions of (4.7) to those of the VCON model as follows: Fix a point $x_0 \neq 0$ and consider the solution of (4.7), $x(t)$, such that $x(0) = x_0$. Since $F(x) > 0$, the solution $x(t)$ increases and approaches equilibrium $2\pi \equiv 0$ as $t \to +\infty$. Let $\theta(t)$ be a solution of the VCON model such that $\theta(0) = x_0$. It also increases and approaches the equilibrium. Then, we define h on the semicircle $[x_0, 2\pi]$ as a transformation that maps $x(t)$ to $\theta(t)$ for every fixed t; see Figure 4.4.

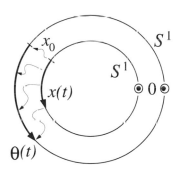

Figure 4.4. The solution $x(t)$ of (4.7) is mapped to the solution $\theta(t)$ of the VCON model.

Now consider the solutions $x(t)$ and $\theta(t)$ for negative t. Both solutions decrease and approach $0 \equiv 2\pi$ as $t \to -\infty$. Again, we define the function h on the semicircle $[0, x_0]$ as the mapping of $x(t)$ to $\theta(t)$ for every $t \le 0$. It is easy to see that such a function is well-defined and continuous, since both $x(t)$ and $\theta(t)$ approach 0 for $t \to \pm\infty$. This fact might not be valid when we consider perturbations of (4.7), as we show in Chapter 8.

According to the way we constructed the function h, it satisfies $\theta(t) = h(x(t))$ for every t if $\theta(0) = h(x(0))$. Therefore, the VCON model is a canonical model for (4.7). The considerations above can be applied to any pair of dynamical systems of the form (4.7) to show that one of them is a canonical model of the other. The VCON is the simplest one in the sense that 1 and $\cos\theta$ are the first two terms of the Fourier series of F.

4.1.7 Example: Phase Oscillators

Consider a family of dynamical systems of the form

$$\dot{x} = F(x)\,, \qquad x \in \mathbb{R}^m\,, \tag{4.8}$$

having an exponentially stable limit cycle $\gamma \subset \mathbb{R}^m$ with the period $T = 2\pi$. Let us show that the dynamical system

$$\dot{\theta} = 1\,, \qquad \theta \in \mathbb{S}^1\,, \tag{4.9}$$

is a local canonical model for all such oscillators. For this we find a neighborhood W of γ and a function $h : W \to \mathbb{S}^1$ such that $\theta(t) - h(x(t))$ is a solution of (4.9).

Let $x(0) \in \gamma$ be a point on the limit cycle. Consider all initial conditions $z(0) \in \mathbb{R}^m$ such that the distance between $z(t)$ and $x(t)$ approaches zero as $t \to \infty$. Winfree (1974) calls the set

$$M_{x(0)} = \{z(0) \in \mathbb{R}^m \mid \|x(t) - z(t)\| \to 0 \text{ as } t \to \infty\}$$

of such points the *isochron* of $x(0)$. Isochrons play crucial roles in the qualitative analysis of oscillators, since the eventual behavior of points on the same isochron look almost the same. Guckenheimer (1975) pointed out the relationship between isochrons and stable manifolds, and proved Winfree's conjecture that there is a neighborhood of a limit cycle that is foliated by its isochrons.

Let W be an open tubular neighborhood of γ so small that each isochron intersects the frontier of W transversally (see Figure 4.5). Since the limit cycle is stable, we can choose W such that if $z(0) \in W$, then $z(t) \in W$ for all $t > 0$. The neighborhood W is invariantly foliated by the isochrons M_x, $x \in \gamma$, in the sense that the flow maps isochrons to isochrons. Let $h_1 : W \to \gamma$ be a function sending $M_x \cap W$ to $x \in \gamma$. It is easy to see that if $z(0) \in M_{x(0)} \cap W$, then $x(t) = h_1(z(t))$ for all $t \geq 0$. Therefore, the dynamical system

$$\dot{y} = F(y), \qquad y \in \gamma\,, \tag{4.10}$$

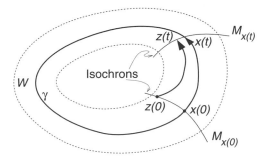

Figure 4.5. A small neighborhood W of a limit cycle γ is invariantly fibered by isochrons. If $x(0)$ and $z(0)$ lie on the isochron $M_{x(0)}$, then $x(t)$ and $z(t)$ lie on the isochron $M_{x(t)}$.

is a model of the original system, but it differs from the original system in that the variable y is defined only on γ. The consideration above can be applied (with some modifications) to many invariant manifolds, not necessarily circles. We do this later in this chapter where we prove that restrictions to the normally hyperbolic invariant manifolds are local models.

Fix $y(0) \in \gamma$ and consider the solution $y(t)$ of the above system. Define a function $h_2 : \gamma \to \mathbb{S}^1$ by $h_2(y(\theta)) = \theta$ for all $\theta \in [0, 2\pi)$. This function is the inverse of $y : \mathbb{S}^1 \to \gamma$. Obviously, h_2 is well-defined and continuous. It associates to each point on the limit γ, parametrized by $y(\theta)$, $\theta \in [0, 2\pi)$, its phase θ. Indeed, for any point $x \in \gamma$, there is precisely one value of $\theta \in [0, 2\pi)$ such that $x = y(\theta)$. Such a value θ is the phase deviation between the solutions starting from $y(0)$ and x.

Now consider equation (4.9). Since $\theta(t) = t = h_2(y(t))$, we have $\theta(t) = h(x(t))$ for any initial condition $x(0) \in \gamma$, where $\theta(0) = h_2(x(0))$, and hence equation (4.9) is a model of (4.10). If we define $h : W \to \mathbb{S}^1$ to be the composition of h_1 and h_2, $h = h_2 \circ h_1$, then equation (4.9) is a model of the original system. Since it does not depend on F or γ, it is a canonical model for a family of dynamical systems having a hyperbolic 2π-periodic solution. If the period $T \neq 2\pi$, then the canonical model is

$$\dot{\theta} = \omega \, ,$$

where $\omega = 2\pi/T$ is the frequency. If we consider the relaxed definition of the model (that is, if we are not interested in the parametrization by time), then (4.9) is a local canonical model for any system having a hyperbolic limit cycle, no matter what the period is.

Notice that system (4.10) is a local canonical model, since the observation of it (h) is defined in some neighborhood W of γ. We cannot define it globally with a continuous function, since there are points of accumulation of isochrons (Guckenheimer 1975).

In this book we derive canonical models that are less simple than the phase oscillator (4.9), but the example above illustrates a few important points:

- (Simplicity) The canonical model is simpler than any dynamical system of the form (4.8) having a limit cycle attractor.

- (Universality) The canonical model captures some essentials of every member of the family.

- (Importance for neuroscience) Derivation of the canonical model is possible even for those F that are not given explicitly. The only requirement is that (4.8) have a limit cycle attractor. One does not need to know the peculiarities of the dynamics of (4.8).

- (Limitations) The canonical model has restricted value: It provides information about local behavior of (4.8) near the limit cycle attractor, but it does not tell anything about global behavior of (4.8); including the transients.

A powerful method of deriving canonical models is based on normal form theory, which we discuss next.

4.2   Normal Forms

Consider a smooth dynamical system

$$\dot{X} = F(X), \qquad X = (X_1, \ldots, X_m) \in \mathbb{R}^m , \qquad (4.11)$$

for which $X = 0$ is an equilibrium; that is, $F(0) = 0$. Let $\lambda_1, \ldots, \lambda_m$ be the eigenvalues of the Jacobian matrix $L = DF$ at $X = 0$. Without loss of generality we may assume that L is given in Jordan normal form. For the sake of simplicity we assume that the eigenvalues are distinct, so that L is diagonal. Each integer-valued relation of the form

$$\lambda_i = n_1 \lambda_1 + \cdots + n_m \lambda_m , \qquad (4.12)$$

where $n_1, \ldots, n_m$ are nonnegative integers and $\sum n_j \geq 2$, is called a *resonance*. With each resonance we can associate a *resonant monomial*

$$v_i X_1^{n_1} \cdots X_m^{n_m} ,$$

where $v_i \in \mathbb{R}^m$ is the ith eigenvector of L corresponding to λ_i.

The *Poincaré-Dulac* theorem (Arnold 1982) asserts that there is a near identity change of variables

$$X = Y + P(Y) , \qquad P(Y) = 0 , \ DP(0) = 0,$$

that transforms the dynamical system (4.11) to

$$\dot{Y} = LY + W(Y), \tag{4.13}$$

where the nonlinear vector-function W consists of only resonant monomials. Such a system is called a *normal form*. In particular, if there are no resonances, then the dynamical system can be transformed to its linearization $\dot{Y} = LY$. We see that it is possible to find a change of variables that removes all nonresonant monomials and transforms the nonlinear system into a simpler form – a property desirable for mathematical neuroscience applications.

Below we outline the idea of the proof of the Poincaré-Dulac theorem. Let us write the dynamical system (4.11) in the form

$$\dot{X} = F(X) = LX + V(X),$$

where the vector-valued function $V(X) = F(X) - LX$ denotes the nonlinear terms of $F(X)$. Let $X = Y + P(Y)$ be the near identity transformation, where P is specified below. Differentiating it with respect to t yields

$$\dot{X} = (I + DP(Y))\dot{Y},$$

where I is the identity $n \times n$ matrix. On the other hand,

$$\dot{X} = F(Y + P(Y)) = LY + V(Y) + LP(Y) + \dots,$$

where $\dots$ denotes the higher-order terms in Y. Since

$$(I + DP(Y))^{-1} = I - DP(Y) + \dots,$$

we have

$$
\begin{aligned}
\dot{Y} &= (I - DP(Y) + \dots)(LY + V(Y) + LP(Y) + \dots) \\
&= LY + V(Y) + LP(Y) - DP(Y)LY + \dots \\
&= LY + V(Y) + \mathrm{ad}_L P + \dots, \tag{4.14}
\end{aligned}
$$

where

$$\mathrm{ad}_Q P = [Q, P] = DQ\,P - DP\,Q$$

is the *Poisson bracket* operation, which is linear in P for fixed Q. In (4.14) $Q = LY$ is linear. If we denote by H_k the linear space of all monomials of degree k, then the Poisson bracket ad_L is a linear operator from H_k to H_k. As with any other linear operator, it has a kernel, $\ker \mathrm{ad}_L$, and an image, $\mathrm{ad}_L H_k$.

When L is diagonal with eigenvalues $\lambda_1, \dots, \lambda_m$, the ith component of the Poisson bracket can be written in the form

$$(\mathrm{ad}_L P)_i = \lambda_i P_i(Y) - \sum_{j=1}^{m} \frac{\partial P_i}{\partial Y_j} \lambda_j Y_j.$$

It is easy to check that the kernel and the image of such a linear operator are spanned by resonant and nonresonant monomials, respectively. Indeed, let $P_i(Y) = Y_1^{n_1} \cdots Y_m^{n_m}$; then

$$(\mathrm{ad}_L P_i)_i = \lambda_i P_i - \sum_{j=1}^{m} n_j \lambda_j P_i = (\lambda_i - \sum_{j=1}^{m} n_j \lambda_j) P_i ,$$

which is zero if there is a resonance (4.12) and nonzero otherwise. We see that when L is diagonal, the linear space of all nonlinear monomials $H = H_2 + H_3 + \cdots$ is the direct sum of the kernel and the image, that is there is a splitting $H = \ker \mathrm{ad}_L \oplus \mathrm{ad}_L H$. The part of $V(Y)$ in (4.14) that lies in $\mathrm{ad}_L H$ can be removed by an appropriate change of variables. This suggests the definition for the normal form:

A dynamical system $\dot{Y} = LY + W(Y)$ is said to be in *normal form* if

$$W(Y) \cap \mathrm{ad}_L H = \{0\} ,$$

that is, the nonlinear part W lies in the complement of $\mathrm{ad}_L H$. This definition also applies to general (not necessarily diagonal) matrices L.

The following procedure is employed to convert a dynamical system $\dot{Y} = LY + V(Y)$ to its normal form: The linear space of monomials of degree k, H_k, is represented as a sum of the image of ad_L and its complement G_k; that is, $H_k = G_k \oplus \mathrm{ad}_L H_k$. Then a suitable choice of $P_k(Y) \in H_k$ can remove the part of $V(Y)$ that lies in the image $\mathrm{ad}_L H_k$. First, $\mathrm{ad}_L H_2$ is removed. For this we write $V(Y) = W_2(Y) + \tilde{V}_2(Y) + V_3(Y)$, where $W_2(Y) \in G_2$, $\tilde{V}_2(Y) \in \mathrm{ad}_L H_2$, and $V_3(Y)$ denotes terms of order 3 and higher. Then there is a monomial $P_2(Y) \in H_2$ satisfying

$$\tilde{V}_2(Y) + \mathrm{ad}_L P_2 = 0 ,$$

so that the dynamical system has the form $\dot{Y} = LY + W_2(Y) + V_3(Y)$. Writing $V_3 = W_3 + \tilde{V}_3 + V_4$ and repeating the procedure above transforms the system to $\dot{Y} = LY + W_2(Y) + W_3(Y) + V_4(Y)$. After $k - 1$ such steps the dynamical system has the form

$$\dot{Y} = LY + W_2(Y) + \cdots + W_k(Y) + \mathcal{O}(Y^{k+1}) , \qquad (4.15)$$

where each $W_i(Y)$ consists only of resonant monomials of degree i. More details on this procedure can be found in Arnold (1982), Sanders and Verhulst (1985), and Guckenheimer and Holmes (1983).

When we discard the nonlinear terms of order $k + 1$, we obtain the *truncated normal form*. The truncated normal form is called the *topological normal form* when it is topologically conjugate to any system of the form (4.15). That is, adding, changing, or removing any term of order $k + 1$ or higher does not change the local phase portrait of the system. This is usually hard to prove in applications.

4.2.1   Example: Normal Form for Andronov-Hopf Bifurcation

Let us illustrate the theory outlined above. In this book we study oscillators near an Andronov-Hopf bifurcation point. In this case the dynamical system can be written in the form

$$\begin{pmatrix} \dot{x} \\ \dot{y} \end{pmatrix} = \begin{pmatrix} 0 & -\Omega \\ \Omega & 0 \end{pmatrix} \begin{pmatrix} x \\ y \end{pmatrix} + \begin{pmatrix} f(x,y) \\ g(x,y) \end{pmatrix}.$$

The eigenvalues of the Jacobian matrix at the origin are $\pm i\Omega \in \mathbb{C}$, and the corresponding eigenvectors are $(1, \mp i)^\top \in \mathbb{C}^2$. The linear change of variables

$$\begin{pmatrix} x \\ y \end{pmatrix} = \begin{pmatrix} 1 \\ -i \end{pmatrix} z + \begin{pmatrix} 1 \\ i \end{pmatrix} \bar{z}, \qquad z \in \mathbb{C},$$

transforms the dynamical system into another one having diagonal linear part:

$$\begin{cases} \dot{z} &= +i\Omega z + q(z, \bar{z}) \\ \dot{\bar{z}} &= -i\Omega \bar{z} + \bar{q}(z, \bar{z}) \end{cases},$$

where q is a nonlinear function. The Jacobian matrix

$$\begin{pmatrix} +i\Omega & 0 \\ 0 & -i\Omega \end{pmatrix}$$

is diagonal. Its eigenvectors are $v_1 = (1, 0)^\top \in \mathbb{C}^2$ and $v_2 = (0, 1)^\top \in \mathbb{C}^2$.

It is easy to see that there is an infinite number of resonances

$$i\Omega = (k+1)i\Omega + k(-i\Omega) \quad \text{and} \quad -i\Omega = ki\Omega + (k+1)(-i\Omega),$$

where $k = 1, 2, \ldots$. Therefore, the resonant monomials have the form

$$\begin{pmatrix} 1 \\ 0 \end{pmatrix} z^{k+1}\bar{z}^k \quad \text{and} \quad \begin{pmatrix} 0 \\ 1 \end{pmatrix} z^k \bar{z}^{k+1}, \qquad k = 1, 2, \ldots$$

The Poincaré-Dulac theorem guarantees the existence of a change of variables that removes all nonresonant monomials, so that in the new variables the dynamical system has the form

$$\begin{pmatrix} \dot{\xi} \\ \dot{\bar{\xi}} \end{pmatrix} = \begin{pmatrix} i\Omega\xi \\ -i\Omega\bar{\xi} \end{pmatrix} + d \begin{pmatrix} 1 \\ 0 \end{pmatrix} \xi^2\bar{\xi} + \bar{d} \begin{pmatrix} 0 \\ 1 \end{pmatrix} \xi\bar{\xi}^2 + \mathcal{O}(|\xi|^5),$$

where $d \in \mathbb{C}$ is a constant that depends on the function q. It is convenient to rewrite the above system as

$$\begin{cases} \dot{\xi} &= +i\Omega\xi + d\xi|\xi|^2 + \mathcal{O}(|\xi|^5) \\ \dot{\bar{\xi}} &= -i\Omega\bar{\xi} + \bar{d}\bar{\xi}|\xi|^2 + \mathcal{O}(|\xi|^5) \end{cases}$$

and study only the first equation, since the second one is its complex conjugate.

The real part of the coefficient d indicates whether the bifurcation is subcritical ($\operatorname{Re} d > 0$) or supercritical ($\operatorname{Re} d < 0$). This can be determined by evaluating the expression

$$
\begin{aligned}
\operatorname{Re} d \;=\; & \frac{1}{16}\left(f_{xxx} + f_{xyy} + g_{xxy} + g_{yyy}\right) + \frac{1}{16\Omega}\left\{f_{xy}\left(f_{xx} + f_{yy}\right)\right. \\
& \left. - g_{xy}\left(g_{xx} + g_{yy}\right) - f_{xx}\, g_{xx} + f_{yy}\, g_{yy}\right\}
\end{aligned}
$$

(Guckenheimer and Holmes 1983, p. 155).

4.2.2 Normal Forms Are Canonical Models

Let us fix the set of resonances of the form (4.12) and consider a family $\mathcal{F}$ of all dynamical systems having those resonances. For any system $F \in \mathcal{F}$ there is a change of variables h putting F into the form (4.13) with the same set of resonant monomials having possibly different coefficients. Therefore, the normal form is a canonical model for the family $\mathcal{F}$. Since the change of variables is usually local, the normal form is a local canonical model.

4.3 Normally Hyperbolic Invariant Manifolds

In this book we use extensively the fact that restrictions to some invariant manifolds are local models. For example, when we study a saddle-node bifurcation on a limit cycle, we first reduce a multidimensional case to an invariant circle and then utilize other techniques to simplify the dynamical system further. Similarly, restrictions to the limit cycle attractor allow us to simplify analysis of the dynamics of oscillatory neurons significantly and obtain phase equations. In both cases we use the fact that the restricted system is a model of the original multidimensional system. In this section we prove this fact for flows. Its generalization for mappings is straightforward.

4.3.1 Invariant Manifolds

The theory of invariant manifolds is at the border between smooth dynamical system theory and differential topology. Since we cannot afford to provide all definitions and properties of all results we use in this book, we refer the reader to other textbooks on this subject. We find particularly useful and readable the following books: Chillingworth (1976) *Differential Topology With a View to Applications*, Ruelle (1989) *Elements of Differentiable Dynamics and Bifurcation Theory*, Wiggins (1994) *Normally Hyperbolic Invariant Manifolds in Dynamical Systems*, and the article Fenichel (1971). Instead of providing strict definitions from differential topology, we rather illustrate them below.

Consider a dynamical system

$$\dot{X} = F(X), \qquad X \in \mathbb{R}^m,$$

and suppose there is a set $M \subset \mathbb{R}^m$ that is invariant for the solutions of the system. That is, if $X(0) \in M$, then $X(t) \in M$ for all $t > 0$. It is frequently the case that the invariant set M is a manifold. Even though we discuss a general theory of invariant manifolds below, we apply the theory in this book only to circles $\mathbb{S}^1$ and tori $\mathbb{T}^n$. Thus, whenever we say that M is a manifold, the reader may think of M as being an n-torus $\mathbb{T}^n = \mathbb{S}^1 \times \cdots \times \mathbb{S}^1$, where $n = 1$ is not excluded. An example of an invariant manifold is a limit cycle attractor $\gamma \subset \mathbb{R}^m$, which is homeomorphic to the circle $\mathbb{S}^1$. An equilibrium for $\dot{X} = F(X)$ and the whole space $\mathbb{R}^m$ are also invariant manifolds. Equilibria, circles, and tori are compact manifolds; the space $\mathbb{R}^m$ is obviously not.

4.3.2 Normal Hyperbolicity

First, we give an intuitive illustration of normal hyperbolicity; then we provide strict definitions. A stable invariant manifold M is said to be *normally hyperbolic* if contractions of vectors orthogonal to the manifold are sharper than those along the manifold. When the manifold is an equilibrium, this definition coincides with the usual definition of hyperbolicity: The Jacobian matrix does not have eigenvalues with zero real part. When the manifold is a limit cycle, then it is normally hyperbolic when its Floquet multipliers are not on the unit circle. Such a limit cycle is said to be hyperbolic.

For more general manifolds (tori, spheres, etc.) normal hyperbolicity can be characterized in terms of generalized Liapunov-type numbers. Instead of providing strict definitions, we rather illustrate geometrically the notion of normal hyperbolicity in Figure 4.6a, which we adapted from Fenichel's

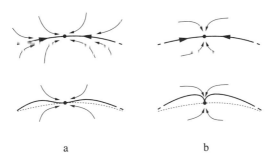

a   b

Figure 4.6. a. Normally hyperbolic invariant manifold (bold curve) and its perturbation. b. The invariant manifold is not normally hyperbolic. A cusp may develop under perturbations.

paper (1971). Since the normal rate of approach to the normally hyperbolic stable invariant manifold is greater than the tangential one, the flow around the manifold is tangent to it. This smoothes out any perturbations, even those near equilibria on the manifold. On the contrary, some perturbations cannot be smoothed when the manifold is not normally hyperbolic, as we depict in Figure 4.6b.

Formal Definition

Next, we study the dynamics of $\dot{X} = F(X)$ in a neighborhood of a compact invariant manifold M. There are two linear mutually orthogonal spaces associated with each point $X \in M$: the *tangent space*, denoted by $T_X M$, and the *normal space*, denoted by $N_X M$. The tangent space can be thought of as the collection of all lines tangent to M at X; see the illustration in Figure 4.7.

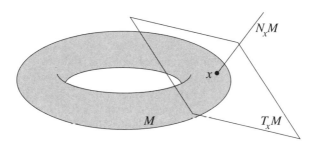

Figure 4.7. Tangent $T_x M$ and normal $N_x M$ linear subspaces to manifold M.

Let

$$\Pi : T_X M \rightarrow N_X M$$

be the orthogonal projection to the normal subspace $N_X M$. To define the notion of contractions of vectors at M, we consider the linear part, $D\Phi_t(X)$, of the flow $\Phi_t(X)$ at the invariant manifold M. Let

$$
\begin{aligned}
v(t) &= D\Phi_t(X)v(0) , & v(0) &\in T_X M , \\
w(t) &= \Pi D\Phi_t(X)w(0) , & w(0) &\in N_X M .
\end{aligned}
$$

The invariant manifold is *stable* when

$$\lim_{t \to \infty} |w(t)| = 0 \qquad \text{(stability)}$$

for all $X \in M$ and all vectors $w(0) \in N_X M$. It is said to be *normally hyperbolic* if

$$\lim_{t \to \infty} \frac{|w(t)|}{|v(t)|} = 0 \qquad \text{(normal hyperbolicity)}$$

for all $X \in M$ and all nonzero vectors $w \in N_X M$ and $v \in T_X M$. This means that the rate of normal contraction to the manifold is larger than the tangential one; see the illustration in Figure 4.8.

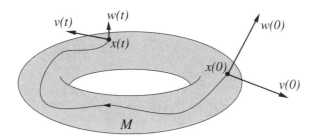

Figure 4.8. Normal contraction to the manifold is sharper than the tangential one.

Persistence

All normally hyperbolic smooth compact invariant manifolds possess a useful property: *They are persistent under perturbations* (Fenichel 1971 and Hirsch et al. 1977).

Theorem 4.1 (Persistence of Normally Hyperbolic Invariant Manifolds) *Consider a dynamical system $\dot{X} = F(X)$, $X \in \mathbb{R}^m$, having an attractive normally hyperbolic compact invariant manifold $M \subset \mathbb{R}^m$. Then the perturbed dynamical system*

$$\dot{X} = F(X) + \varepsilon G(X, \varepsilon), \qquad X \in \mathbb{R}^m, \tag{4.16}$$

has an attractive normally hyperbolic compact invariant manifold $M_\varepsilon \subset \mathbb{R}^m$, which is an ε-perturbation of M.

The proof of this result is not reproduced here.

The perturbed invariant manifold M_ε has normal and tangent subspaces that are ε-close to those of M. Thus, not only the manifold itself, but all other topological objects associated with it are persistent. Moreover, normal hyperbolicity is a necessary and sufficient condition for such persistence.

It can be difficult to verify normal hyperbolicity in applications. In this book we study dynamics of general neural networks having invariant circles and tori. We do not analyze their normal hyperbolicity; we simply assume that they *are* normally hyperbolic.

4.3.3   Stable Submanifolds

Let us generalize the notion of the isochron of a limit cycle attractor that we used in Section 4.1.7 for an arbitrary stable normally hyperbolic invariant manifold M.

Let W be a small open neighborhood of M. In analogy with isochrons, one is tempted to associate with each point $X \in M$ the following set:

$$\{Y \in W \mid \lim_{t \to \infty} |\Phi_t(X) - \Phi_t(Y)| = 0\} \ .$$

Even though this definition works well for limit cycle attractors, it fails when the dynamical system has nontrivial dynamics on M. Indeed, consider as an example a saddle-node bifurcation on a limit cycle. The invariant manifold $M \cong \mathbb{S}^1$ can contain a stable equilibrium, say $p \in M$. Then all $\Phi_t(X) \to p$, and therefore the stable submanifold V_X of every point is the whole W. Obviously, such a definition is not useful in this case. Nevertheless, it provides good intuition about what is needed to generalize the notion of isochron.

Since M is stable, any solution $\Phi_t(Y)$ starting sufficiently close to M approaches the manifold. Fix an initial point $X \in M$ and consider the set of all initial conditions Y from an open neighborhood W of M that approach $X(t) = \Phi_t(X)$ faster than any other solution $Z(t) = \Phi_t(Z)$, $Z \neq X$, starting on M; see Figure 4.9. This set is referred to as a *stable submanifold*

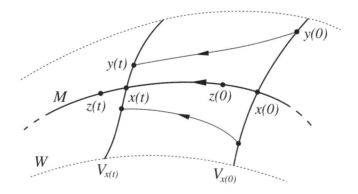

Figure 4.9. To accompany the definition of stable submanifold V_X, $X \in M$.

of $X \in M$ and is denoted by

$$V_X = \left\{Y \in W \mid \text{if } Z \in M, Z \neq X, \text{ then } \lim_{t \to \infty} \frac{|\Phi_t(X) - \Phi_t(Y)|}{|\Phi_t(X) - \Phi_t(Z)|} = 0\right\}.$$

Each stable submanifold V_X of the normally hyperbolic compact invariant manifold M has the same dimension as the normal subspace $N_X M$. When

the invariant manifold M is a limit cycle attractor, the stable submanifold
of each point is its isochron, as one expects.

4.3.4 Invariant Foliation

Although many facts have been proved for stable normally hyperbolic
compact invariant manifolds, we use here only the one stating that *a
small neighborhood W of M is continuously foliated by stable submanifolds*
V_X, $X \in M$; see Figure 4.10. That is,

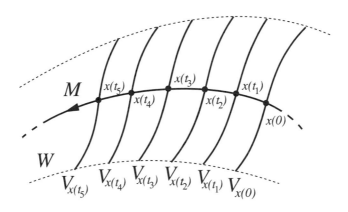

Figure 4.10. A small neighborhood W of normally the hyperbolic invariant manifold M is invariantly foliated by the submanifolds V_X, $X \in M$.

$$W = \bigcup_{X \in M} V_X \quad \text{and} \quad V_X \cap V_Y = \emptyset \text{ for } Y \neq X,$$

where $\emptyset$ is the empty set. The foliation is invariant in the following sense: If
$Y(0) \in V_{X(0)}$, then $Y(t) \in V_{X(t)}$, as depicted in Figure 4.9. Thus, the flow
$\Phi_t(\mathbb{R}^m)$ of the dynamical system maps stable submanifolds to themselves;
that is, $\Phi_t(V_X) \subset V_{\Phi_t(X)}$. The foliation depends continuously on $X \in M$,
but it may or may not be smooth.

4.3.5 Asymptotic Phase

Let us generalize the notion of phase for the case of general M. In analogy
with the limit cycle attractor, we assume that there is a parametrization of
the compact invariant manifold M, and the phase of each point X on the
manifold M is defined to be X. Thus, when $M \cong \mathbb{S}^1$, the phase X is the

angle that parametrizes $\mathbb{S}^1$. When $M \cong \mathbb{T}^n$, the phase is the n-dimensional vector of angles; etc.

Now consider the mapping $h : W \to M$ defined as follows: $h(Y) = X$ if and only if $Y \in V_X$. Thus, h sends each leaf V_X of the foliation of W to $X \in M$. Since the foliation is continuous, the mapping h is continuous. Since the foliation is invariant, h satisfies $h(\Phi_t(X)) = \Phi_t(h(X))$ for all $X \in W$ and all $t \geq 0$. Such a mapping is said to be the *asymptotic phase*. With this definition we prove the following result:

Theorem 4.2 (Restrictions to Invariant Manifolds Are Local Models) *The restriction of a dynamical system*

$$\dot{X} = F(X), \qquad X \in \mathbb{R}^m ,$$

to an attractive normally hyperbolic compact invariant manifold M,

$$\dot{y} = F(y), \qquad y \in M ,$$

is a local model. That is, there is a local continuous mapping $h : W \to M$ defined in a neighborhood W of M that maps solutions of the former system to those of the latter.

Proof. Take h to be the asymptotic phase. $\square$

It is still an open question whether the result is valid if we drop the assumption of stability and/or normal hyperbolicity. A basis for optimism is that the theorem requires only a continuous invariant foliation of a small neighborhood of M. Such a foliation exists in many cases, including those when M is not even hyperbolic.

Due to the persistence of normally hyperbolic manifolds, we can prove the following result:

Theorem 4.3 (Restrictions to Perturbed Invariant Manifolds Are Local Models) *Suppose a dynamical system*

$$\dot{X} = F(X), \qquad X \in \mathbb{R}^m ,$$

has an attractive normally hyperbolic compact invariant manifold $M \subset \mathbb{R}^m$. Then its ε-perturbation ($\varepsilon \ll 1$)

$$\dot{X} = F(X) + \varepsilon G(X, \varepsilon), \qquad X \in \mathbb{R}^m , \qquad (4.17)$$

has an attractive normally hyperbolic compact invariant manifold M_ε in a small neighborhood of M. Let $W \subset \mathbb{R}^m$ be a small open neighborhood containing both M and M_ε.

- *The dynamical system (4.17) has a local model*

$$\dot{y} = F(y) + \varepsilon G(y, \varepsilon), \qquad y \in M_\varepsilon . \qquad (4.18)$$

  That is, there is a local transformation $h_\varepsilon : W \to M_\varepsilon$ that maps solutions of (4.17) to those of (4.18).

- *The dynamical system (4.17) has another local model, which is defined on the unperturbed invariant manifold M:*

$$\dot{x} = F(x) + \varepsilon \tilde{G}(x, \varepsilon), \qquad x \in M , \qquad (4.19)$$

where

$$\tilde{G}(x, 0) = G(x, 0) + \mathrm{ad}_F\, P(x, 0) , \quad \text{where } \mathrm{ad}_F\, P = DF\, P - DP\, F$$

and the vector-valued function P can be determined from the condition

$$G(x, 0) + \mathrm{ad}_F\, P(x, 0) \in T_x M \qquad \text{for all } x \in M.$$

See Figure 4.11 for a summary.

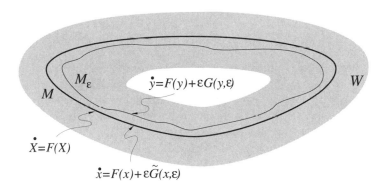

Figure 4.11. To accompany Theorem 4.3.

Proof. Since M is normally hyperbolic, it is persistent under small perturbations (Theorem 4.1). Therefore, (4.17) has a stable normally hyperbolic compact invariant manifold M_ε that is ε-close to M in some metric. Now we apply Theorem 4.2 to conclude that (4.17) has a local model (4.18).

Now let us prove that (4.19) is a local model too. First, notice that each point $y \in M_\varepsilon$ can be represented as

$$y = x + \varepsilon P(x, \varepsilon) , \qquad x \in M , \qquad (4.20)$$

where the small function $\varepsilon P(x, \varepsilon)$ denotes the ε-deviation from M. The representation (4.20) can be made unique if we require that x be the projection of y onto M. In this case the vector-function $\varepsilon P(x, \varepsilon)$ is orthogonal to M at each x.

Let us find the equations for $\dot{x}$. For this we differentiate (4.20) with respect to t to obtain

$$\dot{y} = (I + \varepsilon DP(x, \varepsilon))\dot{x} ,$$

where I is the identity matrix, and $DP(x, \varepsilon)$ denotes the matrix of partial derivatives of P with respect to $X \in \mathbb{R}^m$. It is easy to check that

$$(I + \varepsilon DP(x, \varepsilon))^{-1} = I - \varepsilon DP(x, 0) + \mathcal{O}(\varepsilon^2) .$$

It is also easy to see that

$$
\begin{aligned}
F(y) + \varepsilon G(y, \varepsilon) &= F(x + \varepsilon P(x, \varepsilon)) + \varepsilon G(x + \varepsilon P(x, \varepsilon), \varepsilon) \\
&= F(x) + \varepsilon \left(DF(x) P(x, 0) + G(x, 0) \right) + \mathcal{O}(\varepsilon^2) .
\end{aligned}
$$

Therefore,

$$
\begin{aligned}
\dot{x} &= \left(I - \varepsilon DP(x, 0) + \mathcal{O}(\varepsilon^2) \right) \dot{y} \\
&= F(x) + \varepsilon \left(DF(x) P(x, 0) - DP(x, 0) F(x) + G(x, 0) \right) + \mathcal{O}(\varepsilon^2) \\
&= F(x) + \varepsilon \left(\mathrm{ad}_F P + G(x, 0) \right) + \mathcal{O}(\varepsilon^2) ,
\end{aligned}
\tag{4.21}
$$

where

$$\mathrm{ad}_F P = [F, P] = DF(x) P(x, 0) - DP(x, 0) F(x)$$

is the *Poisson bracket* of F and P, which is a linear operator for fixed F. Our manipulations are similar to those in the normal form theory (Section 4.2) so far. In the normal form theory, the vector-function P is determined that removes nonresonant monomials. In our case, P is given by (4.20) due to the existence and uniqueness of the perturbed invariant manifold M_ε. If we relax the condition that the vector-function P be orthogonal to M, then the function P is not unique.

It is difficult to find such $P(x, \varepsilon)$ in applications. A reasonable way to avoid this problem is to determine $P(x, 0)$ from the condition

$$\mathrm{ad}_F P + G \subset TM \qquad \text{on } M ,
\tag{4.22}$$

where TM is the tangent bundle of M. That is, the vector $\mathrm{ad}_F P(x, 0) + G(x, 0)$ must be tangent to the manifold M at each $x \in M$. This guarantees that x does not leave M, at least up to order ε^2. Condition (4.22) points out the relationship between the invariant manifold reduction and the Lie algebras, which we do not study in this book. We just notice that the solution P to (4.22) is not unique. For example, $P + aF$ is a solution too, since $\mathrm{ad}_F aF = 0$ for any constant a. We could try to find P satisfying (4.22) that transforms the function $\tilde{G} = \mathrm{ad}_F P + G$ to the "simplest" form. This points out the relationship between the invariant manifold and the normal form reductions, which we do not study in this book.

To prove that (4.19) is a local model of (4.17), we should present a function $\tilde{h}_\varepsilon : W \to M$ that maps solutions of the latter system to those of the former. For this, notice that (4.20) is a nearly identity transformation, and therefore it is nonsingular and has an inverse

$$x = y + \varepsilon Q(y, \varepsilon) , \qquad y \in M_\varepsilon ,$$

for some function Q. Then, we take the function $\tilde{h}_\varepsilon : W \to M$ to be the composition of $h_\varepsilon : W \to M_\varepsilon$ defined earlier and the nearly identity transformation from M_ε to M defined above. $\square$

Notice that systems (4.18) and (4.19) are conjugate (topologically equivalent), since there is a homeomorphism (4.20) that transforms one of them to the other. Each system has its own advantages and drawbacks: The system (4.18) has the same form as (4.17), but the state variable y is defined on the perturbed invariant manifold M_ε, which is difficult to find in applications. The state variable x of the system (4.19) is defined on the unperturbed invariant manifold M, but the function $\tilde{G}$ has yet to be determined, which is also a challenging problem. When M is a torus, we can use the Malkin theorem (Section 9.6) to determine $\tilde{G}$.

4.3.6 Center Manifold Restrictions Are Local Models

An analogue of the theorem above can be proved for center manifold restrictions of a dynamical system $\dot{X} = F(X)$, $X \in \mathbb{R}^m$, at a nonhyperbolic equilibrium $X_b \in \mathbb{R}^m$, see the definitions in Section 2.1.1. The center manifold in this case is not compact and is only locally invariant. Nevertheless, one can apply the "bump" function technique (Wiggins 1994, Section 7.2) to modify the center manifold outside a small neighborhood of X_b so that it becomes invariant and normally hyperbolic. Then the theorem above can be used.

Using various other techniques, one can show (Kirchgraber and Palmer 1990) that there is a continuous invariant foliation of a neighborhood of the center manifold similar to the one we discussed above. Then, one can construct a continuous function h that sends each leaf of the foliation to a point on the center manifold. The function possesses the desired properties and can be used to prove that center manifold restrictions are local models. We consider center manifold reductions for weakly connected neural networks in Section 5.2.1.

4.3.7 Direct Product of Dynamical Systems

Consider an uncoupled system

$$\dot{X}_i = F_i(X_i) \,, \qquad X_i \in \mathbb{R}^m, \quad i = 1, \dots, n \,, \qquad (4.23)$$

and suppose that each equation (fixed i) has an attracting normally hyperbolic compact invariant manifold $M_i \subset \mathbb{R}^m$. The dynamical system (4.23) considered for all i simultaneously is the direct product of such equations. It has an invariant manifold M that is the direct product of the M_i. Let

$$\dot{x}_i = F_i(x_i) \,, \qquad x_i \in M_i, \quad i = 1, \dots, n \,, \qquad (4.24)$$

be the restriction of (4.23) to the M.

From Theorem 4.2 it follows that the restriction of each equation in (4.23) onto its invariant manifold M_i is a local model. Let $h_i : W_i \to M_i$ be the local observation satisfying $x_i(t) = h_i(X_i(t))$ for all $t \geq 0$. Then

$$h = (h_1, \ldots, h_n) : W_1 \times \cdots \times W_n \to M$$

is a local observation for the system (4.23). Since the system is uncoupled, it is easy to verify that $x(t) = h(X(t))$ for all $t \geq 0$, where $x(t) = (x_1(t), \ldots, x_n(t)) \in M$ is a solution to (4.24). Therefore, (4.24) is a local model for (4.23). This can be summarized briefly as the following:

Proposition 4.4 *Consider a finite set of dynamical systems each having a model. Then the direct product of the dynamical systems has a model that is the direct product of the models.*

The invariant manifold M of the system (4.23) is attractive and compact, since it is the direct product of a finite number of compact and attractive manifolds M_i. *It is not necessarily normally hyperbolic.* Indeed, the normal hyperbolicity of each M_i guarantees that the rate of normal contractions at each M_i is greater than that of tangential contractions at the same M_i. Both rates may differ substantially for different manifolds, and it could happen that the rate of normal contraction at M_i is smaller than that of tangential contraction at M_j, $j \neq i$.

If the direct product of normally hyperbolic invariant manifolds *is* a normally hyperbolic invariant manifold, then we can apply Theorem 4.2 directly to (4.23) to prove the proposition above. This is the case when each equation in (4.23) has a hyperbolic limit cycle attractor or a saddle-node bifurcation on a stable limit cycle.

Lemma 4.5 *Suppose each equation*

$$\dot{X}_i = F_i(X_i) , \qquad X_i \in \mathbb{R}^m , \quad i = 1, \ldots, n , \qquad (4.25)$$

has a hyperbolic limit cycle attractor $\gamma_i \subset \mathbb{R}^m$. Then (4.25), considered as a system, has a normally hyperbolic invariant manifold $M = \gamma_1 \times \cdots \times \gamma_n \subset \mathbb{R}^{mn}$.

Proof. Let us first show that hyperbolicity of a limit cycle attractor γ_i implies its normal hyperbolicity. Let $X_i(t)$ be the periodic solution of the ith equation in (4.25), and let $A_i(t) = DF_i(X_i(t))$ be the linearization of F_i. Any solution of the equation $\dot{y} = A_i(t)y$ can be represented in the Floquet form $y(t) = P_i(t)e^{B_i t}y(0)$, where $P_i(t)$ is a periodic matrix and B_i is a constant matrix. In particular, $v_i(t) = P_i(t)e^{B_i t}v_i(0)$ and $w_i(t) = \Pi P_i(t)e^{B_i t}w_i(0)$, where vectors $v_i(0) \in T\gamma_i$ and $w_i(0) \in N\gamma_i$, $i = 1, \ldots, n$. Since γ_i is stable and hyperbolic, $m - 1$ Floquet multipliers (eigenvalues of B_i) corresponding to $N\gamma_i$ are inside the unit circle, and one eigenvalue corresponding to $T\gamma_i$ has modulus one. Therefore, $w_i(t) \in N\gamma_i$ vanishes exponentially, and $v_i(t) \in T\gamma_i$ oscillates. Hence

$$\lim_{t\to\infty} \frac{|w_i(t)|}{|v_i(t)|} = 0 \,,$$

meaning that γ_i is normally hyperbolic.

Now consider the system (4.25). Since $M = \gamma_1 \times \cdots \times \gamma_n$, we have $TM = T\gamma_1 \times \cdots \times T\gamma_n$ and $NM = N\gamma_1 \times \cdots \times N\gamma_n$. Therefore, $v \in TM$ can be represented as $v = v_1 + \cdots + v_n$, where $v_i \in T\gamma_i$. Similarly, $w = w_1 + \cdots + w_n$ for $w_i \in N\gamma_i$, $i = 1, \ldots, n$. Since each $w_i(t)$ vanishes exponentially, then so does $w(t)$. Since each $v_i(t)$ oscillates outside some ball, say of radius $\nu > 0$ around the origin, then so does $v(t)$. Therefore,

$$\lim_{t\to\infty} \frac{|w(t)|}{|v(t)|} = 0 \,,$$

which completes the proof. $\square$

We use this result in Chapter 9, where we study weakly connected networks of limit cycle oscillators.

The case of a saddle-node bifurcation on the limit cycle is considered similarly:

Lemma 4.6 *Suppose each equation*

$$\dot{X}_i = F_i(X_i) \,, \qquad X_i \in \mathbb{R}^m \,, \quad i = 1, \ldots, n \,, \tag{4.26}$$

is at the saddle-node bifurcation point $p_i \in \mathbb{R}^m$ on a stable limit cycle γ_i. Then (4.25), considered as a system, has a normally hyperbolic invariant manifold $M - \gamma_1 \times \cdots \times \gamma_n \subset \mathbb{R}^{mn}$.

Proof. The point $p - (p_1, \ldots, p_n) \in \mathbb{R}^{mn}$ is a nonhyperbolic equilibrium of system (4.26). Any solution $X(t)$ starting on M approaches p. Asymptotic behavior of vectors $v(t) \in TM$ and $w(t) \in NM$ in the definition of normal hyperbolicity is determined by the local dynamics at p. Since $v(t)$ approaches the center subspace when $X(t) \to p$, it vanishes at a rate that is slower than exponential. In contrast, $w(t)$ vanishes at an exponential rate, since it is orthogonal to the center subspace. Therefore, the ratio $|w(t)|/|v(t)|$ approaches zero as $t \to \infty$, meaning that M is normally hyperbolic. $\square$

We use the lemma above when we study weakly connected networks of neurons having Class 1 neural excitability. In this case there is a saddle-node bifurcation on a limit cycle, as we depicted in Figure 2.29.

4.3.8 Weakly Connected Systems

Now we are ready to state the major theorem of this section, which we use together with the lemmas above frequently in this book. We treat a weakly connected system as an ε-perturbation of an uncoupled system.

Theorem 4.7 (Invariant Manifold Reduction for Weakly Connected Systems) *Suppose each equation in the uncoupled system ($\varepsilon = 0$)*

$$\dot{X}_i = F_i(X_i)\,, \qquad X_i \in \mathbb{R}^m, \qquad i = 1,\dots,n\,, \qquad (4.27)$$

has a stable normally hyperbolic compact invariant manifold $M_i \subset \mathbb{R}^m$ such that their direct product $M = M_1 \times \cdots \times M_n$ is normally hyperbolic. Then the weakly connected system

$$\dot{X}_i = F_i(X_i) + \varepsilon G_i(X,\varepsilon)\,, \qquad X = (X_1,\dots,X_n) \in \mathbb{R}^{mn}\,, \qquad (4.28)$$

has a local model

$$\dot{x}_i = F_i(x_i) + \varepsilon \tilde{G}_i(x,\varepsilon)\,, \qquad x_i \in M_i\,, \qquad (4.29)$$

where $\tilde{G}_i$ are some functions, $i = 1,\dots,n$. That is, there is an open neighborhood W of M and a continuous function $h_\varepsilon : W \to M$ that maps solutions of (4.28) to those of (4.29).

Generalization of this theorem for weakly connected systems depending on parameters other than ε is straightforward. Studying mutations of the invariant manifold M_ε and its disappearance as ε increases goes beyond the scope of this book.

Proof. The proof follows directly from Theorem 4.3. Indeed, each function $\tilde{G}_i(x,\varepsilon)$ has the form

$$\begin{aligned}
\tilde{G}_i(x,\varepsilon) &= G_i(x,0) + (\mathrm{ad}_F P)_i + \mathcal{O}(\varepsilon) \\
&= G_i(x,0) + DF_i(x_i)\,P_i(x,0) - DP_i(x,0)\,F(x) + \mathcal{O}(\varepsilon)\,.
\end{aligned}$$

Notice that when $G_i(x,0) \notin T_{x_i} M_i$, the Poisson bracket $\mathrm{ad}_F P_i \neq 0$, and hence $\tilde{G}_i \neq G_i$. $\square$

Suppose $G_i(x,\varepsilon)$ does not depend on the variable x_j. In the context of neuroscience, this means that there are no connections from the jth neuron to the ith one. This issue is closely related to the issue of synaptic organization of the brain. Could $\tilde{G}_i(x,\varepsilon)$ depend on x_j? The answer is YES. In the context of neural networks this means that even when there are no explicit synaptic connections between a pair of neurons, there could be implicit interactions between them via other neurons. The invariant manifold reduction captures such interactions. The dependence, though, is of order ε; that is, $\tilde{G}_i(x,0)$ is independent of x_j. To guarantee this we took the function $P(x, 0)$ satisfying

$$DF_i(x_i) P_i(x, 0) - \sum_{j=1}^{n} D_{x_j} P_i(x, 0) F_j(x_j) + G_i(x, 0) \in T_{x_i} M_i$$

to be independent of x_j. We return to this problem in Section 5.2.1, where we study center manifold reductions of weakly connected systems near equilibria.

Part II

Derivation of Canonical Models

5

Local Analysis of WCNNs

The local activity of a weakly connected neural network (WCNN) is described by the dynamical system

$$\dot{X}_i = F_i(X_i, \lambda) + \varepsilon G_i(X_1, \ldots, X_n, \lambda, \rho, \varepsilon), \qquad i = 1, \ldots, n , \qquad (5.1)$$

where $X_i \in \mathbb{R}^m$, $\lambda \in \Lambda$, and $\rho \in \mathcal{R}$, near an equilibrium point. First we use the Hartman-Grobman theorem to show that the network's local activity is not interesting from the neurocomputational point of view unless the equilibrium corresponds to a bifurcation point. In biological terms such neurons are said to be near a threshold. Then we use the center manifold theorem to prove the fundamental theorem of WCNN theory, which says that neurons should be close to thresholds in order to participate nontrivially in brain dynamics. Using center manifold reduction we can substantially reduce the dimension of the network. After that, we apply suitable changes of variables, rescaling or averaging, to simplify further the WCNN. All these reductions yield simple dynamical systems that are canonical models according to the definition given in Chapter 4. The models depend on the type of bifurcation encountered. In this chapter we derive canonical models for multiple saddle-node, cusp, pitchfork, Andronov-Hopf, and Bogdanov-Takens bifurcations. Other bifurcations and critical regimes are considered in subsequent chapters, and we analyze canonical models in the third part of the book.

5.1 Hyperbolic Equilibria

To avoid awkward formulas, we write the WCNN (5.1) in the form

$$\dot{X} = F(X, \lambda) + \varepsilon G(X, \lambda, \rho, \varepsilon) \,, \tag{5.2}$$

where

$$
\begin{array}{rclcl}
X & = & (X_1, \ldots, X_n)^\top & \in \mathbb{R}^{mn} \,, \\
F & = & (F_1, \ldots, F_n)^\top & : \mathbb{R}^{mn} \times \Lambda \to \mathbb{R}^{mn} \,, \\
G & = & (G_1, \ldots, G_n)^\top & : \mathbb{R}^{mn} \times \Lambda \times \mathcal{R} \times \mathbb{R} \to \mathbb{R}^{mn}.
\end{array}
$$

Thus, $X \in \mathbb{R}^{mn}$ denotes activity of the whole network, and $G(X, \lambda, \rho, \varepsilon)$ denotes connections within the network. Note that $F(X, \lambda)$ has diagonal structure; i.e., $F_i = F_i(X_i)$ for $i = 1, \ldots, n$.

In this book we assume that the functions F and G are as smooth as necessary for our computations. In fact, for most of our analysis it suffices to require that F be three times differentiable with continuous third derivative and that G have continuous first derivative.

Suppose (5.2) has an equilibrium point when $\varepsilon = 0$. Without loss of generality, we may assume that the equilibrium is the origin $X = 0$ for $\lambda = 0$; i.e.,

$$F(0, 0) = 0 \,.$$

Let L_i be the Jacobian $m \times m$ matrix of first partial derivatives of the function $F_i(X_i, 0)$ with respect to X_i at the origin,

$$L_i = D_{X_i} F_i = \left(\frac{\partial F_{ik}}{\partial X_{ij}} \right)_{k,j=1,\ldots,m} .$$

When L_i has no eigenvalues with zero real part, the Jacobian matrix L_i is said to be *hyperbolic*. It is easy to see that the Jacobian matrix $L = D_X F(0, 0)$ of the right hand-side of (5.2) for $\varepsilon = 0$ has the form

$$
L = \begin{pmatrix}
L_1 & 0 & \cdots & 0 \\
0 & L_2 & \cdots & 0 \\
\vdots & \vdots & \ddots & \vdots \\
0 & 0 & \cdots & L_n
\end{pmatrix}. \tag{5.3}
$$

Due to the diagonal structure of L, which is inherited from F, the set of eigenvalues of L is the union of those of L_i, $i = 1, \ldots, n$. Therefore, L is hyperbolic if and only if each L_i is hyperbolic.

We say that a neuron is at a hyperbolic equilibrium if the Jacobian matrix is hyperbolic. We say that neuron is *near* hyperbolic equilibrium when the parameter $|\lambda|$ is small.

Theorem 5.1 (Locally Hyperbolic WCNNs Are Not Interesting)

If the activity of each neuron is near a hyperbolic equilibrium, then a weakly connected network (5.2) of such neurons, the uncoupled network

$$\dot{X}_i = F_i(X_i, \lambda) , \qquad i = 1, \ldots, n , \tag{5.4}$$

and the linear system

$$\dot{X} = LX \tag{5.5}$$

have topologically equivalent local flow structures.

The proof of this theorem is standard. It uses the implicit function and Harman-Grobman theorems (see Theorem 2.1) and is a modification of the proof of the fact that hyperbolic dynamical systems are locally structurally stable.

A similar result is also valid when each X_i is from a Banach manifold. In that case, instead of Jacobian matrices we encounter Fréchet derivatives, and along with hyperbolicity we impose some additional technical conditions (for instance, the Fréchet derivatives must have bounded inverse), which are always met in the finite-dimensional case.

The local activity of (5.2) is the direct product of the single neuron dynamics in this case. The entire neural network is, in this sense, no more complex than a single neuron, which has locally linear dynamics. Therefore, a locally hyperbolic WCNN is not interesting as a brain model. If the nonhyperbolic equilibrium is a unique attractor of (5.4), then even global activity of the WCNN (5.2) is not interesting.

It should be pointed out though that we cannot say anything definite about global dynamics of (5.2) without further assumptions. Even when all neurons have only hyperbolic equilibria, there could be some global bifurcations making the network dynamics interesting. But this requires additional knowledge of the neuron dynamical features.

An interesting question arises: How far can we increase the strength of connections ε so that the network activity continues to be simple and uninteresting? A partial answer is given by MacKay and Sepulchre (1995). They show that there is a critical value ε_0, which depends on the peculiarities of the network, such that for all $\varepsilon < \varepsilon_0$ the network behavior is hyperbolic and simple. When ε is close to ε_0, interesting phenomena can occur. For example, one can observe propagating activity fronts, appearances and annihilations of steady localized structures, etc. Since $\varepsilon_0 = \mathcal{O}(1)$, the network is not weakly connected for ε close to ε_0, and studying such a network goes beyond the scope of this book. On the other hand, some of these phenomena can be observed in weakly connected systems near nonhyperbolic equilibria.

Remark. The only equilibrium points that deserve further discussion are nonhyperbolic ones.

As we discussed in Chapter 2, the most important cases when activity is near nonhyperbolic equilibria correspond to bifurcations of neuron behavior. Biologists say that the membrane potential of such neurons is near a threshold value. The rest of this chapter is devoted to networks of such neurons.

5.2   Nonhyperbolic Equilibrium

We consider the unperturbed (uncoupled) system

$$\dot{X}_i = F_i(X_i, \lambda) , \qquad i = 1, \ldots, n ,$$

and compare its behavior with the behavior of the perturbed (coupled) system (5.1). We seek such changes that endow the WCNN (5.1) with nontrivial neurocomputational abilities. In the previous section it was shown that a necessary condition for a weakly connected neural network (5.1) to exhibit local nonlinear behavior is nonhyperbolicity. Next we consider (5.1) near a nonhyperbolic equilibrium. How near will be explained later.

Suppose that the origin $X = 0$ is a nonhyperbolic equilibrium point of the uncoupled system for $\lambda = 0$. This means that the Jacobian matrix L has some eigenvalues with zero real part. Due to the diagonal structure of L, this is possible only if one or more of the Jacobian matrices L_i has eigenvalues with zero real part.

5.2.1   The Center Manifold Reduction

Without loss of generality we may reorder the system and so assume that only the first k equations in (5.1) have a nonhyperbolic Jacobian matrix L_i. Let us represent the phase space $\mathbb{R}^m$ of the ith neuron as a direct sum,

$$\mathbb{R}^m = E_i^{\mathrm{c}} \oplus E_i^{\mathrm{h}} ,$$

where the *center subspace* E_i^{c} is spanned by the eigenvectors of L_i that correspond to eigenvalues with zero real parts, and the *hyperbolic subspace* E_i^{h} is spanned by the other (generalized) eigenvectors; see Section 2.1. We will use the notation $E^{\mathrm{c}} = E_1^{\mathrm{c}} \times \cdots \times E_k^{\mathrm{c}}$.

Recall (from Chapter 4) that a dynamical system $\dot{x} = f(x)$ is a model of $\dot{X} = F(X)$ if there is a continuous function h (the observation) mapping solutions of the latter to those of the former.

Theorem 5.2 (Center Manifold Reduction for Weakly Connected Systems)   *Consider a WCNN*

$$\dot{X}_i = F_i(X_i, \lambda) + \varepsilon G_i(X, \lambda, \rho, \varepsilon), \qquad i = 1, \ldots, n , \qquad (5.6)$$

at the equilibrium $X = 0$ for $\lambda = 0$, and suppose that each of the first k Jacobian matrices

$$L_i = D_{X_i} F_i$$

is nonhyperbolic. Then the WCNN has a local model

$$\dot{x}_i = f_i(x_i, \lambda) + \varepsilon g_i(x, \lambda, \rho, \varepsilon), \qquad i = 1, \dots, k , \qquad (5.7)$$

where each $x_i \in E_i^c$, *the function* g_i *is a projection (up to second order in* x*) of* G_i *to the center subspace* E_i^c, *and*

$$J_i = D_{x_i} f_i = L_i|_{E_i^c} , \qquad i = 1, \dots, k .$$

In particular, all of the eigenvalues of J_i *have zero real part.*

Notice that (5.7) has only k variables, $x_1, \dots, x_k$, modeling activities of the first k "nonhyperbolic" neurons $X_1, \dots, X_k$. Activities of the other $n - k$ "hyperbolic" neurons, $X_{k+1}, \dots, X_n$, do not affect dynamics of (5.7) and hence they are irrelevant.

Suppose each X_i denotes the activity of a single neuron, not a local population of neurons as we discussed in Section 1.4. Then the case of a nonhyperbolic Jacobian matrix L_i corresponds to the activity of a neuron being sensitive to external influences; i.e., the neuron is near its threshold. The fact that only the first k neurons participate nontrivially in (5.7) motivates the following result:

Corollary 5.3 (The Fundamental Theorem of WCNN Theory) *In order to make a nontrivial contribution to brain dynamics, a neuron must be near threshold.*

Remark. The fundamental theorem of WCNN theory is applicable only to resting neurons. It is not applicable to periodically or chaotically spiking neurons.

The theorem is not totally unexpected for some neurobiologists. It corroborates their belief that in the human brain, which is undoubtedly an extremely complex system, neurons must be near thresholds, or otherwise the brain would not be able to cope with its tasks.

Proof. The proof of Theorem 5.2 is technical and involves many tedious details; therefore, it can be omitted on the first reading. To prove the theorem we first show that system (5.7) is a restriction of WCNN (5.6) to a center manifold. Then the fact that (5.7) is a local model of (5.6) will follow from the fact that restrictions to invariant manifolds are local models (Section 4.3). Below we use the center manifold theorem and the fact that the center manifold has a convenient weakly connected form. Our treatment of the center manifold reduction is based on that of Iooss and Adelmeyer (1992, Section I.2). Basic facts and our notation are summarized in Section 2.1.1.

Let $\Pi_i^c : M \to E_i^c$ denote a projection operator such that

$$\ker \Pi_i^c = E_i^h.$$

Let $x_i = \Pi_i^c X_i \in E_i^c$. We use the notation $x = (x_1, \dots, x_k)^\top \in E^c$. In order to apply the center manifold theorem, we must consider the auxiliary system

$$\begin{cases} \dot{X}_i = F_i(X_i, \lambda) + \varepsilon G_i(X, \lambda, \rho, \varepsilon) \\ \dot{\lambda} = 0 \\ \dot{\rho} = 0 \\ \dot{\varepsilon} = 0 \end{cases} \qquad i = 1, \ldots, n .$$

at the equilibrium point $(X, \lambda, \varepsilon) = (0, 0, 0)$. Its center subspace is

$$E^c \times \Lambda \times \mathcal{R} \times \mathbb{R} = \{(x, \lambda, \rho, \varepsilon) \mid x \in E^c; \ \lambda \in \Lambda; \ \rho \in \mathcal{R}; \ \varepsilon \in \mathbb{R}\} .$$

Applying the center manifold theorem, we obtain the function

$$H = (H_1, \ldots, H_n) \ : \ E^c \times \Lambda \times \mathcal{R} \times \mathbb{R} \ \rightarrow \ E_1^h \times \cdots \times E_k^h \times \mathbb{R}^{m(n-k)}$$

with

$$H(0, 0, \rho, 0) = 0$$

and

$$D_x H(0, 0, \rho, 0) = 0 , \tag{5.8}$$

such that for λ and ε sufficiently small and for bounded $\rho \in \mathcal{R}$ the manifolds

$$M(\lambda, \rho, \varepsilon) = \{x + H(x, \lambda, \rho, \varepsilon) \mid x \in E^c\}$$

are locally invariant with respect to (5.6). Furthermore, x on $M(\lambda, \rho, \varepsilon)$ is governed by a dynamical system of the form

$$\dot{x}_i = \Pi_i^c \left(F_i(x_i + H_i(x, \lambda, \rho, \varepsilon), \lambda) + \varepsilon G_i(x + H(x, \lambda, \rho, \varepsilon), \lambda, \rho, \varepsilon) \right) \tag{5.9}$$

for $i = 1, \ldots, k$.

Let us show that H has a weakly connected form. For this, notice that $M(\lambda, \rho, 0)$ is the center manifold for the *uncoupled* ($\varepsilon = 0$) system

$$\dot{X}_i = F_i(X_i, \lambda) , \qquad i = 1, \ldots, n .$$

The function $H(x, \lambda, \rho, 0)$ has the uncoupled form

$$H_i(x, \lambda, \rho, 0) = \begin{cases} V_i(x_i, \lambda) & \text{for } i = 1, \ldots, k , \\ V_i(\lambda) & \text{for } i = k+1, \ldots, n , \end{cases}$$

where the V_i are some functions. In fact, each

$$M_i = \{x_i + V_i(x_i, \lambda) \mid x_i \in E_i^c\}$$

is the center manifold for the ith "nonhyperbolic" neuron dynamics, and the center manifold $M(\lambda, \rho, 0)$ is a copy of their direct product.

Recall that we assume that all data are as smooth as is necessary for our computations. Therefore, H depends smoothly on ε at $\varepsilon = 0$. We can rewrite H for $\varepsilon \neq 0$ in the weakly connected form

$$H(x, \lambda, \rho, \varepsilon) = \begin{cases} V_i(x_i, \lambda) & + \;\; \varepsilon W_i(x, \lambda, \rho, \varepsilon) \quad \text{for } i = 1, \ldots, k \\ V_i(\lambda) & + \;\; \varepsilon W_i(x, \lambda, \rho, \varepsilon) \quad \text{for } i = k+1, \ldots, n \,, \end{cases}$$

where the $W_i = \mathcal{O}(1)$ are some functions. The functions F_i in (5.9) can also be rewritten in the weakly connected form

$$F_i(x_i + V_i(x_i, \lambda) + \varepsilon W_i(x, \lambda, \rho, \varepsilon), \lambda)$$
$$= F_i(x_i + V_i(x_i, \lambda), \lambda) + \varepsilon \hat{F}_i(x, \lambda, \rho, \varepsilon) \tag{5.10}$$

for some functions $\hat{F}_i$ satisfying

$$\hat{F}_i(x, \lambda, \rho, 0) = D_{X_i} F_i(x_i + V_i(x_i, \lambda), \lambda) \, W_i(x, \lambda, \rho, 0) \,. \tag{5.11}$$

If we write for $i = 1, \ldots, k$

$$f_i(x_i, \lambda) = \Pi_i^c F_i(x_i + V_i(x_i, \lambda), \lambda)$$

and

$$g_i(x, \lambda, \rho, \varepsilon) = \Pi_i^c \left(\hat{F}_i(x, \lambda, \rho, \varepsilon) + G_i(x + H(x, \lambda, \rho, \varepsilon), \lambda, \rho, \varepsilon) \right) \,,$$

then (5.9) can be written as (5.7). From (5.11) it follows that the image of the function $\hat{F}_i(x, \lambda, \rho, 0)$ lies in E_i^{h}, and therefore

$$\Pi_i^c \hat{F}_i(x, \lambda, \rho, 0) = 0 \,.$$

Moreover, from (5.8) it follows that H does not have linear terms in x, and therefore

$$g_i(x, \lambda, \rho, 0) = \Pi_i^c G_i(x, \lambda, \rho, 0) + \mathcal{O}(x^2) \,. \tag{5.12}$$

This expression can be used to compute the linear terms of g_i.

Now note that

$$\begin{aligned} J_i & = D_{x_i} f_i(0, 0) \\ & = D_{x_i} \Pi_i^c F_i(x_i + V_i(x_i, 0), 0)|_{x_i = 0} \\ & = \Pi_i^c D_{X_i} F_i(0, 0) \cdot (I + D_{x_i} V_i(0, 0)) \,, \end{aligned}$$

where I is the unit matrix. From (5.8) it follows that $D_{x_i} V_i(0, 0) = 0$. Recall that $D_{x_i} F_i(0, 0) = L_i$. Hence

$$J_i = \Pi_i^c L_i = L_i \Pi_i^c = L_i|_{E_i^c} \,.$$

In particular, each J_i has all its eigenvalues having zero real part. $\square$

Notice that the center manifold could not be unique. This implies that the local model (5.7) could not be unique. This does not affect our subsequent analysis because we will use an initial portion of the Taylor series of (5.7), which is unique.

Remark. Theorem 5.2 can be restated concisely as follows: *Center manifolds for weakly connected systems have weakly connected form.*

Remark. There could be connections between a pair of neurons in the reduced system (5.7) even when there are no connections between them in the original WCNN (5.6).

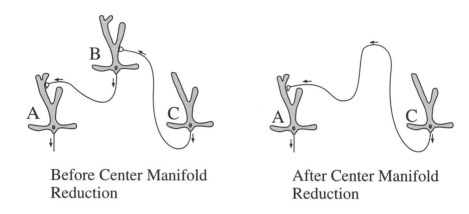

Before Center Manifold
Reduction

After Center Manifold
Reduction

Figure 5.1. Indirect connections between neurons become direct after center manifold reduction. Neurons A and C have nonhyperbolic dynamics and neuron B has hyperbolic dynamics.

That is, the function g_i may depend on the variable x_i even when the function G_i does not depend on X_i. Such curious phenomena can be illustrated as follows: Suppose there are neurons A, B, and C such that C directly affects B, and B directly affects A (see Figure 5.1). If there are no connections between C and A, then there is no direct communication between the two neurons, though there is an indirect communication through the intermediate neuron B. Now suppose A and C are nonhyperbolic and B has hyperbolic local dynamics. Then the center manifold reduction removes B, but it automatically establishes a connection from C to A.

From equation (5.12) it follows that if G_i does not depend on X_j, then g_i does not depend on x_j up to quadratic terms in x. Since the canonical models that we derive in this chapter are affected only by linear terms of g_i, the phenomenon in Figure 5.1 does not play an essential role.

5.3 Multiple Saddle-Node, Pitchfork, and Cusp Bifurcations

Consider the canonical model

$$\dot{x}_i = f_i(x_i, \lambda) + \varepsilon g_i(x, \lambda, \rho, \varepsilon) , \qquad i = 1, \ldots, n , \qquad (5.13)$$

where $x_i \in \mathbb{R}$ is a scalar and

$$\frac{\partial f_i}{\partial x_i} = 0 \qquad (5.14)$$

at $x_i = 0$ and $\lambda = 0$ for all i. Thus, the equilibrium $x = (x_1, \ldots, x_n)^\top = 0 \in \mathbb{R}^n$ of the uncoupled system

$$\dot{x}_i = f_i(x_i, \lambda) , \qquad i = 1, \ldots, n ,$$

for $\lambda = 0$ is nonhyperbolic. We will use various rescalings of variables and parameters to blow up its neighborhood.

A standard approach to the study of such nonhyperbolic problems is to introduce functions $\rho = \rho(\varepsilon)$ and $\lambda = \lambda(\varepsilon)$ such that $\rho(0) = \rho_0$ and

$$\lambda(\varepsilon) = 0 + \varepsilon\lambda_1 + \varepsilon^2\lambda_2 + \mathcal{O}(\varepsilon^3) , \qquad \lambda_1, \lambda_2 \in \Lambda , \qquad (5.15)$$

i.e., $\lambda(\varepsilon)$ is ε-close to the bifurcation value $\lambda = 0$. This is not a restriction on ρ and λ, since we allow ρ_0, λ_1, and λ_2 to assume any values. Thus, a family of curves $(\lambda(\varepsilon), \rho(\varepsilon))$ can cover completely a neighborhood of $(0, \rho_0) \in \Lambda \times \mathcal{R}$.

5.3.1 Adaptation Condition

A technical condition arises in our study. We say that the WCNN satisfies the *adaptation condition* if

$$D_\lambda f_i(0, 0) \cdot \lambda_1 + g_i(0, 0, \rho_0, 0) = 0 \qquad (5.16)$$

for all i. Recall that $\lambda_1 \in \Lambda$ could be a multi- or infinite-dimensional variable. If its dimension is greater than n, then usually there is a linear manifold in Λ of solutions to equation (5.16).

The adaptation condition (5.16) is equivalent to the conditions

$$f_i(0, \lambda(\varepsilon)) + \varepsilon g_i(0, \lambda(\varepsilon), \rho(\varepsilon), \varepsilon) = \mathcal{O}(\varepsilon^2) , \qquad i = 1, \ldots, n . \qquad (5.17)$$

Indeed, we can differentiate equation (5.17) with respect to ε to obtain

$$D_\lambda f_i(0, \lambda(\varepsilon)) \cdot \lambda'(\varepsilon) + g_i(0, \lambda(\varepsilon), \rho(\varepsilon), \varepsilon) + \mathcal{O}(\varepsilon) = \mathcal{O}(\varepsilon)$$

and then substitute $\varepsilon = 0$ to obtain (5.16).

The adaptation condition can be explained in ordinary language: The internal parameter λ counterbalances (up to order ε) the steady-state input onto each neuron from the entire network. In Section 5.3.8 we show that the condition is necessary for some WCNNs to exhibit nontrivial properties.

The adaptation condition can also be restated in terms of the original weakly connected system (5.6). In this case it suffices to demand that λ counterbalance the steady-state input from the network onto each neuron along with the center subspace, E^c, direction.

An important example when the adaptation condition is satisfied corresponds to

$$\lambda = \mathcal{O}(\varepsilon^2) \qquad \text{and} \qquad g_i(0,0,\rho_0,0) = 0 , \quad i = 1,\dots,n .$$

That is, the strength of connections ε between them is the square root of the distance $|\lambda|$ to the threshold, and the steady-state input from the network onto the ith neuron vanishes when the neurons are quiescent; see Figure 5.2. The assumption $g_i(0,0,\rho_0,0) = 0$ arises naturally when we consider

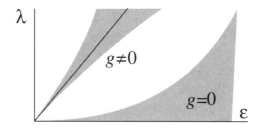

Figure 5.2. The adaptation condition (5.16) is satisfied in the shaded areas.

weakly connected networks of spiking neurons in Chapter 8. It follows from the observation that there is no synaptic transmission when the neurons are silent.

5.3.2 Multiple Saddle-Node Bifurcation

Recall (Section 2.2.1) that a dynamical system $\dot{x}_i = f_i(x_i,\lambda)$, $x_i \in \mathbb{R}$ is near a *saddle-node* bifurcation point $x_i = 0$ for $\lambda = 0$ if

$$\frac{\partial f_i}{\partial x_i} \;=\; 0 \qquad \text{(nonhyperbolicity)},$$

$$p_i \equiv \frac{1}{2}\frac{\partial^2 f_i}{\partial x_i^2} \;\neq\; 0 \qquad \text{(presence of quadratic term)}, \qquad (5.18)$$

$$D_\lambda f_i \;\neq\; 0 \qquad \text{(transversality)},$$

at the point $(x_i,\lambda) = (0,0)$. If all equations in the uncoupled system are near a saddle-node bifurcation, then the bifurcation is called *multiple*.

Theorem 5.4 (Local Canonical Model for Multiple Saddle-Node Bifurcations in WCNNs) *Suppose the WCNN*

$$\dot{x}_i = f_i(x_i, \lambda) + \varepsilon g_i(x, \lambda, \rho, \varepsilon) , \qquad i = 1, \ldots, n ,$$

is near a multiple saddle-node bifurcation point $x = 0$ *for* $\lambda = 0$, *and suppose the external input has the form*

$$\rho(\varepsilon) = \rho_0 + \varepsilon \rho_1 + \mathcal{O}(\varepsilon^2) , \qquad \rho_0, \rho_1 \in \mathcal{R} , \qquad (5.19)$$

and the internal parameter $\lambda(\varepsilon) = \varepsilon \lambda_1 + \mathcal{O}(\varepsilon^2)$ *satisfies the adaptation condition*

$$D_\lambda f_i(0,0) \cdot \lambda_1 + g_i(0,0,\rho_0,0) = 0 \qquad \text{for all } i.$$

Then the invertible change of variables

$$x_i = \varepsilon p_i^{-1} y_i \qquad (5.20)$$

transforms the WCNN to

$$y_i' = r_i + y_i^2 + \sum_{j=1}^{n} c_{ij} y_j + \mathcal{O}(\varepsilon) , \qquad i = 1, \ldots, n , \qquad (5.21)$$

where $' = d/d\tau$, $\tau = \varepsilon t$ *is the slow time.*

Proof. Using (5.15) and (5.19) we can write the initial part of the Taylor series of the WCNN in the form

$$\dot{x}_i = \varepsilon a_i + \varepsilon^2 d_i + \varepsilon \sum_{j=1}^{n} s_{ij} x_j + p_i x_i^2 + \mathcal{O}(|x, \varepsilon|^3) , \qquad (5.22)$$

where

$$a_i = D_\lambda f_i \cdot \lambda_1 + g_i ,$$

$$d_i = D_\lambda f_i \cdot \lambda_2 + (D_\lambda^2 f_i) \cdot (\lambda_1, \lambda_1) + D_\lambda g_i \cdot \lambda_1 + D_\rho g_i \cdot \rho_1 + \frac{\partial g_i}{\partial \varepsilon} ,$$

$$s_{ij} = \frac{\partial g_i}{\partial x_j}$$

for $i \neq j$, and

$$s_{ii} = D_\lambda \frac{\partial f_i}{\partial x_i} \cdot \lambda_1 + \frac{\partial y_i}{\partial x_i} ,$$

where all functions and derivatives are evaluated at the point $(x, \lambda, \rho, \varepsilon) = (0, 0, \rho_0, 0)$. The constants $p_i \neq 0$ were defined in (5.18). The adaptation condition implies $a_i = 0$. If we set

$$r_i = p_i d_i , \qquad c_{ij} = \frac{p_i}{p_j} s_{ij} ,$$

and use the rescaling (5.20), then (5.22) is transformed into (5.21). $\square$

5.3.3 *Multiple Cusp Bifurcation*

Recall (Section 2.2.2) that a dynamical system

$$\dot{x}_i = f_i(x_i, \lambda) , \qquad x_i \in \mathbb{R} ,$$

is near a *cusp bifurcation* $x_i = 0$ for $\lambda = 0$ if the following conditions are satisfied:

$$\frac{\partial f_i}{\partial x_i} \;=\; 0 \qquad \text{(nonhyperbolicity)},$$

$$\frac{\partial^2 f_i}{\partial x_i^2} \;=\; 0 \qquad \text{(absence of quadratic term)},$$

$$q_i \equiv \frac{1}{6}\frac{\partial^3 f_i}{\partial x_i^3} \;\neq\; 0 \qquad \text{(presence of cubic term)}$$

and a transversality condition, which we do not use below.

Theorem 5.5 (Local Canonical Model for Multiple Cusp Bifurcations in WCNNs) *If the WCNN*

$$\dot{x}_i = f_i(x_i, \lambda) + \varepsilon g_i(x, \lambda, \rho, \varepsilon) , \qquad i = 1, \ldots, n ,$$

is near a multiple cusp bifurcation, if the external input has the form

$$\rho(\varepsilon) = \rho_0 + \sqrt{\varepsilon}\rho_{\frac{1}{2}} + \mathcal{O}(\varepsilon) , \qquad \rho_0, \rho_{\frac{1}{2}} \in \mathcal{R} , \tag{5.23}$$

and if the internal parameter $\lambda(\varepsilon) = \varepsilon\lambda_1 + \mathcal{O}(\varepsilon^2)$ satisfies the adaptation condition

$$D_\lambda f_i(0, 0) \cdot \lambda_1 + g_i(0, 0, \rho_0, 0) = 0 \qquad \text{for all } i,$$

then the invertible change of variables

$$x_i = \sqrt{\frac{\varepsilon}{|q_i|}}\, y_i \tag{5.24}$$

transforms the WCNN to

$$y_i' = r_i + \sigma_i y_i^3 + \sum_{j=1}^n c_{ij} y_j + \mathcal{O}(\sqrt{\varepsilon}) , \qquad \sigma_i = \pm 1 , \qquad i = 1, \ldots, n , \tag{5.25}$$

where $' = d/d\tau$, and $\tau = \varepsilon t$ is the slow time.

Proof. The initial portion of the Taylor series of the WCNN has the form

$$\dot{x}_i = \varepsilon\sqrt{\varepsilon}d_i + \varepsilon\sum_{j=1}^n s_{ij}x_j + q_i x_i^3 + \mathcal{O}(|x, \sqrt{\varepsilon}|^4) , \tag{5.26}$$

where s_{ij} and q_i were defined earlier, and

$$d_i = D_\rho g_i(0, 0, \rho_0, 0) \cdot \rho_{\frac{1}{2}}.$$

If we set

$$c_{ij} = \sqrt{\frac{|q_i|}{|q_i|}}\, s_{ij}\,, \quad r_i = \sqrt{|q_i|}\, d_i\,, \quad \sigma_i = \operatorname{sign} q_i$$

and rescale using (5.24), then (5.26) transforms into (5.25). $\square$

5.3.4 Multiple Pitchfork Bifurcation

In Theorem 5.5 we demanded that the deviation of $\rho(\varepsilon)$ from ρ_0 be of order $\sqrt{\varepsilon}$ (formula (5.23)). If we require that $\rho(\varepsilon)$ have the form as in (5.19), i.e., $\rho_{\frac{1}{2}} = 0$, then $r_i = 0$ in (5.25), and (5.25) has the form

$$y_i' = \sigma_i y_i^3 + \sum_{j=1}^{n} c_{ij} y_j + \mathcal{O}(\sqrt{\varepsilon})\,, \quad \sigma_i = \pm 1\,, \quad i = 1, \ldots, n\,. \tag{5.27}$$

Dynamical system (5.27) is a local canonical model for a multiple *pitchfork* bifurcation in WCNNs because (5.27) is exactly what one receives if one considers (5.13) with functions f_i and g_i that are $\mathbb{Z}_2$-equivariant (under the reflection $x_i \to -x_i$). In this case the adaptation condition (5.16) is satisfied automatically.

We study the canonical models for multiple cusp and pitchfork bifurcations in Chapter 11. In particular, we show that they can operate as multiple attractor (MA) or globally asymptotically stable (GAS) type neural networks.

In canonical models (5.25) and (5.27), the choice $\sigma_i = +1$ ($\sigma_i = -1$) corresponds to a subcritical (respectively supercritical) bifurcation in the ith neuron dynamics. The parameter c_{ii} is a bifurcation parameter, since it depends on λ. To stress this fact we sometimes denote it by b_i and write the canonical models in the form

$$\begin{aligned}
y_i' &= r_i + b_i y_i + y_i^2 + \sum_{j=1}^{n} c_{ij} y_j & \text{(saddle-node)}, \\
y_i' &= r_i + b_i y_i \pm y_i^3 + \sum_{j=1}^{n} c_{ij} y_j & \text{(cusp)}, \\
y_i' &= b_i y_i \pm y_i^3 + \sum_{j=1}^{n} c_{ij} y_j & \text{(pitchfork)},
\end{aligned}$$

where $c_{ii} = 0$ for all i.

5.3.5 Example: The Additive Model

Consider a weakly connected, weakly forced network of additive neurons in the form

$$\dot{x}_i = -x_i + S\left(\rho_i^0 + \varepsilon \rho_i + c_{ii} x_i + \varepsilon \sum_{j=1}^{n} c_{ij} x_j\right)\,, \quad i = 1, \ldots, n\,. \tag{5.28}$$

We treat $\lambda = (\rho_1^0, c_{11}, \rho_2^0, c_{22}, \ldots, \rho_n^0, c_{nn}) \in \mathbb{R}^{2n}$ as a multidimensional bifurcation parameter, and $\varepsilon\rho_i$ as a weak external input to the ith neuron. This system can be rewritten in the weakly connected form

$$\dot{x}_i = -x_i + S(\rho_i^0 + c_{ii}x_i) + \varepsilon S'(\rho_i^0 + c_{ii}x_i)\left(\rho_i + \sum_{j=1}^n c_{ij}x_j\right) + \mathcal{O}(\varepsilon^2) \, .$$

Hyperbolic Equilibrium

Suppose that each additive neuron described by the dynamical system

$$\dot{x}_i = -x_i + S(\rho_i^0 + c_{ii}x_i) \tag{5.29}$$

is at a stable hyperbolic equilibrium $x_i \in \mathbb{R}$, namely

$$L_i = -1 + c_{ii}S'(\rho_i^0 + c_{ii}x_i) < 0$$

for all i. Note that (5.29) regarded as a system for $i = 1, .., n$ describes the uncoupled ($\varepsilon = 0$) additive neural network near a stable hyperbolic equilibrium $x = (x_1, .., x_n) \in \mathbb{R}^n$.

The weakly connected additive neural network (5.28) can be treated as an ε-perturbation of system (5.29). Its linearization at the equilibrium has the form

$$\dot{x}_i = L_ix_i + \mathcal{O}(\varepsilon) \, , \qquad L_i < 0 \, , \quad i = 1, .., n \, .$$

The fact that such a system does not acquire any new local qualitative features follows from the Hartman-Grobman theorem and is illustrated in Fig.5.3. Obviously, any small perturbations of the neuron dynamics cannot

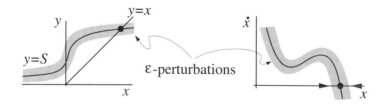

Figure 5.3. Additive neurons having hyperbolic equilibria are not sensitive to ε-perturbations.

destroy the hyperbolic equilibrium or make it unstable. No matter what the activities of the other neurons, each neuron continues to have a stable hyperbolic equilibrium in an ε-neighborhood of the old one. Once the neuron activity $x_i(t)$ is close to the equilibrium, it stays in its ε-neighborhood regardless of the dynamical processes taking place in other neurons. Nothing can draw it away from the neighborhood.

Nonhyperbolic Equilibria

Since the weakly connected additive network cannot have interesting local dynamics near hyperbolic equilibria, one might think that interesting nonlinear features appear as soon as some or all of the neurons are near nonhyperbolic equilibria. That is not necessarily true. Nonhyperbolicity is a necessary but not sufficient condition for a weakly connected systems to have interesting nonlinear local behavior. We illustrate this issue here using the weakly connected additive neural network (5.28).

Suppose each neuron is near a nonhyperbolic equilibrium $x_i^0 \in \mathbb{R}$ for $\varepsilon = 0$ and for some c_{ii}^0 and ρ_i^0. That is

$$-x_i^0 + S(\rho_i^0 + c_{ii}^0 x_i^0) = 0 , \qquad \text{(equilibrium)},$$
$$-1 + c_{ii}^0 S'(\rho_i^0 + c_{ii}^0 x_i^0) = 0 , \qquad \text{(nonhyperbolicity)},$$

for all i. We can use Figure 2.6 to determine such values for ρ_i^0 and c_{ii}^0. We can describe the local dynamics of the network in an ε-neighborhood of the feedback parameter c_{ii}^0 by assuming that $c_{ii} = c_{ii}^0 + \varepsilon b_i$, where b_i is a rescaled perturbation of c_{ii}^0. Let x_i be a local coordinate at the nonhyperbolic equilibrium x_i^0. Thus, the activity of the ith neuron is $x_i^0 + x_i$, and the initial portion of the Taylor series of the additive model has the form

$$\dot{x}_i = 0 + 0x_i + \varepsilon a_i + p_i x_i^2 + \varepsilon \sum_{j=1}^{n} s_{ij} x_j + \varepsilon^2 r_i + \ldots , \qquad (5.30)$$

where a_i, p_i, s_{ij}, and r_i are some constants, which are computable. In particular,

$$a_i = S'(\rho_i^0 + c_{ii}^0 x_i^0) \left(\rho_i + b_i x_i^0 + \sum_{j-1}^{n} c_{ij} x_j^0 \right) .$$

If $a_i \neq 0$, then the rescaling $x_i = \sqrt{\varepsilon} y_i$, $\tau = \sqrt{\varepsilon} t$ transforms (5.30) to

$$y_i' = a_i + p_i y_i^2 + \mathcal{O}(\sqrt{\varepsilon}) .$$

When the nonhyperbolic equilibria correspond to saddle-node bifurcations, the parameters $p_i \neq 0$ and the system above has simple dynamics:

- If $a_i p_i < 0$, then the ith neuron dynamics has a stable equilibrium $y_i = \sqrt{|a_i/p_i|}$

- If $a_i p_i > 0$, then the neuron activity $y_i(\tau)$ leaves a neighborhood of the origin.

In both cases the activity of each additive neuron does not depend on the activities of the other neurons, and the network's local behavior is trivial.

Adaptation Condition

Now suppose the coefficients a_i vanish. Then the network dynamics governed by (5.30) can be transformed to

$$y_i' = r_i + p_i y_i^2 + \sum_{j=1}^{n} s_{ij} y_j + \mathcal{O}(\varepsilon) \tag{5.31}$$

by the change of variables $x_i = \varepsilon y_i$, $\tau = \varepsilon t$. In this case, activity of each neuron *does* depend on the activities of the other neurons provided that the synaptic coefficients s_{ij} are nonzero.

We see that the condition $a_i = 0$, which we call the *adaptation condition*, is important. Since the S-shaped function S satisfies $S' > 0$ everywhere, a_i vanishes only when

$$\rho_i + b_i x_i^0 + \sum_{j=1}^{n} c_{ij} x_j^0 = 0 \ .$$

Therefore, the adaptation condition can be rewritten in the form

$$b_i x_i^0 = -\left(\rho_i + \sum_{j=1}^{n} c_{ij} x_j^0 \right),$$

which means that the internal parameter b_i counterbalances the overall input $\rho_i + \sum c_{ij} x_j^0$ from the external receptors and from other neurons of the network to the ith neuron.

When the adaptation condition is violated, the local activity of the additive neuron depends only on the sign of a_i. Since a_i depends on the external input ρ_i, we can conclude that an additive neuron reacts to the input from receptors but not to the activities of the other neurons. Therefore, the behavior of the network is predetermined by the input ρ. Such a network does not even have memory. When the adaptation condition is satisfied, the neuron reacts to both the external input (the parameter r_i depends on it) and to activities of the other neurons. A network of such neurons can have memories encoded in the synaptic coefficients s_{ij}.

In summary, we have found that local activity of the additive model near a hyperbolic equilibrium is not interesting. In order to be interesting, the equilibrium must be nonhyperbolic and the adaptation condition must be satisfied. In this case local activity of the additive neural network is governed by the canonical model (5.31). In order to study neurocomputational properties of the network in a neighborhood of the equilibrium it suffices to study the canonical model (5.31), which is simpler. Besides, studying the canonical model will shed some light on the neurocomputational properties of *all* weakly connected networks near the bifurcation, not only on the additive one.

5.3.6 Time-Dependent External Input ρ

Suppose the WCNN

$$\dot{x}_i = f_i(x_i, \lambda) + \varepsilon g_i(x, \lambda, \rho(\varepsilon, t), \varepsilon) \tag{5.32}$$

has external input $\rho(\varepsilon, t)$ that depends on time t. The derivation of canonical models in this case involves averaging theory. The canonical models remain the same, though the parameters in the models have different formulae. Below we present an analysis of a multiple saddle-node bifurcation. Analysis of multiple pitchfork and cusp bifurcations is similar and hence omitted.

When the external input depends on time, the adaptation condition takes the form

$$D_\lambda f_i(0,0) \cdot \lambda_1 + \frac{1}{T} \int_0^T g_i(0, 0, \rho(0, t), 0) \, dt = \mathcal{O}(\frac{1}{T}) \,, \tag{5.33}$$

where $T = \mathcal{O}(1/\varepsilon)$. It is easy to see that this equation is equivalent to the assumption that there is $\lambda_1 \in \Lambda$ such that the integral

$$\int_0^T T \, D_\lambda f_i(0,0) \cdot \lambda_1 + g_i(0, 0, \rho(0, t), 0) \, dt$$

is bounded for $T \to \infty$.

Theorem 5.6 *A WCNN (5.32) satisfying the adaptation condition (5.33) at a multiple saddle-node bifurcation point has a local canonical model (5.21).*

Proof. The initial part of the Taylor series of (5.32) can be written in the form

$$\dot{x}_i = \varepsilon a_i(t) + \varepsilon^2 d_i(t) + p_i x_i^2 + \varepsilon \sum_{j=1}^{n} s_{ij}(t) x_j + \mathcal{O}(|x, \varepsilon|^3) \,, \tag{5.34}$$

where

$$
\begin{aligned}
a_i(t) &= D_\lambda f_i(0,0) \cdot \lambda_1 + g_i(0, 0, \rho(0, t), 0) \,, \\
d_i(t) &= D_\lambda f_i(0,0) \cdot \lambda_2 + D_\lambda^2 f_i(0,0) \cdot (\lambda_1, \lambda_1) \\
&\quad + D_\lambda g_i(0, 0, \rho(0, t), 0) \cdot \lambda_1 + D_\rho g_i(0, 0, \rho(0, t), 0) \cdot \frac{\partial \rho}{\partial \varepsilon}(0, t) \\
&\quad + \frac{\partial g_i}{\partial \varepsilon}(0, 0, \rho(0, t), 0), \\
s_{ij}(t) &= \frac{\partial g_i}{\partial x_j}(0, 0, \rho(0, t), 0) \,, \qquad i \neq j \,, \\
s_{ii}(t) &= D_\lambda \frac{\partial f_i}{\partial x_i}(0, 0) \cdot \lambda_1 + \frac{\partial g_i}{\partial x_i}(0, 0, \rho(0, t), 0) \,, \\
p_i &= \frac{1}{2} \frac{\partial^2 f_i}{\partial x_i^2}(0, 0) \neq 0 \,.
\end{aligned}
$$

Notice that the origin $x = 0$ is an equilibrium of (5.34) up to order ε. Now we use the change of variables $x_i = y_i + \varepsilon h_i(t)$, $i = 1, \ldots, n$, to remove the term $\varepsilon a_i(t)$ from (5.34), that is, to make the origin an equilibrium up to order ε^2. To find suitable $h_i(t)$, we plug $x = y + \varepsilon h$ into (5.34) and set $y = 0$. Simple computations show that

$$\dot{h}_i = a_i(t) , \qquad i = 1, \ldots, n ,$$

which has the solution

$$h_i(t) = \int_0^t a_i(s) \, ds, \qquad i = 1, \ldots, n .$$

Now let us check that $h(t)$ is bounded at least for $t = \mathcal{O}(1/\varepsilon)$. For this we use the adaptation condition (5.33), which we write in the form

$$T D_\lambda f_i(0,0) \cdot \lambda_1 + \int_0^T g_i(0,0,\rho(0,t),0) \, dt = \mathcal{O}(1) .$$

Obviously, this is equivalent to

$$\int_0^T D_\lambda f_i(0,0) \cdot \lambda_1 + g_i(0,0,\rho(0,t),0) \, dt = \int_0^T a_i(t) \, dt = \mathcal{O}(1) ,$$

meaning that $h(T)$ is bounded. System (5.34) in the new variables has the form

$$\dot{y}_i = \varepsilon^2 \tilde{d}_i(t) + p_i y_i^2 + \varepsilon \sum_{j=1}^n \tilde{s}_{ij}(t) y_j \; + \; \mathcal{O}(|y,\varepsilon|^3) ,$$

where

$$\begin{aligned}
\tilde{d}_i(t) &= d_i(t) + p_i h_i(t)^2 + \sum_{j=1}^n s_{ij} h_j(t) , \\
\tilde{s}_{ij}(t) &= s_{ij}(t) , \qquad i \neq j , \\
\tilde{s}_{ii}(t) &= s_{ii}(t) + 2 p_i h_i(t) .
\end{aligned}$$

Now we blow up a neighborhood of the origin using the rescaling $y_i = \varepsilon x_i / p_i$, $i = 1, \ldots, n$, so that the system has the form

$$\dot{x}_i = \varepsilon \left(p_i \tilde{d}_i(t) + x_i^2 + \sum_{j=1}^n p_i \tilde{s}_{ij}(t) p_j^{-1} x_j \right) \; + \; \mathcal{O}(|\varepsilon|^2) .$$

After averaging, we obtain

$$\dot{x}_i = \varepsilon \left(r_i + x_i^2 + \sum_{j=1}^n c_{ij} x_j \right) \; + \; \mathcal{O}(|\varepsilon|^2) ,$$

where

$$r_i = \lim_{T \to \infty} \frac{1}{T} \int_0^T p_i d_i(t) \, dt \ ,$$

$$c_{ij} = \lim_{T \to \infty} \frac{1}{T} \int_0^T p_i s_{ij}(t) p_j^{-1} \, dt \ ,$$

which can be written as (5.21) when we use slow time $\tau = \varepsilon t$. $\square$

The proof can be applicable (with obvious modifications) to the multiple cusp and pitchfork bifurcations in WCNNs (5.32) to show that systems (5.25) and (5.27) are also canonical models in the case of time-dependent external input ρ.

Remark. The values of the synaptic coefficients

$$c_{ij} = \frac{p_i}{p_j} \left(\lim_{T \to \infty} \frac{1}{T} \int_0^T \frac{\partial g_i}{\partial x_j}(0, 0, \rho(0, t), 0) \, dt \right)$$

depend crucially on the external input $\rho(\varepsilon, t)$. For some functions g_i the external input could be chosen in such a way that c_{ij} acquires positive, negative, or zero values. If Dale's principle is satisfied, the function $\partial g_i / \partial x_j$ does not change its sign for any value of ρ, and the synaptic coefficient c_{ij} cannot change sign but can only change its amplitude.

The coefficient c_{ij} determines how the jth neuron affects dynamics of the ith neuron. The fact that c_{ij} may change its value depending on an external input ρ is curious from both the mathematical and neurophysiological points of view. We discuss this issue in Section 5.4.5 where we study multiple Andronov-Hopf bifurcations in WCNNs. There we show how profound the changes in the network behavior can be.

5.3.7 Variables and Parameters

It is customary in the neural network literature to refer to the matrix $C = (c_{ij})$ as the *synaptic* (or connection) matrix. It is believed that changes in C can describe memorization of information by the brain. We will show that the same is true for canonical models in the sense that they have interesting neurocomputational properties that depend on C. In particular, they can perform pattern recognition and recall tasks by association – a basic property of the human brain. Each parameter r_i depends on λ and ρ and has the meaning of rescaled external input on the ith neuron. It is convenient to think of $\rho_0 \in \mathcal{R}$ in the expression

$$\rho(\varepsilon) = \rho_0 + \varepsilon \rho_1 + \mathcal{O}(\varepsilon^2)$$

as a parameter describing an environment in which the WCNN (5.6) does something, and $\rho_1 \in \mathcal{R}$ as describing an input pattern from this environment. For example, ρ_0 may parametrize a degree of illumination and ρ_1

represent the shape of an object to be recognized. Notice that the synaptic coefficients c_{ij} computed in the proof of Theorem 5.2 depend on the environment ρ_0, but not on the input ρ_1.

We can also ascribe meaning to each term in the expansion

$$\lambda(\varepsilon) = 0 + \varepsilon\lambda_1 + \mathcal{O}(\varepsilon^2) , \quad \lambda_1 \in \Lambda .$$

The first term, postulated to be 0 (but it could have been any $\lambda_0 \in \Lambda$), means that a WCNN is near a nonhyperbolic equilibrium point. We showed that this was a necessary condition for a WCNN to exhibit any local nonlinear behavior. The second term (which is $\varepsilon\lambda_1$) is of order ε. The fact that coefficient $\lambda_1 \in \Lambda$ satisfies (5.16) means that the network (5.6) is adapted to the particular environment $\rho_0 \in \mathcal{R}$ in which it works.

The choices $\lambda = \mathcal{O}(\varepsilon)$ and $\rho = \rho_0 + \mathcal{O}(\varepsilon)$ for multiple saddle-node and pitchfork bifurcations and $\rho = \rho_0 + \mathcal{O}(\sqrt{\varepsilon})$ for the cusp bifurcation are the most interesting. Indeed, we could consider a multiple saddle-node bifurcation when

$$\lambda = \varepsilon^\alpha\lambda_1 + \varepsilon^{2\alpha}\lambda_2 + o(\varepsilon^{2\alpha}) \quad \text{and} \quad \rho = \rho_0 + \varepsilon^\beta\rho_1 + o(\varepsilon^\beta)$$

for some positive constants α and β. Then it is possible to show that almost all choices of α and β produce uninteresting (locally uncoupled) dynamics except when $\alpha = 1$ or $\alpha \geq 2$ and $\beta \geq 1$ (for saddle-node and pitchfork) or $\beta \geq 1/2$ (for cusp).

5.3.8 Adaptation Condition

Suppose the adaptation condition (5.16) is violated. Consider, for example, the multiple saddle-node bifurcation. The other bifurcations may be considered similarly. The rescaling

$$x_i \to \varepsilon^{\frac{1}{2}}p_i^{-1}x_i , \qquad a_i \to p_i^{-1}a_i , \qquad t \to \varepsilon^{-\frac{1}{2}}t$$

transforms (5.22) to

$$x_i' = a_i + x_i^2 + \sqrt{\varepsilon}\sum_{j=1}^{n} c_{ij}x_j + \mathcal{O}(\varepsilon) , \qquad i = 1,\ldots,n , \qquad (5.35)$$

where the synaptic coefficients c_{ij} were defined earlier.

Suppose all $a_i \neq 0$; that is, the adaptation condition is violated uniformly. Then the system above is a weakly connected network of hyperbolic neurons. The fundamental theorem of WCNN theory prohibits such a system from having any interesting behavior. Indeed, the behavior of this system is predetermined by the environment ρ_0 and the internal parameter λ (a_i depends upon them). The activity of (5.35) is hyperbolic for $\varepsilon = 0$ and hence locally linear. System (5.35) does not have memory and cannot perform the recognition task because it is insensitive to the input pattern ρ_1. It can react only to ρ_0, and its reaction is locally trivial.

Phase portraits of a network of two such neurons is depicted in Figure 5.4. The solution to each condition $a_i = 0$ in (5.16) is a linear manifold in Λ.

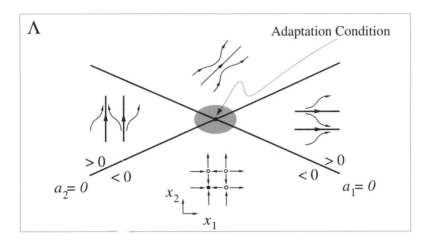

Figure 5.4. Dynamics of a weakly connected neural network is interesting in an ε-neighborhood of intersection of linear manifolds $a_1 = 0$ and $a_2 = 0$. Outside the intersection the activity is locally trivial.

Their intersection corresponds to the solution of the adaptation equation. The activity of such system is trivial outside some ε-neighborhood of the intersection.

Now suppose the adaptation condition is violated nonuniformly. That is, there are some zero and some nonzero coefficients a_i in (5.35). Without loss of generality we take $a_1 = \cdots = a_k = 0$ and $a_i \neq 0$ for $i = k+1, \ldots, n$ for some positive $k < n$. Then the system is

$$\begin{cases} x_i' = r_i^2 + \sqrt{\varepsilon} \sum_{j=1}^n c_{ij} r_j + \mathcal{O}(\varepsilon) , & i = 1, \ldots, k , \\ x_i' = a_i + x_i^2 + \mathcal{O}(\sqrt{\varepsilon}) , & i = k+1, \ldots, n . \end{cases}$$

The second set of equations describes a weakly connected network of hyperbolic neurons. We can apply the center manifold theorem (Theorem 5.2) directly to this system to extract the dynamics of the first k locally nonhyperbolic neurons and discard the dynamics of the other $n - k$ locally hyperbolic neurons.

One can treat the system (5.35) for $\varepsilon = 0$ as the canonical model (5.21) with zero connection matrix $C = (c_{ij})$. In subsequent chapters we show that the connection matrix C is responsible for memorization and recognition by association. Therefore, system (5.35) cannot have such properties.

Similar considerations for a multiple cusp bifurcation shows that the canonical model (5.25) in that case is also uncoupled. This proves the following result.

Proposition 5.7 *The adaptation condition (5.16) is necessary for weakly connected neural network near multiple saddle-node or cusp bifurcations to have local neurocomputational properties.*

We see that the adaptation condition (5.16) is important, at least for local dynamics to have interesting features. Without it recognition is impossible. That is exactly what we expect from our experience. Indeed, when we come into a dark room from bright sunshine, we cannot see anything until we adapt to the new environment. The human brain is an extremely flexible system. It has many subsystems with different time scales that help it to adapt to any new environment. Their functions are not known completely yet. The fact that we did not make any assumptions about the Banach spaces $\mathcal{R}$ and Λ gives us freedom in interpretation of our results. For example, we do not have to specify where exactly the adaptation takes place – in the cortex, in the retina, or in the pupil of the eye. It does take place. And there is a mechanism responsible for that.

5.4 Multiple Andronov-Hopf Bifurcation

A dynamical system

$$\dot{x}_i = f_i(x_i, \lambda)$$

is near an *Andronov-Hopf* bifurcation if the Jacobian matrix

$$J_i = D_{x_i} f_i$$

has a simple pair of purely imaginary eigenvalues. Using the center manifold reduction for WCNN (Theorem 5.2) we assume without loss of generality that J_i is a 2×2 matrix. Let $\pm i\Omega_i$ be the eigenvalues of J_i, where

$$\Omega_i = \sqrt{\det J_i}\ .$$

Let $v_i \in \mathbb{C}^2$ and $\bar{v}_i \in \mathbb{C}^2$ denote the (column) eigenvectors of J_i corresponding to the eigenvalues $i\Omega_i$ and $-i\Omega_i$, respectively. Let w_i and $\bar{w}_i$ be dual (row) vectors to v_i and $\bar{v}_i$, respectively. Let V_i be the matrix whose columns are v_i and $\bar{v}_i$; i.e.,

$$V_i = (v_i, \bar{v}_i)\ .$$

Notice that V_i^{-1} has w_i and $\bar{w}_i$ as its rows. Moreover, it is easy to check that

$$V_i^{-1} J_i V_i = \begin{pmatrix} i\Omega_i & 0 \\ 0 & -i\Omega_i \end{pmatrix}. \tag{5.36}$$

We use this fact in the proof of the following theorem.

Theorem 5.8 (Local Canonical Model for Multiple Andronov-Hopf Bifurcations in WCNNs) *If the WCNN*

$$\dot{x}_i = f_i(x_i, \lambda) + \varepsilon g_i(x, \lambda, \rho, \varepsilon) \,, \quad x_i \in \mathbb{R}^2 \,, \ i = 1, \ldots, n \,,$$

is near a multiple Andronov-Hopf bifurcation and $\lambda(\varepsilon) = \varepsilon \lambda_1 + \mathcal{O}(\varepsilon^2)$, then there is a change of variables

$$x_i(t) = \sqrt{\varepsilon} V_i \begin{pmatrix} e^{i\Omega_i t} z_i(\tau) \\ e^{-i\Omega_i t} \bar{z}_i(\tau) \end{pmatrix} + \mathcal{O}(\varepsilon) \,, \tag{5.37}$$

where $\tau = \varepsilon t$ is a slow time that transforms the WCNN to

$$z_i' = b_i z_i + d_i z_i |z_i|^2 + \sum_{\substack{j \neq i}}^{n} c_{ij} z_j + \mathcal{O}(\sqrt{\varepsilon}) \,, \tag{5.38}$$

where $' = d/d\tau$; $b_i, d_i, z_i \in \mathbb{C}$; and the (synaptic) coefficients $c_{ij} \in \mathbb{C}$ are given by

$$c_{ij} = \begin{cases} w_i \cdot D_{x_j} g_i \cdot v_j & \text{if } \Omega_i = \Omega_j \,, \\ 0 & \text{if } \Omega_i \neq \Omega_j \,. \end{cases} \tag{5.39}$$

System (5.38) is a canonical model for a WCNN near a multiple Andronov-Hopf bifurcation.

Proof. Consider the uncoupled ($\varepsilon = 0$) system

$$\dot{x}_i = f_i(x_i, 0) = J_i x_i + N_i(x_i) \,,$$

where each function N_i accounts for nonlinear terms in x_i. The change of variables

$$x_i = V_i \begin{pmatrix} z_i \\ \bar{z}_i \end{pmatrix} \tag{5.40}$$

transforms the system to

$$V_i \begin{pmatrix} \dot{z}_i \\ \dot{\bar{z}}_i \end{pmatrix} = J_i V_i \begin{pmatrix} z_i \\ \bar{z}_i \end{pmatrix} + N_i \left(V_i \begin{pmatrix} z_i \\ \bar{z}_i \end{pmatrix} \right) .$$

Multiplying both sides by V_i^{-1} and taking into account (5.36), we obtain

$$\begin{pmatrix} \dot{z}_i \\ \dot{\bar{z}}_i \end{pmatrix} = \begin{pmatrix} i\Omega_i & 0 \\ 0 & -i\Omega_i \end{pmatrix} \begin{pmatrix} z_i \\ \bar{z}_i \end{pmatrix} + V_i^{-1} N_i \left(V_i \begin{pmatrix} z_i \\ \bar{z}_i \end{pmatrix} \right) ,$$

which we rewrite in the form

$$\begin{cases} \dot{z}_i &= \ i\Omega_i z_i + h_i(z_i, \bar{z}_i) \\ \dot{\bar{z}}_i &= \ -i\Omega_i \bar{z}_i + \bar{h}_i(z_i, \bar{z}_i) \end{cases} ,$$

where each function h_i accounts for all nonlinear terms in z_i and $\bar{z}_i$. Since the second equation is the complex conjugate of the first one, it suffices to consider only the first equation

$$\dot{z}_i = \mathrm{i}\Omega_i z_i + h_i(z_i, \bar{z}_i) \, . \tag{5.41}$$

Next, we use the well-known fact from normal form theory (see Section 4.2) that there is a near identity change of variables

$$z_i = \tilde{z}_i + q_i(\tilde{z}_i, \bar{\tilde{z}}_i) \tag{5.42}$$

that transforms (5.41) to the truncated normal form

$$\dot{z}_i = \mathrm{i}\Omega_i z_i + d_i z_i |z_i|^2 + \mathcal{O}(|z_i|^5)$$

for some $d_i \in \mathbb{C}$, where we have erased the $\tilde{}$. This is the standard procedure for transforming a dynamical system having a pair of purely imaginary eigenvalues $\pm \mathrm{i}\Omega_i$ to its normal form.

Now suppose $\varepsilon \neq 0$ and the parameter $\lambda(\varepsilon) = \varepsilon\lambda_1 + \mathcal{O}(\varepsilon^2)$. Since the Jacobian matrix of the uncoupled system

$$J = \begin{pmatrix} J_1 & 0 & \cdots & 0 \\ 0 & J_2 & \cdots & 0 \\ \vdots & \vdots & \ddots & \vdots \\ 0 & 0 & \cdots & J_n \end{pmatrix}$$

is nonsingular, the implicit function theorem guarantees the existence of a family of equilibria $x^\star(\varepsilon) = (x_1^\star(\varepsilon), \ldots, x_n^\star(\varepsilon))^\top \in \mathbb{R}^{2n}$ in the coupled system. It is easy to check that

$$x_i^\star(\varepsilon) = -\varepsilon J_i^{-1} g_i(0, 0, \rho_0, 0) + \mathcal{O}(\varepsilon^2)$$

for all $i = 1, \ldots, n$. Introducing local coordinates at $x^\star(\varepsilon)$, we can rewrite the WCNN in the form

$$\dot{x}_i = J_i x_i + N_i(x_i) + \varepsilon \sum_{j=1}^{n} S_{ij} x_j + \mathcal{O}(\varepsilon|x|^2, \varepsilon^2|x|) \, , \tag{5.43}$$

where the functions N_i account for the nonlinear terms as before, the matrices S_{ij} are given by

$$S_{ij} = D_{x_j} g_i \tag{5.44}$$

for $i \neq j$, and

$$S_{ii} = D_\lambda(D_{x_i} f_i) \cdot \lambda_1 + D_{x_i} g_i - 2(D_{x_i}^2 N) J_i^{-1} g_i \, . \tag{5.45}$$

We apply (5.40) to transform system (5.43) to the complex form

$$\dot{z}_i = \mathrm{i}\Omega_i z_i + h_i(z_i, \bar{z}_i) + \varepsilon \sum_{j=1}^{n} (s_{ij} z_j + e_{ij} \bar{z}_j) + \mathcal{O}(|z_i|^5, \varepsilon|z|^2, \varepsilon^2|z|) \, ,$$

where the complex coefficients s_{ij} and e_{ij} are entries of the 2×2 matrix $V_i^{-1} S_{ij} V_j$; that is,

$$V_i^{-1} S_{ij} V_j = \begin{pmatrix} s_{ij} & e_{ij} \\ \bar{e}_{ij} & \bar{s}_{ij} \end{pmatrix} . \tag{5.46}$$

Now we apply (5.42) to obtain

$$\dot{\tilde{z}}_i = \mathrm{i}\Omega_i \tilde{z}_i + d_i \tilde{z}_i |\tilde{z}_i|^2 + \varepsilon \sum_{j=1}^{n} (s_{ij}\tilde{z}_j + e_{ij}\bar{\tilde{z}}_j) + \mathcal{O}(|\tilde{z}_i|^5, \varepsilon|\tilde{z}|^2, \varepsilon^2|\tilde{z}|) ,$$

Introducing slow time $\tau = \varepsilon t$ and changing variables

$$\tilde{z}_i(t) = \sqrt{\varepsilon} e^{\mathrm{i}\frac{\Omega_i}{\varepsilon}\tau} z_i(\tau) , \tag{5.47}$$

we transform the system to

$$z_i' = d_i z_i |z_i|^2 + \sum_{j=1}^{n} \left(e^{\mathrm{i}\frac{\Omega_j - \Omega_i}{\varepsilon}\tau} s_{ij} z_j + e^{\mathrm{i}\frac{-\Omega_j - \Omega_i}{\varepsilon}\tau} e_{ij}\bar{z}_j \right) + \mathcal{O}(\sqrt{\varepsilon}) . \tag{5.48}$$

Notice that each

$$e^{\mathrm{i}\frac{-\Omega_j - \Omega_i}{\varepsilon}\tau}$$

is a high-frequency oscillatory term as $\varepsilon \to 0$. The same is true for

$$e^{\mathrm{i}\frac{\Omega_j - \Omega_i}{\varepsilon}\tau}$$

unless $\Omega_j = \Omega_i$. After averaging, all terms that have the factor $e^{\mathrm{i}\frac{\delta}{\varepsilon}\tau}$ for $\delta \neq 0$ vanish, and we obtain (5.38), where $b_i = s_{ii}$ and

$$c_{ij} = \begin{cases} s_{ij} & \text{if } \Omega_i = \Omega_j , \\ 0 & \text{if } \Omega_i \neq \Omega_j . \end{cases}$$

It is easy to see that (5.37) is a composition of (5.40), (5.42), and (5.47). Equation (5.39) follows from (5.44), (5.46), and the fact that the first column of V_j is the eigenvector v_j and the first row of V_i^{-1} is the dual vector w_i. ⊔

Alternative proofs of this theorem can be found, for example, in Aronson et al. (1990); Schuster and Wagner (1990); Ermentrout and Kopell (1992), or Hoppensteadt and Izhikevich (1996a). Multiple Andronov-Hopf bifurcations in weakly connected systems were also studied by Malkin (1949, 1956) and Blechman (1971), though they called the oscillators the Liapunov's systems.

The chart in Figure 5.5 summarizes the results obtained in this and the previous sections.

There is a remarkable resemblance of (5.27) and (5.38). The latter has the same form as the former except that all variables are complex. It highlights the fact that there is a pitchfork bifurcation in the amplitude of oscillation at the Andronov-Hopf bifurcation.

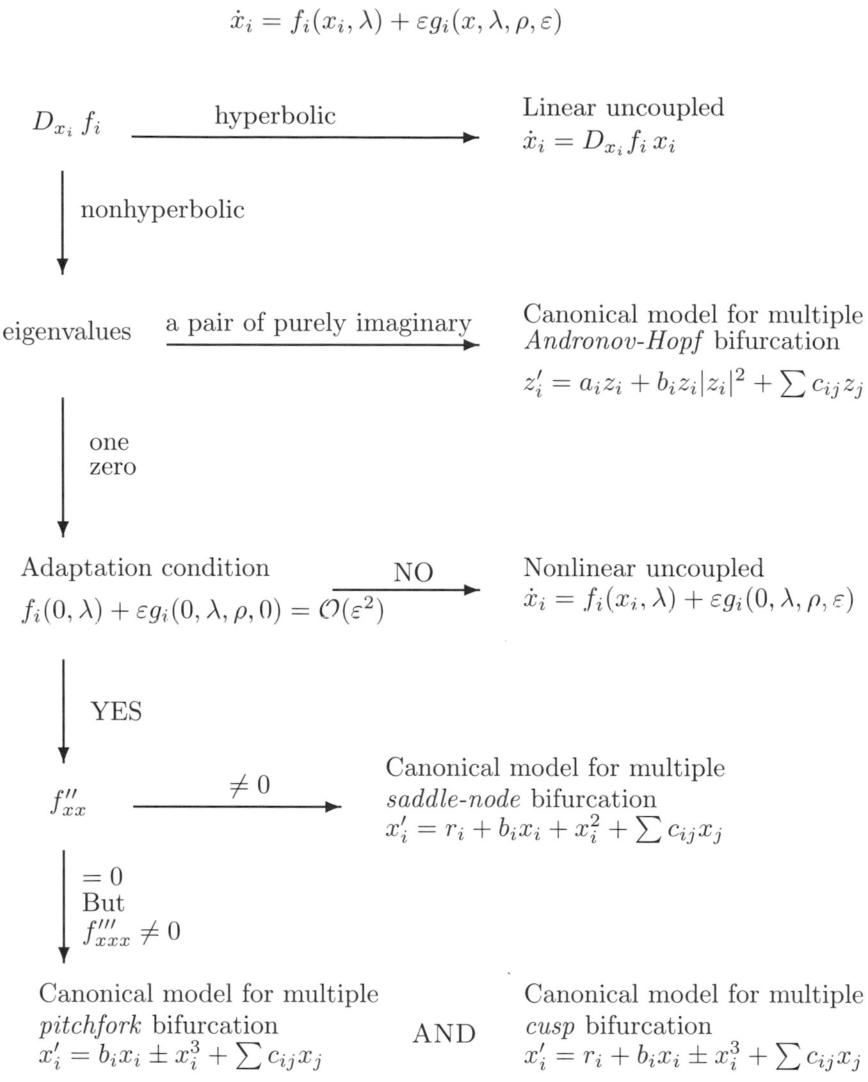

$$\dot{x}_i = f_i(x_i, \lambda) + \varepsilon g_i(x, \lambda, \rho, \varepsilon)$$

$D_{x_i} f_i$ $\xrightarrow{\text{hyperbolic}}$ Linear uncoupled
$\dot{x}_i = D_{x_i} f_i \, x_i$

$\downarrow$ nonhyperbolic

eigenvalues $\xrightarrow{\text{a pair of purely imaginary}}$ Canonical model for multiple
Andronov-Hopf bifurcation

$$z_i' = a_i z_i + b_i z_i |z_i|^2 + \sum c_{ij} z_j$$

$\downarrow$ one zero

Adaptation condition $\xrightarrow{\text{NO}}$ Nonlinear uncoupled
$f_i(0, \lambda) + \varepsilon g_i(0, \lambda, \rho, 0) = \mathcal{O}(\varepsilon^2)$ $\dot{x}_i = f_i(x_i, \lambda) + \varepsilon g_i(0, \lambda, \rho, \varepsilon)$

$\downarrow$ YES

f_{xx}'' $\xrightarrow{\neq 0}$ Canonical model for multiple
saddle-node bifurcation
$x_i' = r_i + b_i x_i + x_i^2 + \sum c_{ij} x_j$

$\downarrow$ $= 0$
But
$f_{xxx}''' \neq 0$

Canonical model for multiple
pitchfork bifurcation
$x_i' = b_i x_i \pm x_i^3 + \sum c_{ij} x_j$

AND

Canonical model for multiple
cusp bifurcation
$x_i' = r_i + b_i x_i \pm x_i^3 + \sum c_{ij} x_j$

Figure 5.5. Summary of bifurcations and canonical models for weakly connected neural networks.

5.4.1 Ginzburg-Landau (Kuramoto-Tsuzuki) Equation

A system similar to the canonical model (5.38) can be derived for a non-linear reaction-diffusion system

$$\dot{X} = F(X) + \varepsilon G \bigtriangleup X \ ,$$

where G is a matrix of diffusion coefficients and $\bigtriangleup$ is the diffusion operator. When the linear operator $DF + \varepsilon G \bigtriangleup$ has a pair of purely imaginary eigenvalues, the reaction-diffusion system can be transformed to (Kuramoto 1984)

$$z' = bz + dz|z|^2 + c \bigtriangleup z + \mathcal{O}(\sqrt{\varepsilon}) \ , \tag{5.49}$$

which is called the complex Ginzburg-Landau (or the Kuramoto-Tsuzuki) equation. The canonical model (5.38) can be viewed as being a discretization of (5.49) where we take the connection matrix $C = (c_{ij})$ to be

$$c_{i,i-1} = c_{i,i+1} = \frac{c}{h^2} \quad \text{and} \quad c_{ii} = -2\frac{c}{h^2} \ ,$$

and $h \ll 1$ is the discretization step, $i = 2, \ldots, n - 1$. Even for this simple choice of C, the system can exhibit complicated dynamical behavior, including spatio-temporal chaos (Kuramoto 1984).

5.4.2 Equality of Frequencies and Interactions

It is customary to call Ω_i the natural frequency of the ith neuron. Since $\lambda(\varepsilon) = \mathcal{O}(\varepsilon)$, each such oscillator has natural frequency $\Omega_i + \mathcal{O}(\varepsilon)$, which is ε-close to Ω_i. Therefore, when we consider a pair of oscillators having equal frequencies Ω_i, we mean up to terms of $\mathcal{O}(\varepsilon)$. One direct consequence of Theorem 5.8 is the following result.

Corollary 5.9 *All neural oscillators can be divided into groups, or pools, according to their natural frequencies. Oscillators from different pools have different natural frequencies, and interactions between them are negligible (see Figure 5.6).*

The proof of this corollary follows from (5.39). This result needs an adjustment when the external forcing $\rho = \rho(t)$ is time-dependent; see Section 5.4.5 below. We prove a similar result for weakly connected limit cycle oscillators (Corollary 9.13), where instead of equality of frequencies we require them to be commensurable. Such frequency dependent linking can also be observed in excitable media (Keener 1989).

Oscillators from different pools work independently of each other even when they have nonzero synaptic contacts $S_{ij} = D_{x_j} g_i$. We say that such pools have a *frequency gap*. One neuron can "feel" another one only when they have equal natural frequencies; that is, when there is no frequency gap. In terms of equation (5.38), this result implies that the mathematical

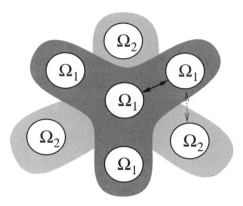

Figure 5.6. Synaptic connections between neurons having different natural frequencies are functionally insignificant. Therefore, the network can be divided into relatively independent ensembles (pools) of oscillators having equal or ε-close natural frequencies. Solid line – functionally significant connections, dashed line – functionally insignificant connections. For simplicity only two connections are shown.

synaptic coefficients c_{ij} and c_{ji} between two oscillators z_i and z_j having different frequencies vanish even when there *are* physiological synaptic connections between them. Although physiologically present and active, those synaptic connections are functionally insignificant and do not play any role in the network dynamics. They could be altered without changing the network behavior. Therefore, when it is necessary to establish communication between two neural oscillators, it does not suffice only to grow synaptic connections from one of them to the other: They must also establish a common frequency of transmission to remove the frequency gap. It is reasonable to speculate that the brain has mechanisms to regulate the natural frequencies Ω_i of its neurons so that the neurons can be entrained into different pools at different times simply by adjusting the Ω_i.

There is a temptation to identify the mechanism of linking and separating of oscillators with the Dynamic Link Architecture (von der Malsburg 1995), but there is a difference between them. The latter is based on phase deviations (phase gaps) and the former on the frequency deviations (frequency gaps). The difference however is not one of principle, since both mechanisms use oscillations to dynamically link or separate ensembles of neurons.

The mechanism for selective tuning that we described above is similar to the one used in tuning the FM (frequency modulation) radio. Since the frequency controls the "channel" of communication, we reach a conclusion that the neural information transmitted via the channel is encoded as phase deviations (timings). Thus, the *neural code* of the brain is a timing of the signal, not the mean firing rate. We continue to study this issue in Section 5.4.5 and Chapters 9, 10, and 13.

5.4.3 Distinct Frequencies

Are there any interactions between oscillators having different frequencies? The answer is YES. A pair of oscillators having $\Omega_i \neq \Omega_j$ can interact, but these interactions are of smaller order, hidden in the term $\mathcal{O}(\sqrt{\varepsilon})$ in the canonical model

$$z_i' = b_i z_i + d_i z_i |z_i|^2 + \mathcal{O}(\sqrt{\varepsilon}) , \qquad z_i \in \mathbb{C} .$$

They are noticeable only on time scales of order $\mathcal{O}(1/\sqrt{\varepsilon})$ (for slow time τ, or $\mathcal{O}(1/\varepsilon\sqrt{\varepsilon})$ for normal time t) and are negligible on shorter time scales.

For example, if all oscillators have different natural frequencies, then the canonical model has the form

$$z_i' = b_i z_i + d_i z_i |z_i|^2 + \sqrt{\varepsilon} p_i(z_1, \bar{z}_1, \dots, z_n, \bar{z}_n, \varepsilon, \tau) , \qquad z_i \in \mathbb{C} , \quad (5.50)$$

for some functions p_i. Obviously, such a system is weakly connected, since $\sqrt{\varepsilon} \ll 1$, and it can be studied the same way as (5.6). If all $\operatorname{Re} b_i < 0$, then the uncoupled ($\sqrt{\varepsilon} = 0$) system has an asymptotically stable hyperbolic equilibrium $z_1 = \cdots = z_n = 0$, and the activity of the coupled ($\sqrt{\varepsilon} \neq 0$) system is locally trivial. When $\operatorname{Re} b_i > 0$ for at least two different i's and the corresponding $\operatorname{Re} d_i < 0$, then (5.50) is a weakly connected network of limit cycle oscillators. Due to the special form of this network, it is possible to simplify it further using averaging theory or normal form theory.

A Network of Two Oscillators

Consider, for the sake of illustration, a network comprising two oscillators having frequencies Ω_1 and Ω_2:

$$\begin{cases} \dot{z}_1 = i\Omega_1 z_1 + h_1(z_1, \bar{z}_1) + \varepsilon p_1(z_1, \bar{z}_1, z_2, \bar{z}_2, \varepsilon) \\ \dot{z}_2 = i\Omega_2 z_2 + h_2(z_2, \bar{z}_2) + \varepsilon p_2(z_1, \bar{z}_1, z_2, \bar{z}_2, \varepsilon) \end{cases} ,$$

where we omitted the equations for $\dot{\bar{z}}_1$ and $\dot{\bar{z}}_2$, since they are complex conjugate to those above.

If the frequencies are equal, then the network is described by the canonical model (5.38) for $n = 2$. When $\Omega_1 \neq \Omega_2$, we can still apply the composition of variable changes (5.42) and (5.47). This results in the system

$$\begin{cases} z_1' = b_1 z_1 + d_1 z_1 |z_1|^2 + \sqrt{\varepsilon} p_1(z_1, \bar{z}_1, z_2, \bar{z}_2, \sqrt{\varepsilon}, \tau) \\ z_2' = b_2 z_2 + d_2 z_2 |z_2|^2 + \sqrt{\varepsilon} p_2(z_1, \bar{z}_1, z_2, \bar{z}_2, \sqrt{\varepsilon}, \tau) \end{cases}$$

for some nonlinear functions p_1 and p_2.

Suppose the frequencies are commensurable, say $\Omega_1 = 2\Omega_2$. Then the original uncoupled system

$$\begin{cases} \dot{z}_1 = i\Omega_1 z_1 + h_1(z_1, \bar{z}_1) \\ \dot{z}_2 = i\Omega_2 z_2 + h_2(z_2, \bar{z}_2) \end{cases}$$

has eigenvalues $\pm i\Omega_1$ and $\pm i\Omega_2$ (recall that there are four equations). It is easy to check that the eigenvalues satisfy resonant conditions. The first four of them are

$$\begin{aligned} i\Omega_1 &= 2i\Omega_2, \\ i\Omega_2 &= i\Omega_1 - i\Omega_2 \\ -i\Omega_1 &= -2i\Omega_2, \\ -i\Omega_2 &= -i\Omega_1 + i\Omega_2, \end{aligned}$$

where the last two are obtained from the first two by conjugation. In fact, if we add $k(i\Omega_1 - 2i\Omega_2), k \in \mathbb{Z}$, to any of the resonances above, then we obtain new ones. Therefore, there is an infinite number of them.

Using normal form theory we can remove all nonlinear terms in the functions p_1 and p_2 except the resonant ones. The resonant monomials that have lowest power and correspond to the relations above are

$$z_2^2, \quad z_1\bar{z}_2, \quad \bar{z}_2^2, \quad \bar{z}_1 z_2,$$

where the last two are complex conjugate to the first two. These terms persist in the functions $p_1, p_2, \bar{p}_1$, and $\bar{p}_2$, respectively. However, they have order $\sqrt{\varepsilon}$, and the canonical model in this case has the form

$$\begin{cases} z_1' = b_1 z_1 + d_1 z_1 |z_1|^2 + \sqrt{\varepsilon} c_{12} z_2^2 + \mathcal{O}(\varepsilon) \\ z_2' = b_2 z_2 + d_2 z_2 |z_2|^2 + \sqrt{\varepsilon} c_{21} z_1 \bar{z}_2 + \mathcal{O}(\varepsilon) \end{cases} \tag{5.51}$$

Obviously, the resonant nonlinear terms are negligible if we study the system on the slow time τ scale up to order $\mathcal{O}(1/\sqrt{\varepsilon})$, which corresponds to the original time t scale up to $\mathcal{O}(\varepsilon^{-3/2})$. Similar canonical models can be obtained for the frequency relationship of the form $\Omega_2 = 2\Omega_1$. In both cases the canonical models are weakly connected networks of limit cycle oscillators.

As another example, suppose the relationship between frequencies is $\Omega_1 = 3\Omega_2$; then the low-order resonances are

$$i\Omega_1 = 3i\Omega_2 \quad \text{and} \quad i\Omega_2 = i\Omega_1 - 2i\Omega_2$$

plus their complex conjugates. The resonant monomials are z_2^3 and $z_1\bar{z}_2^2$, and the canonical model for this case is

$$\begin{cases} z_1' = b_1 z_1 + d_1 z_1 |z_1|^2 + \varepsilon(c_{12} z_2^3 + w_1 z_1 |z_1|^4) + \mathcal{O}(\varepsilon\sqrt{\varepsilon}) \\ z_2' = b_2 z_2 + d_2 z_2 |z_2|^2 + \varepsilon(c_{21} z_1 \bar{z}_2^2 + w_2 z_2 |z_2|^4) + \mathcal{O}(\varepsilon\sqrt{\varepsilon}) \end{cases} \tag{5.52}$$

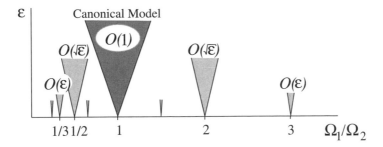

Figure 5.7. The canonical model is strongly connected when $\Omega_1 = \Omega_2$. It is described by the $\sqrt{\varepsilon}$-weakly connected system (5.51) when $\Omega_1 = 2\Omega_2$ or $\Omega_2 = 2\Omega_1$, and the ε-weakly connected system (5.52) when $\Omega_1 = 3\Omega_2$ or $\Omega_2 = 3\Omega_1$, etc.

for some constants $w_1, w_2 \in \mathbb{C}$. Again, the resonant terms can play a role only on large time scales and hence are negligible.

We see that blowing up (rescaling) a neighborhood of a nonhyperbolic equilibrium corresponding to a multiple Andronov-Hopf bifurcation can produce various results depending on the relationship between oscillator frequencies. In one case (when $\Omega_1 = \Omega_2$) we obtain a strongly connected canonical model (5.38). In other cases (such as $\Omega_1 = 2\Omega_2$ or $\Omega_1 = 3\Omega_2$) we obtain weakly connected canonical models (such as (5.51) and (5.52); see Figure 5.7). When $\mathrm{Re}\, b_i > 0$ and $\mathrm{Re}\, d_i < 0$ for $i = 1, 2$, then each equation

$$z_i' = b_i z_i + d_i z_i |z_i|^2$$

describes an oscillator having a stable limit cycle as an attractor. Then the canonical models for oscillators having distinct frequencies describe a weakly connected network of such limit cycle oscillators, which we study in Chapter 9.

General Case

In general, any resonant relationship among oscillator frequencies of the form

$$\Omega_s = m_1 \Omega_1 + \cdots + m_n \Omega_n, \qquad m_1, \ldots, m_n \in \mathbb{Z}, \qquad \sum |m_i| \geq 2 \,,$$

produces a resonance among eigenvalues of the uncoupled system. The corresponding nonremovable resonant term has exponent $\sum |m_i|$ and participates in the canonical model with a coefficient of order $\mathcal{O}(\varepsilon^{(\sum |m_i|-1)/2})$. Therefore, it is negligible on time scales of order up to $\mathcal{O}(1/\varepsilon^{(\sum |m_i|-1)/2})$.

Suppose the whole network is divided into two or more subnetworks (pools) of oscillators each having equal frequencies. Then it is natural to study the dynamics of one such pool first, and then study how these pools interact. In order to study the first problem we may assume that $\Omega_1 = \cdots = \Omega_n = \Omega$; i.e., the whole network is one such pool. We use this assumption in Chapters 10 and 13. In order to study the interactions between the pools, we have to consider a dynamical system of the form

$$X_i' = F_i(X_i) + \sqrt{\varepsilon}\, \varepsilon G_i(X_1, \ldots, X_k, \sqrt{\varepsilon}, \tau) , \quad i = 1, \ldots, k , \qquad (5.53)$$

where $X_i = (z_{j_1}, \ldots, z_{j_{m_i}})$ describes the activity of the ith pool and k is the number of pools. Such a system is weakly connected. It coincides with (5.6) and hence can be studied by the bifurcation methods discussed in this book.

For example, suppose there are only two distinct frequencies in the network, say Ω_1 and Ω_2, such that

$$\Omega_1 = 2\Omega_2.$$

Without loss of generality we may assume that the first n_1 oscillators have frequency Ω_1 and the remaining $n - n_1$ have frequency Ω_2. Then the network is described by the canonical model

$$z_i' = b_i z_i + d_i z_i |z_i|^2 + \sum_{j=1}^{n_1} c_{ij} z_j + \sqrt{\varepsilon} \sum_{j=n_1+1}^{n} c_{ij} z_j^2 + \mathcal{O}(\varepsilon) ,$$

for $i = 1, \ldots, n_1$, and

$$z_i' = b_i z_i + d_i z_i |z_i|^2 + \sum_{j=n_1+1}^{n} c_{ij} z_j + \sqrt{\varepsilon} z_i \sum_{j=1}^{n_1} c_{ij} \bar{z}_j + \mathcal{O}(\varepsilon),$$

for $i = n_1 + 1, \ldots, n$. This system can be rewritten as (5.53) for $k = 2$, where $X_1 = (z_1, \ldots, z_{n_1}) \in \mathbb{C}^{n_1}$ and $X_2 = (z_{n_1+1}, \ldots, z_n) \in \mathbb{C}^{n-n_1}$ describe the activities of the first and the second pools, respectively. We see that the system above is weakly connected. Therefore, it is reasonable to study the uncoupled ($\sqrt{\varepsilon} = 0$) case first and then to study how such pools of oscillators interact. In the rest of the book we study networks of oscillators having a common frequency; that is, we study a single pool.

5.4.4 Distance to an Andronov-Hopf bifurcation

In the proof of Theorem 5.8 we assumed that the bifurcation parameter has the form

$$\lambda(\varepsilon) = \varepsilon \lambda_1 + \mathcal{O}(\varepsilon^2) ;$$

that is, the distance to the bifurcation value $\lambda = 0$ is of order ε. What happens when the distance is not $\mathcal{O}(\varepsilon)$? Suppose $\lambda(\varepsilon) = \mathcal{O}(\varepsilon^\alpha)$ for some value $\alpha > 0$.

When $\alpha > 1$ the canonical model (5.38) is not changed. Indeed, this case corresponds to $\lambda_1 = 0$ in the equation above. This affects only the value S_{ii} defined in (5.45), which may affect the parameter b_i in the canonical model. Values of b_i do not change the form of the canonical models, even when all of them vanish.

The case $\alpha < 1$ does change the canonical model, but not much. Let $\delta = \varepsilon^{\alpha}$ be a new small parameter. Then $\lambda(\delta) = \delta\lambda_1 + o(\delta)$ and $\varepsilon = \delta^{1/\alpha}$, which is smaller than δ. Then the canonical model (5.38) has the form

$$z_i' = b_i z_i + d_i z_i |z_i|^2 + \delta^{\frac{1}{\alpha}-1} \sum_{j \neq i}^{n} c_{ij} z_j + \mathcal{O}(\sqrt{\delta}) , \qquad \delta \ll 1,$$

which resembles that of (5.50). If all $\operatorname{Re} b_i < 0$, then the uncoupled ($\delta = 0$) system has a stable hyperbolic equilibrium $z_1 = \cdots = z_n = 0$. If $\operatorname{Re} b_i > 0$ and $\operatorname{Re} d_i < 0$ for some i, then the canonical model is a (linearly) weakly connected network of limit cycle oscillators, which we study in Chapter 9.

5.4.5 Time-Dependent External Input ρ

Consider a weakly connected network

$$\dot{x}_i = f_i(x_i, \lambda) + \varepsilon g_i(x, \lambda, \rho(t), \varepsilon) , \qquad x_i \in \mathbb{R}^2 , \qquad (5.54)$$

near a multiple Andronov-Hopf bifurcation when the external input ρ depends on time t. As before, we assume $\lambda = \varepsilon\lambda_1 + \mathcal{O}(\varepsilon^2)$ and

$$\rho(t) = \rho_0(t) + \varepsilon\rho_1(t) + \mathcal{O}(\varepsilon^2) ,$$

where $\rho_0, \rho_1 : \mathbb{R} \to \mathcal{R}$ are some "nice" functions, i.e., bounded, sufficiently smooth, etc. We do not assume here that they are periodic.

Systems of the form (5.54) arise when one studies neural networks receiving periodic or chaotic input from other brain structures. For example, each x_i can describe a cortex column, and the multidimensional input $\rho(t)$ can describe influences from subcortical structures such as the thalamus. Another example is the hippocampus forced by the medial septum; see Section 2.3.

We use the same notation as in the beginning of Section 5.4. Let $x_i \in \mathbb{R}^2$ be the activity of the ith neuron and assume that each $J_i = D_{x_i} f_i(0,0)$ has a pair of purely imaginary eigenvalues $\pm i\Omega_i$. Let $v_i, \bar{v}_i$ and $w_i, \bar{w}_i \in \mathbb{C}^2$, be the eigenvectors and dual vectors of J_i, respectively. We say that the input is *nonresonant* to Ω_i if the integral

$$\int_0^T w_i g_i(0, 0, \rho_0(t), 0) e^{-i\Omega_i t} dt$$

is bounded as $T \to \infty$ for all i. This is always the case when $\rho_0(t) = \text{const}$, or when $\rho_0(t)$ is a periodic function with the period not an integer multiple of $2\pi/\Omega_i$ (see below).

Theorem 5.10 *The WCNN (5.54) with a nonresonant input $\rho(t)$ near a multiple Andronov-Hopf bifurcation has a canonical model*

$$z_i' = d_i z_i |z_i|^2 + \sum_{j=1}^{n}(c_{ij}z_j + e_{ij}\bar{z}_j) + \mathcal{O}(\sqrt{\varepsilon}) , \qquad (5.55)$$

where $' = d/d\tau$, $\tau = \varepsilon t$ is slow time, $d_i, z_i, c_{ij}, e_{ij} \in \mathbb{C}$. In particular, the synaptic coefficients c_{ij}, e_{ij} are given by

$$c_{ij} = \lim_{T \to \infty} \frac{1}{T} \int_0^T w_i D_{x_j} g_i(0, 0, \rho_0(t), 0) v_j e^{\mathrm{i}(+\Omega_j - \Omega_i)t} dt , \quad (5.56)$$

$$e_{ij} = \lim_{T \to \infty} \frac{1}{T} \int_0^T w_i D_{x_j} g_i(0, 0, \rho_0(t), 0) \bar{v}_j e^{\mathrm{i}(-\Omega_j - \Omega_i)t} dt , \quad (5.57)$$

provided that these limits exist.

What a canonical model for WCNN (5.54) is when the external input is resonant or when the limits above do not exist is still an open question.

Proof. Below we use the same notations as in the proof of Theorem 5.8. Consider WCNN (5.54) written in the form

$$\dot{x}_i = J_i x_i + N_i(x_i) + \varepsilon a_i(t) + \varepsilon \sum_{j=1}^{n} S_{ij}(t)x_j + \mathcal{O}(\varepsilon^2, \varepsilon x^2) ,$$

where

$$\begin{aligned}
a_i(t) &= D_\lambda f_i \cdot \lambda_1 + g_i(0, 0, \rho_0(t), 0) , &(5.58)\\
S_{ij}(t) &= D_{x_j} g_i(0, 0, \rho_0(t), 0) , \quad i \neq j , \\
S_{ii}(t) &= D_\lambda(D_{x_i} f_i) \cdot \lambda_1 + D_{x_i} g_i(0, 0, \rho_0(t), 0) ,
\end{aligned}$$

and $N_i(x_i) = f_i(x_i, 0) - J_i x_i$ accounts for a nonlinear terms in x_i. Using the change of variables

$$x_i = V_i \begin{pmatrix} z_i \\ \bar{z}_i \end{pmatrix}$$

and then multiplying both sides by V_i^{-1}, we rewrite the WCNN in the form

$$\dot{z}_i = \mathrm{i}\Omega_i z_i + h_i(z_i, \bar{z}_i) + \varepsilon w_i a_i(t) + \varepsilon \sum_{j=1}^{n} w_i S_{ij}(t) V_j \begin{pmatrix} z_j \\ \bar{z}_j \end{pmatrix} + \mathcal{O}(\varepsilon^2, \varepsilon |z|^2) ,$$

$$(5.59)$$

where we omitted the (complex-conjugate) equation for $\dot{\bar{z}}_i$. Here w_i is the first row of the matrix V_i^{-1}.

Now we use normal form theory to obtain a near identity transformation $z_i = \tilde{z}_i + q_i(\tilde{z}_i, \bar{\tilde{z}}_i)$ that converts each uncoupled equation

$$\dot{z}_i = \mathrm{i}\Omega_i z_i + h_i(z_i, \bar{z}_i) , \qquad i = 1, \ldots, n ,$$

to the normal form

$$\dot{\tilde{z}}_i = i\Omega_i\tilde{z}_i + d_i\tilde{z}_i|\tilde{z}_i|^2 + \mathcal{O}(|\tilde{z}_i|^5), \qquad i = 1,\ldots,n.$$

Application of these transformations to the coupled system (5.59) yields

$$\dot{\tilde{z}}_i = i\Omega_i\tilde{z}_i + d_i\tilde{z}_i|\tilde{z}_i|^2 + \varepsilon w_i a_i(t) + \varepsilon\sum_{j=1}^{n} w_i S_{ij}(t)V_j\left(\begin{array}{c}\tilde{z}_j \\ \bar{\tilde{z}}_j\end{array}\right) + \mathcal{O}(\varepsilon^2, \varepsilon|\tilde{z}|^2, |\tilde{z}|^5).$$

Up to this point our proof is similar to the one for Theorem 5.8. But they differ in that we cannot apply the implicit function arguments as we did on page 166.

The origin $\tilde{z} = 0$ is not an equilibrium of the system above. Let us find a time-dependent shift $\tilde{z} = \hat{z} + \varepsilon p(t, \varepsilon)$ such that $\hat{z} = 0$ is the equilibrium. For this we substitute $\hat{z} + \varepsilon p(t, \varepsilon)$ into the equation and then set $\hat{z} = 0$ to obtain

$$\dot{p}_i(t, \varepsilon) = i\Omega_i p_i(t, \varepsilon) + w_i a_i(t) + \mathcal{O}(\varepsilon),$$

which we can solve explicitly. The solution is

$$p_i(t, \varepsilon) = e^{i\Omega_i t}\int_0^t w_i a_i(s)e^{-i\Omega_i s}ds + \mathcal{O}(\varepsilon t).$$

Let us check that $p(t, \varepsilon)$ is bounded, at least for $t = \mathcal{O}(1/\varepsilon)$. For this we must check that the integral is bounded. Using (5.58) we see that it is the sum of two terms

$$w_i D_\lambda f_i \cdot \lambda_1 \int_0^t e^{-i\Omega_i s}ds \quad \text{and} \quad \int_0^t w_i g_i(0,0,\rho_0(s),0)e^{-i\Omega_i s}ds.$$

The first one is obviously bounded, and the second one is bounded because of the nonresonance condition. Thus, the function $p(t, \varepsilon) = \mathcal{O}(1)$ for $t = \mathcal{O}(1/\varepsilon)$, and the WCNN in the new variables has the form

$$\dot{\hat{z}}_i = i\Omega_i\hat{z}_i + d_i\hat{z}_i|\hat{z}_i|^2 + \varepsilon\hat{b}_i(t)\left(\begin{array}{c}\hat{z}_i \\ \bar{\hat{z}}_i\end{array}\right)$$
$$+ \varepsilon\sum_{j=1}^{n} w_i S_{ij}(t)V_j\left(\begin{array}{c}\hat{z}_j \\ \bar{\hat{z}}_j\end{array}\right) + \mathcal{O}(\varepsilon|z|^2, |z|^5),$$

where

$$\hat{b}_i(t) = d_i\left(\begin{array}{c}2|p_i(t,0)|^2 \\ p_i(t,0)^2\end{array}\right) \in \mathbb{C}^2$$

arises as a result of the time-dependent shift $\tilde{z} = \hat{z} + \varepsilon p(t, \varepsilon)$. Now the origin is always an equilibrium.

Next, we use the change of variables

$$\hat{z}_i = \sqrt{\varepsilon}e^{i\Omega_i t}z_i, \qquad i = 1,\ldots,n,$$

that transforms the system above to

$$\dot{z}_i = \varepsilon d_i z_i |z_i|^2 + \varepsilon \hat{b}_i(t) \begin{pmatrix} z_i \\ e^{-2i\Omega_i t} \bar{z}_i \end{pmatrix}$$

$$+ \varepsilon \sum_{j=1}^{n} w_i S_{ij}(t)(v_j e^{i(\Omega_j - \Omega_i)t} z_j + \bar{v}_j e^{i(-\Omega_j - \Omega_i)t} \bar{z}_j) + \mathcal{O}(\varepsilon\sqrt{\varepsilon}) .$$

Averaging this system produces

$$\dot{z}_i = \varepsilon d_i z_i |z_i|^2 + \varepsilon \sum_{j=1}^{n} (c_{ij} z_j + e_{ij} \bar{z}_j) + \mathcal{O}(\varepsilon\sqrt{\varepsilon}) ,$$

where c_{ij} and e_{ij} for $i \neq j$ are given by (5.56) and (5.57), respectively, and

$$c_{ii} = \lim_{T \to \infty} \frac{1}{T} \int_0^T 2d_i |p_i(t,0)|^2 + w_i S_{ii}(t) v_i \, dt ,$$

$$e_{ii} = \lim_{T \to \infty} \frac{1}{T} \int_0^T (d_i p_i(t,0)^2 + w_i S_{ii}(t) \bar{v}_i) e^{-2i\Omega_i t} dt .$$

Introduction of slow time $\tau = \varepsilon t$ yields (5.55). $\square$

Many parameters in the canonical model (5.55) are given in the form

$$\lim_{T \to \infty} \frac{1}{T} \int_0^T q(t) \, dt$$

for some function $q(t)$. The limit above might not exist even when $q(t)$ has nice properties. For example, the integral

$$\frac{1}{T} \int_0^T \sin(\ln t) \, dt = (\sin \ln T - \cos \ln T)/2$$

does not converge as $T \to \infty$ even though $q(t) = \sin \ln t$ is smooth and bounded. Obviously, the averaging procedure that we used at the end of the proof is not valid in this case, so one must find alternative ways to study the WCNN (5.54).

Frequency Gap

Corollary 5.9 requires modification when the external input ρ is time-dependent. Even though the oscillators can still be divided into pools or ensembles according to their natural frequencies Ω_i, the interactions between oscillators from different pools may not be negligible; that is, the synaptic coefficients c_{ij} or e_{ij} from the jth to ith oscillator may be nonzero when $\Omega_i \neq \Omega_j$. We elaborate below.

Constant Forcing

First, let us check that $c_{ij} = e_{ij} = 0$ if $\Omega_i \neq \Omega_j$ when the external input $\rho_0(t) = \rho_0 = \text{const}$. Indeed, consider c_{ij} given by (5.56) (e_{ij} given by (5.57) may be considered likewise). The integral

$$
\begin{aligned}
c_{ij} &= \lim_{T \to \infty} \frac{1}{T} \int_0^T w_i D_{x_j} g_i(0, 0, \rho_0, 0) v_j e^{\mathrm{i}(\Omega_j - \Omega_i)t} dt \\
&= w_i D_{x_j} g_i(0, 0, \rho_0, 0) v_j \lim_{T \to \infty} \frac{1}{T} \int_0^T e^{\mathrm{i}(\Omega_j - \Omega_i)t} dt = 0
\end{aligned}
$$

as in Theorem (5.8), and we say that there is a *frequency gap* between the jth and ith oscillators, which obstructs interactions between them.

Periodic Forcing

The integral above can vanish also when $\rho_0(t)$ is not a constant, but, for example, a periodic function with a period not a multiple of $2\pi/(\Omega_j - \Omega_i)$. Indeed, let L be the period of $\rho_0(t)$. Then its frequency is $\omega = 2\pi/L$, and our assumption in terms of frequencies means that

$$
\Omega_j - \Omega_i \neq \omega m \tag{5.60}
$$

for any $m \in \mathbb{Z}$. We can use the Fourier series

$$
w_i D_{x_j} g_i(0, 0, \rho_0(t), 0) v_i = \sum_{m=-\infty}^{\infty} a_m e^{\mathrm{i}\omega m t}
$$

to represent this function, where $a_m \in \mathbb{C}$ are the Fourier coefficients satisfying

$$
\sum_{m=-\infty}^{\infty} |a_m| < \infty .
$$

The integral in (5.56) is an infinite sum of the form

$$
a_m \int_0^T e^{\mathrm{i}(\Omega_j - \Omega_i + \omega m)t} = \frac{a_m}{\mathrm{i}(\Omega_j - \Omega_i + \omega m)} \left(e^{\mathrm{i}(\Omega_j - \Omega_i + \omega m)T} - 1 \right) .
$$

Since m is an integer, we can use (5.60) to conclude that

$$
\min_m |\Omega_j - \Omega_i - \omega m| = \mu > 0 .
$$

Then the integral in (5.56) is bounded, since it can be majorized:

$$
\begin{aligned}
\left| \sum_{m=-\infty}^{\infty} a_m \int_0^T e^{\mathrm{i}(\Omega_j - \Omega_i + \omega m)t} \right| &\leq \sum_{m=-\infty}^{\infty} \left| a_m \int_0^T e^{\mathrm{i}(\Omega_j - \Omega_i + \omega m)t} \right| \\
&\leq \frac{2}{\mu} \sum_{m=-\infty}^{\infty} |a_m| < \infty .
\end{aligned}
$$

Therefore,

$$\lim_{T \to \infty} \frac{1}{T} \left| \sum_{m=-\infty}^{\infty} a_m \int_0^T e^{\mathrm{i}(\Omega_j - \Omega_i + \omega m)t} \right| = 0 \,,$$

meaning $c_{ij} = 0$. We see that even time-dependent forcing $\rho_0(t)$ sometimes cannot fill the frequency gap between two oscillators, and interactions between them are still negligible in that case.

If the frequency of forcing is an integer fraction of the frequency gap $\Omega_j - \Omega_i$, then the coefficient c_{ij} is nonzero, provided that the corresponding Fourier coefficient a_m is nonzero. Indeed, in this case the only term that survives the integration and limit is a_m, where $m = (\Omega_j - \Omega_i)/\omega \in \mathbb{Z}$, and the synaptic coefficient has the form

$$c_{ij} = a_m = \frac{1}{L} \int_0^L w_i D_{x_j} g_i(0, 0, \rho_0(t), 0) v_i e^{-\mathrm{i}\omega m t} \, dt \neq 0 \,.$$

In this case we say that the forcing $\rho_0(t)$ fills the frequency gap between the oscillators, thereby enabling the interactions between them.

Finally, the periodic forcing $\rho_0(t)$ can impede interactions between neurons transmitting on the same frequency. This happens when the integral

$$c_{ij} = \lim_{T \to \infty} \frac{1}{T} \int_0^T w_i D_{x_j} g_i(0, 0, \rho_0(t), 0) v_i \, dt = 0 \,.$$

A necessary condition for this is that the real and complex parts of the complex-valued function $w_i D_{x_j} g_i(0, 0, \rho, 0) v_i$ change signs, but this is not always the case.

The considerations above lead to the following result.

Corollary 5.11 *External forcing $\rho_0(t)$ can establish interactions between oscillators transmitting on different frequencies or it can disrupt interactions between oscillators transmitting on the same frequency.*

This result can be illustrated in terms of FM radio as follows: Even though our receiver is tuned to a particular radio station, an external forcing can make us hear other stations when the forcing is resonant with the frequency gaps between the stations.

A periodic forcing can effectively link a pair of oscillators having frequency gaps that are in certain integer relations with the period of forcing. When it is necessary to link many pairs of oscillators having various incommensurable frequency gaps, the periodic signal might not be able to cope with this task. This leads to forcing $\rho_0(t)$ having many incommensurable frequencies in its spectrum, and hence to essentially chaotic dynamics. If so, this might be another mechanism by which the brain uses chaos: Chaotic dynamics might be an effective tool for establishing or disrupting interactions between oscillatory neurons. We return to this issue in Section 9.5.

Corollary 5.11 might revolutionize our interpretation of many neurophysiological phenomena, such as theta rhythms in the hippocampus induced by external forcing from the medial septum and oscillations in neocortex induced (or modified) by external forcing from the thalamus. In all these cases a weak forcing can dramatically change the way in which neurons of those brain structures interact.

5.5   Multiple Bogdanov-Takens Bifurcations

Consider a WCNN of the form

$$\dot{x}_i = f_i(x_i, \lambda) + \varepsilon g_i(x, \lambda, \rho, \varepsilon), \qquad x_i \in \mathbb{R}^2, \quad i = 1, \dots, n,$$

and suppose that each equation in the uncoupled system

$$\dot{x}_i = f_i(x_i, \lambda), \qquad x_i \in \mathbb{R}^2, \quad i = 1, \dots, n,$$

has a nonhyperbolic equilibrium $x_i = 0 \in \mathbb{R}^2$ for $\lambda = 0$. Suppose that the Jacobian matrix $J_i = D_{x_i} f_i$ at the equilibrium has a double zero eigenvalue. Generically, J_i has the Jordan canonical form

$$\begin{pmatrix} 0 & 1 \\ 0 & 0 \end{pmatrix},$$

and without loss of generality we may assume that the Jacobian matrices $J_i, i = 1, \dots, n$, are of this form, so that the WCNN is

$$\dot{x}_i = \begin{pmatrix} 0 & 1 \\ 0 & 0 \end{pmatrix} x_i + \hat{f}_i(x_i, \lambda) + \varepsilon g_i(x, \lambda, \rho, \varepsilon), \tag{5.61}$$

where the nonlinear function $\hat{f}_i$ does not have linear terms in x_i. We assume that weakly connected system (5.61) is near *multiple Bogdanov-Takens* bifurcation, which implies certain transversality conditions, which we do not present here. As usual, we assume that

$$\lambda(\varepsilon) = \varepsilon\lambda_1 + \varepsilon^2\lambda_2 + \mathcal{O}(\varepsilon^3), \qquad \lambda_1, \lambda_2 \in \Lambda$$

and

$$\rho(\varepsilon) = \rho_0 + \varepsilon\rho_1 + \mathcal{O}(\varepsilon^2), \qquad \rho_0, \rho_1 \in \mathcal{R}.$$

Unlike saddle-node, cusp or Andronov-Hopf bifurcations, other choices of $\lambda(\varepsilon)$ and $\rho(\varepsilon)$ for multiple Bogdanov-Takens bifurcations do produce interesting results. We discuss them later.

The activity of the system (5.61) depends crucially on the values of the two-dimensional parameters

$$a_i = D_\lambda f_i \cdot \lambda_1 + g_i \in \mathbb{R}^2; \tag{5.62}$$

in particular, on their second components. In analogy with the multiple saddle-node and cusp bifurcations, we say that the parameter λ satisfies the *adaptation condition* near a multiple Bogdanov-Takens bifurcation if

$$a_i = \begin{pmatrix} a_{i1} \\ 0 \end{pmatrix} \in \mathbb{R}^2$$

for some $a_{i1} \in \mathbb{R}$.

With this we have the following result.

Theorem 5.12 (Local Canonical Model for Multiple Bogdanov-Takens Bifurcations in WCNNs) *Suppose a WCNN is near a multiple Bogdanov-Takens bifurcation and the internal parameter λ satisfies the adaptation condition. Then the WCNN has a local canonical model*

$$\begin{cases} u_i' = v_i + \mathcal{O}(\varepsilon) \\ v_i' = r_i + u_i^2 + \sum_{j=1}^n c_{ij} u_j + \mathcal{O}(\sqrt{\varepsilon}) \end{cases}, \qquad (5.63)$$

where $' = d/d\tau$ and $\tau = \sqrt{\varepsilon} t$ is the slow time.

Proof. Any WCNN of the form (5.61) satisfying the adaptation condition near Bogdanov-Takens bifurcation point can be transformed to a form

$$\begin{cases} \dot{x}_{i1} = x_{i2} + \mathcal{O}(|\varepsilon, x|^3) \\ \dot{x}_{i2} = \varepsilon^2 r_i + \varepsilon \sum_{j=1}^n S_{ij} x_j + x_{i1}^2 + \sigma_i x_{i1} x_{i2} + \mathcal{O}(|\varepsilon, x|^3) \end{cases}$$

by a suitable change of variables x_{i2}. Here $S_{ij} = (c_{ij}, e_{ij}) \in \mathbb{R}^2$, $r_i \in \mathbb{R}$ and $\sigma_i \in \mathbb{R}$ are some parameters. Rescaling

$$x_{i1} = \varepsilon u_i, \quad x_{i2} = \varepsilon \sqrt{\varepsilon} v_i \quad \text{and} \quad \tau = \sqrt{\varepsilon} t$$

transforms the system above to

$$\begin{cases} u_i' = v_i + \mathcal{O}(\varepsilon) \\ v_i' = r_i + u_i^2 + \sum_{j=1}^n c_{ij} u_j + \sqrt{\varepsilon} \left(\sigma_i u_i v_i + \sum_{j=1}^n e_{ij} v_j \right) + \mathcal{O}(\varepsilon) \end{cases} \qquad (5.64)$$

$\square$

One may need the representation (5.64) when $C = (c_{ij})$ is symmetric or it vanishes. In both cases the canonical model is a Hamiltonian system with the Hamiltonian

$$H(u, v) = \sum_{i=1}^n \left(\frac{v_i^2}{2} - r_i u_i - \frac{u_i^3}{3} - \frac{1}{2} \sum_{j=1}^n c_{ij} u_i u_j \right),$$

and it is sensitive to any, even very small, perturbations, such as those of the form $\sqrt{\varepsilon}(\sigma_i u_i v_i + \sum e_{ij} v_j)$.

5.5.1 Violation of Adaptation Condition

Suppose the adaptation condition for a multiple Bogdanov-Takens bifurcation is violated uniformly, that is, suppose

$$a_i = D_\lambda f_i \cdot \lambda_1 + g_i = \begin{pmatrix} a_{i1} \\ a_{i2} \end{pmatrix} , \quad a_{i2} \neq 0 ,$$

for all i. The WCNN (5.61) can be written in the form

$$\begin{cases} \dot{x}_{i1} = x_{i2} + \mathcal{O}(|\varepsilon, x|^3) \\ \dot{x}_{i2} = \varepsilon a_{i2} + \varepsilon^2 r_i + \varepsilon \sum_{j=1}^{n} S_{ij} x_j + x_{i1}^2 + \sigma_i x_{i1} x_{i2} + \mathcal{O}(|\varepsilon, x|^3) \end{cases}$$

in this case. The rescaling

$$x_{i1} = \varepsilon^{1/2} u_i , \quad x_{i2} = \varepsilon^{3/4} v_i , \quad \tau = \varepsilon^{1/4} t ,$$

transforms the system above to

$$\begin{cases} u_i' = v_i + \mathcal{O}(\varepsilon^{1/4}) \\ v_i' = a_{i2} + u_i^2 + \varepsilon^{1/4} \sigma_i u_i v_i + \mathcal{O}(\varepsilon^{1/2}) \end{cases} , \quad i = 1, \ldots, n . \tag{5.65}$$

In particular, the term $\mathcal{O}(\varepsilon^{1/2})$ in the second equation accounts for the connections between neurons.

Suppose $a_{i2} > 0$ for at least one i. In this case the activity of this system does not have equilibria, and $(u(\tau), v(\tau))$ leaves any small neighborhood of the origin no matter what the initial conditions $(u(0), v(0))$ are. To study subsequent behavior of this system we need global information about the WCNN, which is not available.

Now suppose that all $a_{i2} < 0$; that is, each equation in the uncoupled system has at least one equilibrium (in fact, two, a saddle and a center, which becomes a focus for $\varepsilon > 0$). The system (5.65) is a perturbation of the integrable (uncoupled) Hamiltonian system

$$\begin{cases} u_i' = v_i \\ v_i' = a_{i2} + u_i^2 \end{cases} , \quad i = 1, \ldots, n , \tag{5.66}$$

with the Hamiltonian

$$H(u, v) = \sum_{i=1}^{n} \left(\frac{v_i^2}{2} - a_{i2} u_i - \frac{u_i^3}{3} \right) .$$

The phase portrait of each Hamiltonian subsystem (fixed i, or $n = 1$) is similar to the one depicted in Figure 5.8. It is easy to see that there are closed orbits and a homoclinic orbit (a saddle separatrix loop). Thus, system (5.65) is a weakly connected network of neurons each having a homoclinic orbit. It is reasonable to expect that the perturbations, such as the term $\varepsilon^{1/4} \sigma_i u_i v_i$ or synaptic connections, may destroy the "nice" phase

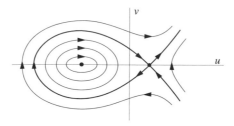

Figure 5.8. Sketch of the phase portrait of Hamiltonian system (5.66) for $n = 1$.

portrait and make dynamics richer. Indeed, they do destroy the homoclinic orbit, but they do not make the dynamics much more interesting.

To study this issue we rewrite a subsystem (fixed i) from (5.65) in the form

$$\begin{cases} u' = v \\ v' = a + bu + u^2 + \sigma uv , \end{cases}$$

where $b = \mathcal{O}(\sqrt{\varepsilon})$ and $\sigma = \mathcal{O}(\sqrt[4]{\varepsilon})$. The homoclinic orbit observable for $\varepsilon = 0$ can persist for $\varepsilon > 0$ when there is a certain relationship between the parameters a and b, in particular, $a = \mathcal{O}(b^2)$ (see Figure 2.14 for reference), which is impossible, since b is small, but a is a nonzero constant in our case. When b vanishes but a does not, the homoclinic orbit disappears for $\varepsilon \neq 0$, and the system's phase portrait is as depicted in Figure 5.9 (for $\sigma > 0$). The

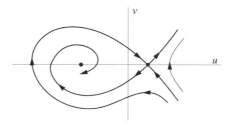

Figure 5.9. A perturbation of the phase portrait depicted in Figure 5.8.

saddle remains a saddle, since it is a hyperbolic equilibrium, but the center becomes a stable focus (unstable for $\sigma < 0$). Linear analysis at the focus shows that it corresponds to an Andronov-Hopf bifurcation. Even though one may observe a stable or unstable periodic orbit for $\varepsilon > 0$, the most interesting dynamical structure, the homoclinic orbit, disappears. We do not give the details here but refer to the exposition of the Bogdanov-Takens bifurcation by Kuznetsov (1995).

Let us return to the weakly connected system (5.65). The homoclinic orbit disappears due to perturbations of the order $\mathcal{O}(\varepsilon^{1/4})$. The WCNN (5.65) is in an $\varepsilon^{1/4}$-neighborhood of a multiple Andronov-Hopf bifurcation and can be analyzed using methods discussed in the previous section. In particular, since the distance to the bifurcation (which is of order $\mathcal{O}(\varepsilon^{1/4})$) is larger than the strength of synaptic connections (which is of order $\mathcal{O}(\varepsilon^{1/2})$), the WCNN either has a locally hyperbolic dynamics or is a weakly connected network of limit cycle oscillators, which are studied in Chapter 9.

Although WCNN activity turns out to be trivial when the adaptation condition is violated, the considerations above motivate another problem for WCNN theory: identifying a canonical model for weakly connected neurons each having a homoclinic orbit, such as the one in Figure 5.8, and finding conditions ensuring rich and interesting local dynamics. This problem might help in analysis of Class 1 neural excitability. It has yet to be solved.

5.5.2 Other Choices of λ and ρ

In the analysis above we assumed that $\lambda = \mathcal{O}(\varepsilon)$ and $\rho = \rho_0 + \mathcal{O}(\varepsilon)$. It is possible to study other choices for λ and ρ. While we do not present details here, since they require many tedious computations, we do summarize some results below.

Suppose

$$\lambda = \varepsilon^\alpha \lambda_1 + \varepsilon^{2\alpha} \lambda_2 + o(\varepsilon^{2\alpha}) \quad \text{and} \quad \rho = \rho_0 + \varepsilon^\beta \rho_1 + o(\varepsilon^\beta)$$

for some positive constants α and β. Then the WCNN (5.6) near a multiple Bogdanov-Takens bifurcation is described by one of the following equations, depending on the values of α and β (see Figure 5.10).

$$(\text{I}) \quad \begin{cases} u_i' = v_i + o(\varepsilon^{\alpha/4}) \\ v_i' = r_i + u_i^2 + \varepsilon^{\alpha/4} \sigma_i u_i v_i + o(\varepsilon^{\alpha/4}) \end{cases}$$

$$(\text{II}) \quad \begin{cases} u_i' = v_i + o(\varepsilon^{1/4}) \\ v_i' = r_i + u_i^2 + \varepsilon^{1/4} \sigma_i u_i v_i + o(\varepsilon^{1/4}) \end{cases}$$

$$(\text{III}) \quad \begin{cases} u_i' = v_i + o(\varepsilon^{\alpha/2}) \\ v_i' = r_i + u_i^2 + \varepsilon^{\alpha/2} (\sigma_i u_i v_i + e_i v_i) + o(\varepsilon^{\alpha/2}) \end{cases}$$

$$(\text{IV}) \quad \begin{cases} u_i' = v_i + o(\varepsilon^{(1+\beta)/4}) \\ v_i' = r_i + u_i^2 + \varepsilon^{(1+\beta)/4} \sigma_i u_i v_i + o(\varepsilon^{(1+\beta)/4}) \end{cases}$$

$$(\text{V}) \quad \begin{cases} u_i' = v_i + o(\varepsilon^{1-\alpha/2}) \\ v_i' = r_i + u_i^2 + \varepsilon^{1-\alpha/2} \sum c_{ij} u_j + o(\varepsilon^{1-\alpha/2}) \end{cases}$$

$$(\text{VI}) \quad \begin{cases} u_i' = v_i + o(\varepsilon^{(1-\beta)/2}) \\ v_i' = r_i + u_i^2 + \varepsilon^{(1-\beta)/2} \sum c_{ij} u_j + o(\varepsilon^{(1-\beta)/2}) \end{cases}$$

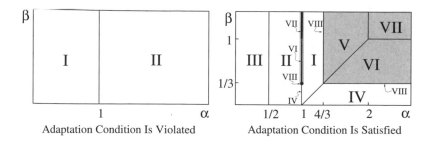

Figure 5.10. A multiple Bogdanov-Takens bifurcation in WCNNs is reduced to various cases depending on parameters α and β. The shaded area is where the activity is nontrivial.

$$
(VII) \quad
\begin{cases}
u_i' = v_i + o(1) \\
v_i' = r_i + u_i^2 + \sum c_{ij} u_j + o(1)
\end{cases}
$$

$$
(VIII) \quad
\begin{cases}
u_i' = v_i + o(\varepsilon^{2/3}) \\
v_i' = r_i + u_i^2 + \varepsilon^{2/3}(\sum c_{ij} u_j + \sigma_i u_i v_i) + o(\varepsilon^{2/3})
\end{cases}
$$

where r_i, σ_i, e_i, and c_{ij} are some constants, $i, j = 1, \ldots, n$. The result depends crucially on the adaptation condition, which we use in the form

$$
\varepsilon^{\alpha} D_\lambda f_i \cdot \lambda_1 + \varepsilon g_i =
\begin{pmatrix}
a_{i1} \\
o(\varepsilon^{\min(1,\alpha)})
\end{pmatrix}.
$$

This definition of the adaptation condition coincides with the one given at the beginning of this section when $\alpha = 1$.

Notice that models I, II, and IV do not produce interesting behavior unless further assumptions are made (e.g., $\sigma_i = 0$). This observation follows from the earlier discussion. The same is true for model III unless there is a certain relationship between r_i, σ_i and e_i. In contrast, models V, VI, VII, and VIII produce interesting behavior. Indeed, canonical model VII is a strongly coupled network of neurons having nonlinear dynamics. Even though models V, VI, and VIII are weakly connected, each uncoupled network is a Hamiltonian system and hence is sensitive to small perturbations such as the weak connections between the neurons. An interesting feature of these models is that each neuron has a homoclinic orbit (saddle separatrix loop), and the synaptic connections determine how the orbit disappears. In this case a local analysis near the homoclinic orbit can provide global information about the behavior of such systems.

5.6   Codimensions of Canonical Models

Let us count how many conditions should be satisfied so that WCNN (5.6) is governed, for example, by the canonical model (5.21). The number of conditions is called the *codimension*.

First, each neuron must be at a saddle-node bifurcation (n conditions). Second, the adaptation condition (5.16) gives another n conditions. Thus, the codimension of (5.21) as a canonical model of (5.6) is $2n$. Similar calculations show that the codimension of the canonical model for cusp and Bogdanov-Takens bifurcations is $3n$, and for an Andronov-Hopf bifurcation it is $2n - 1$, where all neurons have equal frequencies. Thus, there is a natural question: Which of the models is more likely to occur?

Suppose we have two adapted networks of 100 neurons near a multiple saddle-node bifurcation and of 10 neurons near a multiple cusp bifurcation. From a mathematical point of view, the second network is more generic because it requires only 30 conditions instead of the 200 for the first network. A neurobiologist would point out that despite the quantitative differences, many neurons are qualitatively similar. If there is an electrophysiological mechanism that forces one neuron to be near threshold and to be adapted, then the same mechanism should be present in the other 99 neurons in a 100-neuron network. In other words, if two conditions are satisfied for one neuron, then it is physiologically plausible that approximately the same conditions are satisfied for the other neurons. Thus, there are only two dissimilar conditions imposed on the codimension-$2n$ network, whereas there are three conditions in codimension $3n$ networks. We see that there could be a discrepancy between the mathematical and the biological notions of codimension. In fact, there is no contradiction if we take into account the observation that the "dynamical similarity" of neurons is some kind of a symmetry in the system. Systems with symmetry are ubiquitous in nature.

Another aspect of codimension is that when we have a dynamical system and a set of bifurcation parameters, it could be difficult (even numerically) to find a bifurcation point of high codimension. But if we know that there are interesting neurocomputational properties near some multiple bifurcation with high codimension, then it is an easy task to construct a dynamical system that is close to the multiple bifurcation and exhibits the neurocomputational properties.

6
Local Analysis of Singularly Perturbed WCNNs

In this chapter we study the local dynamics of singularly perturbed weakly connected neural networks of the form

$$\begin{cases} \mu X_i' = F_i(X_i, Y_i, \lambda, \mu) + \varepsilon P_i(X, Y, \lambda, \rho, \mu, \varepsilon) \\ Y_i' = G_i(X_i, Y_i, \lambda, \mu) + \varepsilon Q_i(X, Y, \lambda, \rho, \mu, \varepsilon) \end{cases}, \quad \varepsilon, \mu \ll 1, \quad (6.1)$$

where $X_i \in \mathbb{R}^k, Y_i \in \mathbb{R}^m$ are fast and slow variables, respectively; τ is a slow time; and $' = d/d\tau$. The parameters ε and μ are small, representing the strength of synaptic connections and ratio of time scales, respectively. The parameters $\lambda \in \Lambda$ and $\rho \in \mathcal{R}$ have the same meaning as in the previous chapter: They represent a multidimensional bifurcation parameter and external input from sensor organs, respectively. As before, we assume that all functions F_i, G_i, which represent the dynamics of each neuron, and all P_i and Q_i, which represent connections between the neurons, are as smooth as necessary for our computations.

We perform a local analysis of (6.1) at an equilibrium to study how and when the equilibrium loses stability. Global analysis of (6.1) is more complicated and goes beyond the scope of this book. Detailed local and global analysis of a singularly perturbed dynamical system of the form (6.1) for $n = 1$ can be found in Mishchenko et al. (1994). Some global features of dynamics of two coupled relaxation oscillators were studied, e.g., by Belair and Holmes (1984), Chakraborty and Rand (1988), Grasman (1987), Kopell and Somers (1995), Skinner et al. (1994), Somers and Kopell (1993), and Storti and Rand (1986). Terman and Wang (1995) studied some neurocomputational properties of a network of $n > 2$ relaxation oscillators using computer simulation.

6.1   Introduction

We can treat system (6.1) as an ε-perturbation of the uncoupled singularly perturbed system

$$\begin{cases} \mu X_i' = F_i(X_i, Y_i, \lambda, \mu) \\ \quad Y_i' = G_i(X_i, Y_i, \lambda, \mu) \end{cases} , \quad i = 1, \dots, n , \tag{6.2}$$

or as a singular perturbation of the weakly connected system

$$\begin{cases} X_i' = F_i(X_i, Y_i, \lambda, \mu) + \varepsilon P_i(X, Y, \lambda, \rho, \mu, \varepsilon) \\ Y_i' = G_i(X_i, Y_i, \lambda, \mu) + \varepsilon Q_i(X, Y, \lambda, \rho, \mu, \varepsilon) \end{cases} , \quad i = 1, \dots, n.$$

System (6.2) describes an uncoupled network of relaxation neurons (RNs). In this chapter we assume that each RN has an equilibrium $(X_i^\star, Y_i^\star) \in \mathbb{R}^{k+m}$. Then system (6.2) has an equilibrium $(X^\star, Y^\star) \in \mathbb{R}^{n(k+m)}$, and we study the local dynamics of WCNN (6.1) in a neighborhood of the equilibrium.

6.1.1   Motivational Examples

Local analysis can provide some global information about the behavior of a system, especially when the system is near a bifurcation.

Suppose X_i and Y_i in (6.2) are one-dimensional variables and the null-clines $F_i(X_i, Y_i, \lambda^\star, 0) = 0$ and $G_i(X_i, Y_i, \lambda^\star, 0) = 0$ intersect transversally as depicted in Figure 6.1a. Then, each RN has a stable hyperbolic equi-

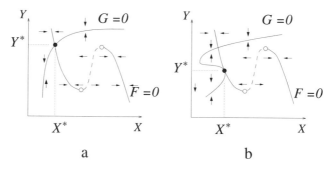

a   b

Figure 6.1. Possible intersections of nullclines of the relaxation neuron described by (6.2) for fixed i.

librium $(X_i^\star, Y_i^\star)$, and hence, the uncoupled system (6.2) as a whole has a stable hyperbolic equilibrium $(X^\star, Y^\star)$. The WCNN (6.1), which is an ε-perturbation of (6.2), also has a stable hyperbolic equilibrium that is, in some neighborhood of $(X^\star, Y^\star)$. Theorem 6.1, presented in Section 6.2, and the fundamental theorem of WCNN theory (Section 5.2.1) ensure that (6.1) does not acquire any nonlinear features that make its activity more interesting than that of (6.2).

Suppose the nullclines intersect nontransversally, as depicted in Figure 6.1b. Obviously, under perturbations, the nonhyperbolic equilibrium $(X_i^\star, Y_i^\star)$ may disappear or transform into a pair of equilibria. Thus, it is reasonable to expect that the WCNN (6.1) might have some nonlinear properties that (6.2) does not have. This case can also be reduced to the one studied in the previous chapter.

These examples show that sometimes the existence of two time scales (slow τ and fast $t = \tau/\mu$) is irrelevant when the network of RNs can be reduced to a network of nonrelaxation neurons (see Section 6.2).

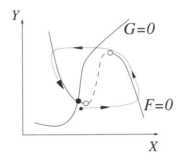

Figure 6.2. An excitable system. There are initial conditions for which the system (6.2), fixed i or $n = 1$, generates an action potential, or spike (dotted line).

When the nullclines intersect, as depicted in Figure 6.2, the equilibrium point is globally attractive. Nevertheless, there are initial conditions, which are relatively close to the equilibrium, such that the activity of the RN leads to large changes in the state variables before the RN activity eventually returns to the equilibrium. This amplified response (the dotted line in Figure 6.2) is called an action potential, or spike. Such systems are called *excitable* (Alexander et al. 1990). The case we depicted in Figure 6.2 corresponds to Class 2 neural excitability.

The excitability can be observed when

$$\frac{\partial F_i}{\partial X_i}(X_i^\star, Y_i^\star, \lambda, \mu) \approx 0, \qquad (6.3)$$

i.e., when the intersection of nullclines, $(X_i^\star, Y_i^\star)$, is near a point for which

$$\frac{\partial F_i}{\partial X_i} = 0.$$

We depict such points as open circles in Figure 6.1 and 6.2. If X_i is a vector, then condition (6.3) should be replaced by the condition that the Jacobian matrix $D_{X_i}F_i$ evaluated at the equilibrium is nonhyperbolic (i.e., has eigenvalues with zero real part). If the parameter λ is close to $\lambda^\star$ and the closeness is compatible with the strength of connections ε, then WCNN (6.1) may have interesting nonlinear properties. The ε-perturbations from the other neurons can force an RN to generate the action potential or to remain silent. Local analysis at the nonhyperbolic equilibrium may often determine which RN is to fire a spike and which is to be quiescent. This is in the spirit of the nonhyperbolic neural network approach, which we discussed in Section 3.1.

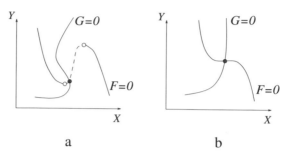

a   b

Figure 6.3. Intersections of nullclines which do not correspond to an excitable system. a. A relaxation oscillator with nonzero amplitude. b. A relaxation oscillator with zero amplitude.

The excitable system is one of the many relaxation systems under consideration. For example, the RNs whose nullclines intersect as depicted in Figure 6.3 are not excitable. Nevertheless, our theory is applicable to them too. In both cases condition (6.3) is satisfied, and a network of such neurons can have interesting nonlinear properties. We start studying networks of such RNs in Section 6.3.

Below we show that if condition (6.3) is violated for all RNs, then the network of such RNs behaves similarly to a network of nonrelaxation neurons, which we studied in the previous chapter.

6.2   Reduction to the Regular Perturbation Problem

One can consider (6.1) as a singular perturbation of the unperturbed ($\mu = 0$) weakly connected system

$$\begin{cases} 0 & = & F_i(X_i, Y_i, \lambda, \mu) + \varepsilon P_i(X, Y, \lambda, \rho, \mu, \varepsilon) \\ Y_i' & = & G_i(X_i, Y_i, \lambda, \mu) + \varepsilon Q_i(X, Y, \lambda, \rho, \mu, \varepsilon) \end{cases}, \quad i = 1, \dots, n. \quad (6.4)$$

The reduced system (6.4) is a quasi-static approximation to (6.1). Then the following questions arise: When do (6.1) and (6.4) have similar dynamical properties? Can (6.4) be further simplified? Partial answers are contained in the following theorem.

Theorem 6.1 *Suppose that (6.2) has an equilibrium point $(X^\star, Y^\star)$ for $\lambda = \lambda^\star$. Suppose that each Jacobian matrix $D_{X_i} F_i$ at the equilibrium is* *hyperbolic. Then the regularly perturbed WCNN*

$$y_i' = g_i(y_i, \lambda, \mu) + \varepsilon q_i(y, \lambda, \rho, \mu, \varepsilon) , \qquad i = 1, \dots, n , \qquad (6.5)$$

where $y = Y + \mathcal{O}(\varepsilon, \mu)$ and g_i and q_i are some functions, is a local model for the singularly perturbed WCNN (6.1).

The proof of the theorem is based on implicit function arguments and singular perturbation techniques. We do not present the proof here because Theorem 6.1 is a corollary of Theorem 6.2, which we prove in the next section.

We see that when Jacobian matrices $D_{X_i} F_i$ are hyperbolic, we can completely eliminate the fast variables X_i and concentrate our efforts on studying slow dynamics described by variables Y_i. We studied regularly perturbed systems of the form (6.5) in the previous chapter, where we showed how they can be further simplified.

In the rest of this chapter we study (6.1) for the case when its reduction to (6.5) is impossible, i.e., when some (or all) of the Jacobian matrices $D_{X_i} F_i$ are nonhyperbolic (have eigenvalues with zero real parts). This occurs when the corresponding RNs are near thresholds and are sensitive to external perturbations.

6.3   Center Manifold Reduction

Consider the fast subsystem of (6.2),

$$\dot{X}_i = F_i(X_i, Y_i, \lambda, \mu) , \qquad i = 1, \dots, n , \qquad (6.6)$$

where $\dot{} = d/dt$ and $t = \tau/\mu$ is fast time. We assume that each $X_i = 0$ is an equilibrium of (6.6) for $(Y_i, \lambda, \mu) = (0, 0, 0)$.

Theorem 6.2   (The Center Manifold Reduction for Singularly Perturbed WCNNs) *Suppose that each of the first n_1 Jacobian matrices $L_i = D_{X_i} F_i$ in (6.6) at the equilibrium are nonhyperbolic, and the other $n - n_1$ Jacobian matrices are hyperbolic. Then (6.1) has a local model*

$$\begin{cases} \mu x_i' &= f_i(x_i, y_i, \lambda, \mu) + \varepsilon p_i(x, y, \lambda, \rho, \mu, \varepsilon) , & i = 1, \dots, n_1, \\ y_i' &= g_i(x_i, y_i, \lambda, \mu) + \varepsilon q_i(x, y, \lambda, \rho, \mu, \varepsilon) , & i = 1, \dots, n_1, \\ y_j' &= g_j(y_j, \lambda, \mu) + \varepsilon q_j(x, y, \lambda, \rho, \mu, \varepsilon) , & j = n_1 + 1, \dots, n. \end{cases}$$

$$(6.7)$$

Here $x_i \in E_i^c$, where E_i^c is the center subspace, that is, the subspace spanned by the (generalized) eigenvectors corresponding to the eigenvalues with zero real parts. Moreover, $J_i = D_{x_i} f_i = L_{i|E_i^c}$ for $i = 1, \dots, n_1$. In particular, the J_i have all eigenvalues with zero real parts.

Proof. Let us rewrite (6.1) as

$$\begin{cases} \dot{X}_i &= F_i(X_i, Y_i, \lambda, \mu) + \varepsilon P_i(X, Y, \lambda, \rho, \mu, \varepsilon) \\ \dot{Y}_i &= \mu\left(G_i(X_i, Y_i, \lambda, \mu) + \varepsilon Q_i(X, Y, \lambda, \rho, \mu, \varepsilon)\right) \end{cases}.$$

Notice that this system is weakly connected and regularly perturbed. For $\varepsilon = \mu = 0$ it is

$$\begin{cases} \dot{X}_i &= F_i(X_i, Y_i, \lambda, 0) \\ \dot{Y}_i &= 0 \end{cases}, \qquad i = 1, \dots, n.$$

One can treat each couple $(X_i, Y_i) \in \mathbb{R}^{k+m}$ as a single multidimensional variable. Its center subspace is $\hat{E}_i^c = E_i^c \times \mathbb{R}^m$ for $i = 1, \dots, n_1$ and $\hat{E}_j^c = \mathbb{R}^m$ for $j = n_1 + 1, \dots, n$. Application of the center manifold reduction for regularly perturbed WCNNs (Theorem 5.2) eliminates variables along the hyperbolic subspaces and produces a local model of the form

$$\begin{cases} \dot{x}_i &= f_i(x_i, y_i, \lambda, \mu) + \varepsilon p_i(x, y, \lambda, \rho, \mu, \varepsilon) \\ \dot{y}_i &= \mu\left(g_i(x_i, y_i, \lambda, \mu) + \varepsilon q_i(x, y, \lambda, \rho, \mu, \varepsilon)\right) \end{cases}$$

for $i = 1, \dots, n_1$ and

$$\dot{y}_j = \mu\left(g_j(y_j, \lambda, \mu) + \varepsilon q_j(x, y, \lambda, \rho, \mu, \varepsilon)\right)$$

for $j = n_1 + 1, \dots, n$, where f_i, g_i, p_i, and q_i are some functions. Now we can use slow time to rewrite the local model above in the form (6.7). $\square$

Corollary 6.3 *Theorem 6.1 follows from Theorem 6.2 when $n_1 = 0$.*

In the proof of the theorem we did not use the assumption that $(X, Y) = (0, 0)$ is an equilibrium of the singularly perturbed WCNN (6.1). What we used is that $X = 0$ is an equilibrium of the reduced fast uncoupled system

$$\dot{X}_i = F_i(X_i, Y_i, \lambda, 0), \qquad i = 1, \dots, n,$$

for some Y_i and λ. We do not need Y to be an equilibrium for the slow uncoupled system

$$\dot{Y}_i = G_i(X_i, Y_i, \lambda, 0), \qquad i = 1, \dots, n.$$

Thus, the center manifold reduction is still valid when functions $G_i = \mathcal{O}(1)$; that is, when X is on the nullcline N_X but off the intersection of N_X and N_Y.

Without loss of generality we assume in the following that all Jacobian matrices L_i, $i = 1, \dots, n_1$, have all eigenvalues with zero real parts. This means that we consider the WCNN (6.1) to be restricted to its center manifold.

6.4   Preparation Lemma

In this section we study the local dynamics of the singularly perturbed WCNN

$$\begin{cases} \mu x_i' &= f_i(x_i, y_i, \lambda, \mu) + \varepsilon p_i(x, y, \lambda, \rho, \mu, \varepsilon) \\ y_i' &= g_i(x_i, y_i, \lambda, \mu) + \varepsilon q_i(x, y, \lambda, \rho, \mu, \varepsilon) \end{cases}, \quad i = 1, \ldots, n, \quad (6.8)$$

at the equilibrium $(x, y) = (0, 0)$ for $(\lambda, \rho, \mu, \varepsilon) = (0, \rho_0, 0, 0)$ for the case when each Jacobian matrix $L_i = D_{x_i} f_i$ has one simple zero eigenvalue. By Theorem 6.2 we may assume that each $x_i \in \mathbb{R}$; i.e., it is a one-dimensional variable, and

$$L_i = \frac{\partial f_i}{\partial x_i} = 0 \ \in \mathbb{R} \ ;$$

i.e., we restricted (6.8) to its center manifold.

Fix $y_i = 0$ and $\mu = \varepsilon = 0$ and consider the reduced fast system

$$\dot{x}_i = f_i(x_i, 0, \lambda, 0) \qquad (6.9)$$

at the equilibrium point $(x_i, \lambda) = (0, 0)$. Condition $\partial f_i / \partial x_i = 0$ means that $x_i = 0$ is a nonhyperbolic equilibrium for (6.9) and a bifurcation might occur. In this case the RN

$$\begin{cases} \mu x_i' &= f_i(x_i, y_i, \lambda, \mu) \\ y_i' &= g_i(x_i, y_i, \lambda, \mu) \end{cases}$$

is at a quasi-static bifurcation. We study here the case when all equations above undergo bifurcations of the same type simultaneously. Such bifurcations are called multiple. We derive canonical models for the simplest and most interesting cases, namely multiple quasi-static saddle-node, pitchfork, cusp, and Andronov-Hopf bifurcations.

Our analysis of the WCNN (6.8) is local in the sense that we study its dynamics in an c neighborhood of the equilibrium point $(x, y, \lambda, \rho, \mu, \varepsilon) = (0, 0, 0, \rho_0, 0, 0)$. In particular, we study the dependence of solutions of (6.8) on parameters $\lambda, \mu, \varepsilon$ by supposing that

$$\begin{aligned} \lambda(\varepsilon) &= 0 + \varepsilon \lambda_1 + \mathcal{O}(\varepsilon^2) \ , & \lambda_1 &\in \Lambda \ , \\ \rho(\varepsilon) &= \rho_0 + \varepsilon \rho_1 + \mathcal{O}(\varepsilon^2) \ , & \rho_0, \rho_1 &\in \mathcal{R} \ , \\ \mu(\varepsilon) &= 0 + \varepsilon \mu_1 + \varepsilon^2 \mu_2 + \mathcal{O}(\varepsilon^3) \ , & \mu_1, \mu_2 &\in \mathbb{R} \ . \end{aligned}$$

The analysis below does not depend crucially on λ_1, ρ_0, or ρ_1. In contrast to this, the values of μ_1 and μ_2 are important, in particular whether or not $\mu_1 = 0$.

We also assume that the following transversality condition

$$D_{y_i} f_i \cdot D_{x_i} g_i \neq 0 \qquad (6.10)$$

is satisfied for all i. If this condition is satisfied, we say that the relationship between x_i and y_i is nondegenerate. This condition is generic in the sense that the set of functions for which this is not true has measure zero. We need (6.10) to exclude many degenerate situations that may complicate our analysis. One of them is depicted in Figure 6.4 and corresponds to the case $D_{x_i} g_i = 0$.

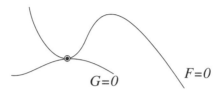

Figure 6.4. An example of violation of the transversality condition (6.10).

Before proceeding to derivations of the canonical models, we prove the following lemma, which transforms the singularly perturbed WCNN (6.8) to a convenient form.

Lemma 6.4 *There is a linear shift $y \to y + \tilde{y}$ such that the initial portion of the Taylor expansion of WCNN (6.8) at $(x, y) = (0, \tilde{y}(\varepsilon))$ has the form*

$$\begin{cases} \dot{x}_i &= A_i y_i + p_i x_i^2 + \varepsilon \sum_{j=1}^n C_{ij} x_j &+ \mathcal{O}(xy, \varepsilon y, y^2, x|\varepsilon, x|^2) \\ \dot{y}_i &= \mu(\varepsilon R_i + B_i x_i + H_i y_i) &+ \mu \mathcal{O}(|\varepsilon, x, y|^2) \end{cases} \quad (6.11)$$

for $i = 1, \ldots, n$. In particular, the equations for $\dot{x}_i$ do not have terms $\varepsilon, \varepsilon^2, \ldots$

Proof. Let $A_i = D_{y_i} f_i \in \mathbb{R}^m$ and $B_i = \partial g_i / \partial x_i \in \mathbb{R}^m$. From the transversality condition $A_i B_i \neq 0$ (eq. (6.10)) it follows that $A_i \neq 0$ for all i. Applying the implicit function theorem to the system of algebraic equations

$$0 = f_i(0, y_i, \lambda(\varepsilon), \mu(\varepsilon)) + \varepsilon p_i(0, y, \lambda(\varepsilon), \rho(\varepsilon), \mu(\varepsilon), \varepsilon) , \quad i = 1, \ldots, n$$

gives a unique smooth function

$$\tilde{y}(\varepsilon) = y(\lambda(\varepsilon), \rho(\varepsilon), \mu(\varepsilon), \varepsilon)$$

such that

$$f_i(0, \tilde{y}_i(\varepsilon), \lambda(\varepsilon), \mu(\varepsilon)) + \varepsilon p_i(0, \tilde{y}(\varepsilon), \lambda(\varepsilon), \rho(\varepsilon), \mu(\varepsilon), \varepsilon) \equiv 0 , \quad i = 1, \ldots, n ,$$

for all sufficiently small ε. Let $(x, y) \in \mathbb{R}^n \times \mathbb{R}^{mn}$ be local coordinates at $(0, \tilde{y}(\varepsilon))$. The equation above guarantees that the Taylor expansions of the equations for x_i' in (6.8) at $(0, \tilde{y}(\varepsilon))$ do not have terms of order $\varepsilon, \varepsilon^2, \ldots$

not multiplied by state variables. Then, the initial portion of the Taylor expansion of (6.8) is defined in (6.11) using the fast time $t = \tau/\mu$, where $A_i = D_{y_i} f_i$, $B_i = \partial g_i / \partial x_i$, $H_i = D_{y_i} g_i$, and $C_{ij} = \partial p_i / \partial x_j \in \mathbb{R}$ for $i \neq j$ and

$$C_{ii} = \frac{\partial}{\partial x_i} \left(D_{y_i} f_i \frac{d\tilde{y}_i(0)}{d\varepsilon} + D_\lambda f_i \lambda_1 + \frac{\partial f_i}{\partial \mu} \mu_1 + p_i \right) \in \mathbb{R} \,.$$

Also,

$$p_i = \frac{1}{2} \frac{\partial^2 f_i}{\partial x_i^2} \in \mathbb{R}$$

and

$$R_i = D_{y_i} g_i \frac{d\tilde{y}_i(0)}{d\varepsilon} + D_\lambda g_i \lambda_1 + \frac{\partial g_i}{\partial \mu} \mu_1 + q_i \,,$$

where all derivatives and functions are evaluated at the equilibrium and $\rho = \rho_0$. $\square$

We use representation (6.11) in our derivation of canonical models.

6.5 The Case $\mu = \mathcal{O}(\varepsilon)$

We first study the case

$$\mu(\varepsilon) = 0 + c\mu_1 + \mathcal{O}(\varepsilon^2) \,, \qquad \mu_1 > 0 \,.$$

According to Lemma 6.4, system (6.8) can be expanded in the Taylor series shown in (6.11), where $B_i = D_{x_i} g_i \in \mathbb{R}^m$ is a vector. Let us represent each variable $y_i \in \mathbb{R}^m$ as

$$y_i = \varepsilon(B_i v_i + w_i) \,, \qquad v_i \in \mathbb{R} \,, \qquad w_i \in \text{span}\,\{B_i\}^\perp \,,$$

where $\text{span}\,\{B_i\}^\perp$ is a subspace of $\mathbb{R}^m$ that is orthogonal to $\text{span}\,\{B_i\}$. Let $x_i \to \sqrt{\varepsilon} x_i$ and $t = \varepsilon^{-1/2}\tau$, then (6.11) in new variables has the form

$$\begin{cases} \sqrt{\varepsilon} x_i' &= a_i v_i + A_i w_i + p_i x_i^2 + \mathcal{O}(\sqrt{\varepsilon}) \\ \sqrt{\varepsilon} v_i' &= \mu_1 x_i + \mathcal{O}(\sqrt{\varepsilon}) \\ w_i' &= \mu_1 (r_i + h_i w_i + e_i x_i^2) + \mathcal{O}(\sqrt{\varepsilon}) \end{cases} \tag{6.12}$$

where $a_i = A_i B_i \neq 0$ because of the transversality condition (6.10). Vectors r_i, e_i and operator h_i are projections of R_i, $1/2\partial^2 g_i/\partial x_i^2$ and H_i, respectively, on the linear subspace $\text{span}\,\{B_i\}^\perp$.

The system above is weakly connected and singularly perturbed. To study it we need to consider the reduced fast uncoupled system

$$\begin{cases} \dot{x}_i &= a_i v_i + A_i w_i + p_i x_i^2 \\ \dot{v}_i &= \mu_1 x_i \end{cases} \tag{6.13}$$

near the equilibrium point $(x_i, v_i) = (0, -A_i w_i/a_i)$, where we treat w_i as a multidimensional parameter. The Jacobian matrix at the equilibrium is

$$L_i = \begin{pmatrix} 0 & a_i \\ \mu_1 & 0 \end{pmatrix}.$$

The equilibrium corresponds to the saddle when $a_i > 0$ and to the center when $a_i < 0$.

Suppose all $a_i > 0$. Since the saddle is a hyperbolic equilibrium, we can apply Theorem 6.1 to the singularly perturbed WCNN (6.12) to exclude the "hyperbolic" variables x_i and v_i. The resulting dynamical system for w_i has the form

$$w'_i = g_i(w_i) + \sqrt{\varepsilon} q_i(w, \sqrt{\varepsilon}), \qquad i = 1, \ldots, n,$$

for some functions g_i and q_i, and it can be studied using methods developed in the previous chapter.

Suppose all $a_i < 0$. Then the reduced fast uncoupled system (6.13) is at a multiple Andronov-Hopf bifurcation and the singularly perturbed WCNN (6.12) is at a multiple quasi-static Andronov-Hopf bifurcation. We study such systems in Section 6.6 below.

If some of the nonzero coefficients a_i, say $a_1, \ldots, a_{n_1}$, are positive and the others are negative, we can apply the center manifold reduction Theorem 6.2 to exclude $x_1, v_1, \ldots, x_{n_1}, v_{n_1}$ from the WCNN (6.12). As a result, we obtain a dynamical system of the form

$$\begin{cases} \sqrt{\varepsilon} \begin{pmatrix} x'_i \\ v'_i \end{pmatrix} &= f_i(x_i, v_i, w_i) + \sqrt{\varepsilon} p_i(x, v, w, \sqrt{\varepsilon}) \\ w'_i &= g_i(x_i, v_i, w_i) + \sqrt{\varepsilon} q_i(x, v, w, \sqrt{\varepsilon}) \\ w'_j &= g_j(w_j) + \sqrt{\varepsilon} q_j(x, v, w, \sqrt{\varepsilon}) \end{cases},$$

where $i = 1, \ldots, n_1$ and $j = n_1 + 1, \ldots, n$ and initial portion of Taylor series of each function f_i has the same form as in (6.13). Such a system is a singularly perturbed WCNN at a multiple quasi-static Andronov-Hopf bifurcation.

6.5.1   The Case $\dim y_i = 1$

When each variable y_i in (6.8) is a scalar, the considerations above lead to the following results.

Theorem 6.5 If $\mu = \varepsilon \mu_1 + \mathcal{O}(\varepsilon^2)$, $\mu_1 > 0$, and all coefficients

$$a_i = \frac{\partial f_i}{\partial y_i} \frac{\partial g_i}{\partial x_i} \in \mathbb{R} \tag{6.14}$$

are positive, then the singularly perturbed WCNN (6.8) and the linear uncoupled system

$$\begin{pmatrix} \dot{x}_i \\ \dot{y}_i \end{pmatrix} = \begin{pmatrix} 0 & a_i \\ \mu_1 & 0 \end{pmatrix} \begin{pmatrix} x_i \\ y_i \end{pmatrix}, \qquad i = 1, \ldots, n,$$

are topologically equivalent.

Proof. The initial portion of the Taylor series of system (6.8) represented in the form (6.11) can be blown up to

$$
\begin{cases}
\dot{x}_i & = & a_i v_i + p_i x_i^2 + \mathcal{O}(\sqrt{\varepsilon}) \\
\dot{v}_i & = & \mu_1 x_i + \mathcal{O}(\sqrt{\varepsilon})
\end{cases}
\tag{6.15}
$$

which is obtained from the system (6.12) by noticing that the set of variables $w_1, \ldots, w_n$ is empty, since each y_i is a scalar. The WCNN above is topologically conjugate to its linearization by Theorem 5.1. $\square$

Thus, the singularly perturbed WCNN (6.8) does not have interesting local nonlinear neurocomputational properties, since it is essentially linear and uncoupled in that case.

Theorem 6.6 *If $\mu = \varepsilon \mu_1 + \mathcal{O}(\varepsilon^2)$, $\mu_1 > 0$, and the coefficients a_i defined by (6.14) satisfy $a_i < 0$ for $i = 1, \ldots, n_1$ and $a_j > 0$ for $j = n_1 + 1, \ldots, n$, then the singularly perturbed WCNN (6.8) has a local canonical model*

$$
z_i' = b_i z_i + d_i z_i |z_i|^2 + \sum_{\substack{j \neq i}}^{n_1} c_{ij} z_j + \mathcal{O}(\varepsilon^{1/4}) , \quad i = 1, \ldots, n_1 ,
$$

where $b_i, c_{ij}, d_i, z_i \in \mathbb{C}$.

Proof. The blown up initial portion of the Taylor series of the singularly perturbed WCNN (6.8) is written in the form (6.15) and has n_1 pairs of purely imaginary eigenvalues. The result follows from the theory presented in the previous chapter (see Section 5.4). $\square$

Thus, when $\mu = \mathcal{O}(\varepsilon)$, the singularly perturbed WCNN (6.8) has either uninteresting local dynamics or it is near a multiple Andronov-Hopf or multiple quasi-static Andronov-Hopf bifurcation.

6.6 Multiple Quasi-Static Andronov-Hopf Bifurcations

In this section we study singularly perturbed WCNNs (6.1) near multiple quasi-static Andronov-Hopf bifurcations. Using the center manifold reduction for singularly perturbed WCNNs (Theorem 6.2) we may assume without loss of generality that each variable $x_i \in \mathbb{R}^2$. Hence the Jacobian 2×2 matrix

$$
L_i = D_{x_i} f_i
$$

at the equilibrium has a pair of purely imaginary nonzero eigenvalues $\pm i\Omega_i$.

Theorem 6.7 (Local Canonical Model for Multiple Quasi-Static Andronov-Hopf Bifurcations in WCNNs) *Suppose that (6.1) is at a multiple quasi-static Andronov-Hopf bifurcation point; then it has a local canonical model*

$$
\begin{cases}
z_i' = (b_i + A_i v_i) z_i + d_i z_i |z_i|^2 + \sum_{j=1}^n c_{ij} z_j + \mathcal{O}(\sqrt{\varepsilon}) \\
v_i' = \hat{\mu}(R_i + S_i |z_i|^2 + T_i v_i) + \mathcal{O}(\hat{\mu}\sqrt{\varepsilon})
\end{cases}
, \qquad (6.16)
$$

where $z_i \in \mathbb{C}$, $v_i \in \mathbb{R}^m$, and $b_i, c_{ij}, d_i \in \mathbb{C}$; $A_i : \mathbb{R}^m \to \mathbb{C}$; $R_i, S_i \in \mathbb{R}^m$; $T_i : \mathbb{R}^m \to \mathbb{R}^m$, and $\hat{\mu} = \mu/\varepsilon$.

We do not present a proof here. The derivation of (6.16) coincides (after obvious modifications) with the derivation of the canonical model for multiple Andronov-Hopf bifurcations in weakly connected networks of regularly perturbed (nonrelaxation) neurons and can be found in Section 5.4.

Notice again that in the canonical model (6.16) (as well as in (5.38)) only those RNs interact that have equal natural frequencies Ω_i. If $\Omega_i \neq \Omega_j$ for some i and j, then the ith and jth RNs do not "feel" each other because $c_{ij} = 0$ and $c_{ji} = 0$ in this case. The ith RN can turn on and off its communications with other RNs simply by adjusting its natural frequency Ω_i.

The canonical model (6.16) is valid not only for $\mu = \mathcal{O}(\varepsilon)$, but also for other choices of μ. Thus, if $\mu = \mathcal{O}(\varepsilon^2)$, the parameter $\hat{\mu} = \mathcal{O}(\varepsilon)$ is small in the canonical model, and it becomes a singularly perturbed strongly connected network. We do not study such systems in this book. In fact, behavior of the canonical model (6.16) even for $\hat{\mu} = \mathcal{O}(1)$ is far from being understood.

Notice that the fast subsystem described by z may be oscillatory and the slow subsystem described by v determines the amplitude of oscillation. We encountered such systems when we discussed bursting mechanisms in Section 2.9. When the Andronov-Hopf bifurcation is supercritical ($\operatorname{Re} d_i < 0$), one can treat (6.16) as a *local* canonical model for a weakly connected network of elliptic bursters. A global canonical model has yet to be found.

6.7   The Case $\mu = \mathcal{O}(\varepsilon^2)$

Below we consider singularly perturbed WCNNs for the case $\mu = \mathcal{O}(\varepsilon^2)$.

6.7.1   Multiple Quasi-Static Saddle-Node Bifurcations

Recall that each equation in the reduced fast system (6.9) being at saddle-node bifurcation implies that for all i

$$
\frac{\partial f_i}{\partial x_i} = 0 \quad \text{and} \quad \frac{\partial^2 f_i}{\partial x_i^2} \neq 0 .
$$

Theorem 6.8 (Local Canonical Model for Multiple Quasi-Static Saddle-Node Bifurcations in WCNNs) *Suppose that the singularly perturbed WCNN (6.8) is at a multiple quasi-static saddle-node bifurcation point, that the condition (6.10) is satisfied, and that*

$$\mu(\varepsilon) = \varepsilon^2 \mu_2 + \mathcal{O}(\varepsilon^3), \qquad \mu_2 > 0.$$

Then the change of variables that transforms (6.8) locally to

$$\begin{cases} x_i' = -y_i + x_i^2 + \sum_{j=1}^{n} c_{ij} x_j + \mathcal{O}(\varepsilon) \\ y_i' = a_i(x_i - r_i) + \mathcal{O}(\varepsilon) \end{cases}, \qquad i = 1, \dots, n, \qquad (6.17)$$

where $x_i, y_i \in \mathbb{R}$ are scalar variables and $a_i \neq 0$. The dynamical system (6.17) is a canonical model for the multiple quasi-static saddle-node bifurcations in singularly perturbed WCNNs.

The theorem asserts, in particular, that any weakly connected network of relaxation oscillators (neurons) having nullclines intersected as in Figure 6.5a is locally equivalent to a strongly linearly connected network of non-relaxation oscillators having nullclines as in Figure 6.5b.

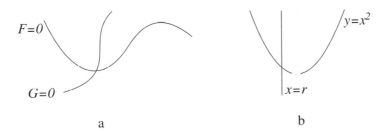

$F=0$

$G=0$

$y=x^2$

$x=r$

a

b

Figure 6.5. A relaxation oscillator with nullclines as in (a) has local canonical model with nullclines as in (b).

Proof. Let $\alpha_i = (1/2)\partial^2 f_i/\partial x_i^2$. If we use

$$\begin{aligned} r_i &= \varepsilon^{-1}\alpha_i x_i, \\ y_i &= -\varepsilon^{-2}\alpha_i A_i y_i, \end{aligned}$$

and rescale the time $\tau = \varepsilon t$, then (6.11) transforms to (6.17), where

$$\begin{aligned} a_i &= -\mu_2 A_i B_i, \\ r_i &= \alpha_i A_i R_i (A_i B_i)^{-1}, \\ c_{ij} &= \alpha_i C_{ij} \alpha_j^{-1}, \end{aligned}$$

for all i and j. Note that $A_i B_i \neq 0$ due to transversality condition (6.10). Hence $a_i \neq 0$, and r_i is finite. $\square$

Note that while $y_i \in \mathbb{R}^m$ is a multidimensional variable, the canonical variable $y_i \in \mathbb{R}$ is a scalar. Therefore, the change of variables cannot be invertible if $m > 1$, and the canonical model describes the dynamics of (6.11) projected into a linear subspace spanned by vectors $A_1, \ldots, A_n$. Thus, we might lose some dynamics taking place along the kernel of the projection.

Remark. We see that in the canonical model (6.17) only connections between the fast variables x_i are significant. The other three types of connections have order ε or ε^2 and hence are negligible. Indeed, (6.17) can be rewritten in the form

$$\begin{cases} x'_i = -y_i + x_i^2 + \sum_{j=1}^n c_{ij} x_j + \varepsilon \sum_{j=1}^n d_{ij} y_j + \mathcal{O}(\varepsilon) \\ y'_i = a_i(x_i - r_i) + \varepsilon \sum_{j=1}^n e_{ij} x_j + \varepsilon^2 \sum_{j=1}^n f_{ij} y_j + \mathcal{O}(\varepsilon) \end{cases} , \quad (6.18)$$

for some $d_{ij}, e_{ij}, f_{ij} \in \mathbb{R}$. We discuss possible applications of this fact in Section 6.9.1.

Using the translations

$$\begin{aligned} \tilde{x}_i &= x_i - r_i \\ \tilde{y}_i &= y_i - r_i^2 - \sum_{j=1}^n c_{ij} r_j \end{aligned} , \quad i = 1, \ldots, n$$

we can rewrite (6.17) in the form (where we erase $\sim$)

$$\begin{cases} x'_i = -y_i + (2r_i + c_{ii})x_i + x_i^2 + \sum_{j \neq i}^n c_{ij} x_j + \mathcal{O}(\varepsilon) \\ y'_i = a_i x_i + \mathcal{O}(\varepsilon) \end{cases} \quad (6.19)$$

which is sometimes more convenient. For, example, it is easy to see that (6.19) has a unique equilibrium $(x, y) = (0, 0)$ up to terms of order $\mathcal{O}(\varepsilon)$, which is asymptotically stable when $r_i \to -\infty$.

With the obvious change of variables, the canonical model (6.17) can be written as

$$y''_i - c_{ii} y'_i - (y'_i)^2 + a_i y_i = \sum_{j \neq i}^n c_{ij} y'_j + \mathcal{O}(\varepsilon) , \quad i = 1, \ldots, n . \quad (6.20)$$

If all $a_i > 0$, then the left-hand side of each equation in (6.20) is the classical model for studying singular Andronov-Hopf bifurcations (Baer and Erneux 1986, 1992; Eckhaus 1983).

6.7.2 Multiple Quasi-Static Pitchfork Bifurcations

Suppose that the conditions $\partial f_i / \partial x_i = \partial^2 f_i / \partial x_i^2 = 0$ and

$$e_i = \frac{1}{6} \frac{\partial^3 f_i}{\partial x_i^3} \neq 0 \quad (6.21)$$

are satisfied for all i. Then the uncoupled system (6.9), which describes fast dynamics of (6.8), is near a multiple cusp singularity. Using Lemma 6.4 one can see that the reduced fast system

$$\mu x_i' = f_i(x_i, y_i, \lambda, \mu) + \varepsilon p_i(x, y, \lambda, \mu, \varepsilon) \quad (6.22)$$

has a family of equilibrium points $(x, y, \lambda, \mu, \varepsilon) = (0, \tilde{y}(\varepsilon), \lambda(\varepsilon), \mu(\varepsilon), \varepsilon)$ for all sufficiently small ε. If we additionally demand that

$$\frac{\partial^2 f_i}{\partial \lambda \partial x} \neq 0 \qquad (6.23)$$

for all i, then (6.22) at $(x, y) = (0, \tilde{y}(\varepsilon))$ has a multiple pitchfork bifurcation. The result presented below is valid even when (6.23) is violated.

Theorem 6.9 (Local Canonical Model for Multiple Quasi-Static Pitchfork Bifurcations in WCNNs)   *Suppose that the singularly perturbed weakly connected system (6.8) is at a multiple quasi-static pitchfork bifurcation point, transversality condition (6.10) is satisfied, and*

$$\mu(\varepsilon) = \varepsilon^2 \mu_2 + \mathcal{O}(\varepsilon^3) , \qquad \mu_2 > 0 .$$

Then there is a change of variables that converts (6.8) to

$$\begin{cases} x_i' = -y_i + \sigma_i x_i^3 + \sum_{j=1}^n c_{ij} x_j + \mathcal{O}(\sqrt{\varepsilon}) \\ y_i' = a_i x_i + \mathcal{O}(\sqrt{\varepsilon}) \end{cases} , \qquad i = 1, \ldots, n , \quad (6.24)$$

where $a_i \neq 0$ and $\sigma_i = \operatorname{sign} e_i = \pm 1$ (the parameters e_i are defined in (6.21)). The dynamical system (6.24) is a canonical model for the multiple quasi-static pitchfork bifurcation.

Thus, a weakly connected network of RNs with nullclines as in Figure 6.6a has a local canonical model (6.24), which is a strongly linearly connected network of oscillators with nullclines as in Figure 6.6b.

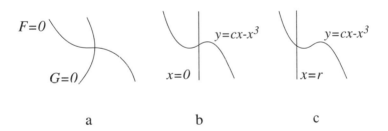

Figure 6.6. A relaxation oscillator depicted in (a) is locally equivalent to an oscillator with the nullclines as in (b) or (c).

Proof. If we use

$$x_i = \sqrt{|e_i|\varepsilon^{-1}} x_i ,$$
$$y_i = -\sqrt{|e_i|\varepsilon^{-3}} A_i y_i ,$$

and rescale time $\tau = \varepsilon t$, then (6.11) transforms to (6.24), where

$$a_i = -\mu_2 A_i B_i ,$$
$$c_{ij} = \sqrt{|e_i|} C_{ij} \sqrt{|e_j|^{-1}} ,$$

for all i and j. Note that $A_i B_i \neq 0$ due to (6.10). Hence $a_i \neq 0$. $\square$

With an obvious change of variables, the canonical model (6.24) can be rewritten as

$$y_i'' - c_{ii}y_i' - \sigma_i(y_i')^3 + a_i y_i = \sum_{j \neq i}^{n} c_{ij} y_j' + \mathcal{O}(\sqrt{\varepsilon}), \qquad i = 1, \ldots, n. \quad (6.25)$$

If all $a_i > 0$ and $\sigma_i = -1$, then the left-hand side of each equation in (6.25) is a Van der Pol's oscillator in Lienard representation. Notice that the oscillators are connected through the derivatives y_j', which are fast variables, not through the slow variables y_j as is frequently assumed (Grasman 1987, Section 3.1).

6.7.3   Multiple Quasi-Static Cusp Bifurcations

Recall that the pitchfork bifurcation is a particular case of the cusp bifurcation. An unfolding of the cusp bifurcation should include the terms x_i^2 in the equations for x_i'. This is equivalent to introducing constant terms in the equations for y_i':

$$\begin{cases} x_i' = -y_i + \sigma_i x_i^3 + \sum_{j=1}^{n} c_{ij} x_j + \mathcal{O}(\sqrt{\varepsilon}) \\ y_i' = a_i(x_i - r_i) + \mathcal{O}(\sqrt{\varepsilon}) \end{cases}, \qquad i = 1, \ldots, n.$$

When $\sigma_i = -1$ and $a_i > 0$, this is a network of connected Bonhoeffer-Van der Pol oscillators.

In fact, the canonical model above could be obtained directly from the singularly perturbed WCNN (6.8) if we assumed that $\rho = \rho_0 + \sqrt{\varepsilon}\rho_{\frac{1}{2}} + \mathcal{O}(\varepsilon)$ for some $\rho_{\frac{1}{2}} > 0$. A similar assumption made in the previous chapter led to a similar result about multiple cusp bifurcations.

The oscillators in the canonical model above have nullclines depicted in Figure 6.6c. Allowing $r_i \neq 0$, we destroy the symmetry observable in the case of the quasi-static pitchfork bifurcation depicted in Figure 6.6b.

6.8   The Case $\mu = \mathcal{O}(\varepsilon^k),\ k > 2$

If

$$\mu(\varepsilon) = 0 + \mathcal{O}(\varepsilon^k), \quad k > 2,$$

then (6.8) can also be reduced to the canonical models (6.17) and (6.24), respectively. But in these cases,

$$a_i = \mathcal{O}(\varepsilon^{k-2}), \quad i = 1, \ldots, n;$$

i.e., each oscillator in the canonical models is a relaxation one.

6.9 Conclusion

We have analyzed the local dynamics of the singularly perturbed WCNN (6.1) at an equilibrium point and showed that in many cases it is governed locally by well-studied systems like regularly perturbed weakly connected systems (6.5) (Theorem 6.1). In other interesting cases the dynamics of (6.1) are governed by the canonical models (6.17), (6.24), or (6.16); see the summary in Figure 6.9. To the best of our knowledge these models have not been analyzed yet. We think they have interesting computational properties. We present some basic analysis of the canonical models (6.17) and (6.24) in Chapter 12.

6.9.1 Synaptic Organizations of the Brain

We saw (remark on p. 202) that in the canonical models only connections between fast variables X_i are significant, and the other three types of connections (fast $\rightarrow$ slow, slow $\rightarrow$ fast, slow $\rightarrow$ slow) are negligible. Even this simple fact has some important biological implications.

Suppose each system (6.2) describes a mechanism of generation of action potentials by a Class 2 neuron. Then the significance of fast $\rightarrow$ fast connections means that the synaptic transmission between neurons is triggered by fast ion channels (in fact, by Na^+ and Ca^{++}; the former are responsible for depolarization, the latter open synaptic vesicles). Since the synaptic transmission mechanism is well studied at present, this observation does not carry much new information, though it would have been valuable say fifty years ago.

Suppose system (6.2) describes the activity of the relaxation neural oscillator, such as Wilson-Cowan's. Then fast $\rightarrow$ fast connections are synaptic connections between excitatory neurons. Thus, one can conclude that information is transmitted from one part of the brain to another through excitatory neurons, while the inhibitory neurons serve only local purposes. And indeed, copious neurophysiological data (Rakic 1976; Shepherd 1976) suggest that excitatory neurons usually have long axons capable of forming distant synaptic contacts, while inhibitory neurons are local-circuit neurons having short axons (or without axons at all). They provide reciprocal inhibition.[1] This division into relay- and interneurons has puzzled biologists for decades (see Rakic 1976). Our analysis shows that even if the inhibitory neurons had long axons, their impact onto other parts of the brain would be negligible. Hence, the long-axon inhibitory neurons are functionally in-

---

[1] There are exceptions. For example, Purkinje cells in the cerebellum are inhibitory, but they have long axons. They transmit information from cerebellar cortex to deep cerebellar nucleus and are believed to be functionally significant. Therefore, our analysis is not applicable to them. This happens because the Purkinje cell activity does not satisfy our prime hypothesis: It is not $\mathcal{O}(\varepsilon^2)$-slow.

$$\begin{cases} \mu X_i' = F_i(X_i, Y_i, \lambda, \mu) + \varepsilon P_i(X, Y, \lambda, \rho, \mu, \varepsilon) \\ Y_i' = G_i(X_i, Y_i, \lambda, \mu) + \varepsilon Q_i(X, Y, \lambda, \rho, \mu, \varepsilon) \end{cases}$$

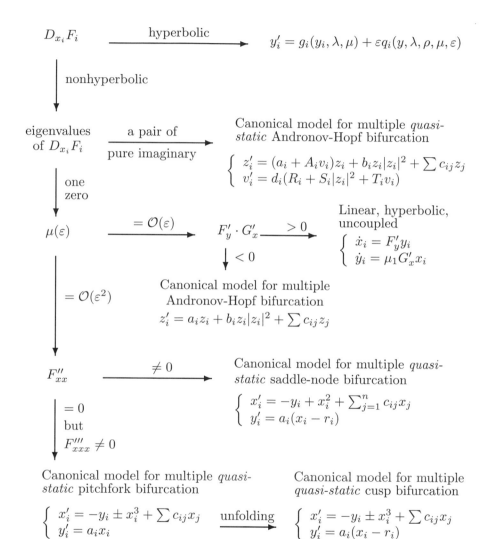

Figure 6.7. Summary of multiple quasi-static bifurcations and canonical models for singularly perturbed weakly connected neural networks.

significant. Of course, this conclusion is valid only when the neural oscillators are of relaxation type and when the ratio of fast and slow time scales μ is of order $\mathcal{O}(\varepsilon^2)$. If this condition is violated or if there are no connections between excitatory neurons, then the other types of connections come to play. In particular, we must consider terms of order $\mathcal{O}(\varepsilon)$ or even $\mathcal{O}(\varepsilon^2)$ in (6.18).

The analysis above attempts to describe why a brain might have the anatomical structure it does. We are still far away from satisfactory explanations of this problem. We continue our study of possible synaptic organizations of the brain in Chapter 13.

7
Local Analysis of Weakly Connected Maps

In previous chapters we studied dynamics of weakly connected networks governed by a system of ordinary differential equations. It is also feasible to consider weakly connected networks of difference equations, or mappings, of the form

$$X_i \mapsto F_i(X_i, \lambda) + \varepsilon G_i(X, \lambda, \rho, \varepsilon) \,, \qquad i = 1, \dots, n, \ \varepsilon \ll 1, \qquad (7.1)$$

where the variables $X_i \in \mathbb{R}^m$, the parameters $\lambda \in \Lambda$, $\rho \in \mathcal{R}$, and the functions F_i and G_i have the same meaning as in previous chapters. The weakly connected mapping (7.1) can be also written in the form

$$X_i^{k+1} = F_i(X_i^k, \lambda) + \varepsilon G_i(X^k, \lambda, \rho, \varepsilon) \,, \qquad i = 1, \dots, n, \ \varepsilon \ll 1,$$

where X^k is the kth iteration of the variable X. In this chapter we use form (7.1) unless we explicitly specify otherwise.

Global analysis of weakly connected mappings having chaotic dynamics goes beyond the scope of this book. Some neurocomputational properties of weakly coupled logistic-like maps have been studied, e.g., by Ishii et al. (1996), Kaneko (1994), and Pitkovski and Kurths (1994).

We study here local dynamics of weakly connected mappings in a neighborhood of a fixed point or a periodic orbit. In analogy with weakly connected flows, the local activity becomes interesting at fixed points corresponding to bifurcations. We derive canonical models for multiple bifurcations and show how they can be reduced to canonical models for flows.

7.1   Weakly Connected Mappings

A weakly connected difference equation (7.1) can arise as a time-T mapping of a periodically forced WCNN. Indeed, consider such a network in the form

$$\dot{X}_i = \tilde{F}_i(X_i, \lambda, Q_1(t)) + \varepsilon \tilde{G}_i(X, \lambda, \rho, Q_2(t), \varepsilon),$$

where $Q_i : \mathbb{R} \to \mathcal{Q}$, $i = 1, 2$, are T-periodic functions. Let $\Phi_t(X, \lambda, \rho, \varepsilon)$ be its flow, which is a continuous function of parameters λ, ρ, and ε. Therefore, it can be written in the weakly connected form

$$\Phi_t(X, \lambda, \rho, \varepsilon) = \Phi_t^0(X, \lambda) + \varepsilon \Psi_t(X, \lambda, \rho, \varepsilon) , \qquad i = 1, \ldots, n ,$$

where Φ_t^0 has the diagonal form

$$\Phi_t^0(X, \lambda) = (\Phi_{1t}^0(X_1, \lambda), \ldots, \Phi_{nt}^0(X_n, \lambda))^\top .$$

Therefore, the time-T mapping induces the weakly connected difference equation (7.1) with

$$F_i(X_i, \lambda) = \Phi_{iT}(X_i, \lambda) \quad \text{and} \quad G_i(X, \lambda, \rho, \varepsilon) = \Psi_{iT}(X, \lambda, \rho, \varepsilon) .$$

Below, we study such weakly connected systems. Recall that our strategy is to compare dynamic behavior of the uncoupled system ($\varepsilon = 0$)

$$X_i \mapsto F_i(X_i, \lambda) , \qquad i = 1, \ldots, n , \tag{7.2}$$

and the coupled ($\varepsilon \neq 0$) system (7.1). Obviously, the uncoupled system (7.2) is not interesting as a model of the brain. We are looking for such regimes and parameter values that endow the coupled system (7.1) with "interesting" neurocomputational properties.

As in previous chapters, we study the dynamics of (7.2) near a fixed point $X^\star = (X_1^\star, \ldots, X_n^\star)^\top$ for some $\lambda^\star \in \Lambda$. Thus, we have

$$X_i^\star = F_i(X_i^\star, \lambda^\star)$$

for all i. Without loss of generality we may assume $X^\star = 0$ for $\lambda^\star = 0$. Recall that the time-T map having a fixed point corresponds to a continuous-time dynamical system having a limit cycle.

Analysis of weakly connected mappings is parallel to analysis of weakly connected systems of flows, which we performed in Chapter 5. As one can expect, local activity near a fixed point is not interesting when the point is hyperbolic (Section 7.2). Therefore, the only points that deserve our attention are nonhyperbolic. In this chapter we study the dynamics of (7.1) in some neighborhood of a nonhyperbolic fixed point corresponding to saddle-node and flip bifurcations (Sections 7.3.1 and 7.3.2). First we derive canonical models, and then we reveal the relationship between them and the canonical models for continuous-time WCNNs (Section 7.4).

7.2   Hyperbolic Fixed Points

A fixed point $X_i^* = 0$ is said to be *hyperbolic* if the Jacobian matrix

$$L_i = D_{X_i} F_i$$

evaluated at the point does not have eigenvalues of unit modulus. Notice that the Jacobian matrix for the uncoupled system (7.2) near the fixed point $X^\star = (X_1^\star, \ldots, X_n^\star)^\top$ has the form

$$L = \begin{pmatrix} L_1 & 0 & \cdots & 0 \\ 0 & L_2 & \cdots & 0 \\ \vdots & \vdots & \ddots & \vdots \\ 0 & 0 & \cdots & L_n \end{pmatrix},$$

and the fixed point $X^\star$ is hyperbolic if and only if each X_i^* is hyperbolic.

Theorem 7.1 *If the activity of each neuron is near a hyperbolic fixed point, then the weakly connected network (7.1) of such neurons, the uncoupled network (7.2), and the linear mapping*

$$X \mapsto LX$$

have topologically equivalent local orbit structures.

In this case the local behavior of the weakly connected system (7.1) is essentially linear and uncoupled.

The proof of the theorem uses the implicit function and Hartman-Grobman theorems and coincides (with the obvious modifications) with the proof of the analogous theorem for flows (Theorem 5.1).

7.3   Nonhyperbolic Fixed Points

It follows from the previous section that the only fixed points requiring further discussion are the nonhyperbolic ones. An important case in which such fixed points occur corresponds to bifurcations in the dynamics of (7.2). In this section we study the simplest cases: saddle-node and flip bifurcations. For each, the Jacobian matrix L_i has only one eigenvalue on the unit circle: $+1$ for the saddle-node bifurcation and -1 for the flip bifurcation.

The center manifold theorem for the maps (analogous to the theorem for flows used in Section 5.2.1) guarantees that there is a locally invariant manifold in $\mathbb{R}^{mn}$ on which the activity of (7.1) is governed by the lower-dimensional system of the form

$$x_i \mapsto f_i(x_i, \lambda) + \varepsilon g_i(x, \lambda, \rho, \varepsilon), \tag{7.3}$$

where each $x_i \in \mathbb{R}$ is a scalar and

$$f_i = 0 , \qquad \frac{\partial f_i}{\partial x_i} = \pm 1$$

at $x^\star = 0$, $\lambda^\star = 0$. As before, we assume that

$$\lambda = \lambda(\varepsilon) = 0 + \varepsilon \lambda_1 + \mathcal{O}(\varepsilon^2)$$

and

$$\rho = \rho(\varepsilon) = \rho_0 + \varepsilon \rho_1 + \mathcal{O}(\varepsilon^2)$$

for some $\lambda_1 \in \Lambda$ and $\rho_0, \rho_1 \in \mathcal{R}$.

7.3.1   Multiple Saddle-Node Bifurcations

Consider the weakly connected system (7.3) near a multiple saddle-node bifurcation point $x^\star = 0$ for $\lambda^\star = 0$. Recall (see Section 2.3) that each equation in the uncoupled $\varepsilon = 0$ system

$$x_i \mapsto f_i(x_i, \lambda)$$

is at the saddle-node bifurcation point if the following are satisfied:

$$f_i = 0 , \qquad \frac{\partial f_i}{\partial x_i} = 1 , \qquad p_i = \frac{1}{2} \frac{\partial^2 f_i}{\partial x_i^2} \neq 0 ,$$

and a transversality condition $D_\lambda f_i \neq 0$, which we do not use below. Using these conditions, we see that the initial portion of the Taylor series of the right-hand side of (7.3) at the origin for small ε has the form

$$x_i + p_i x_i^2 + \varepsilon \Big(a_i + \sum_{j=1}^{n} c_{ij} x_j \Big) + \text{h.o.t.},$$

where $c_{ij} = \partial g_i / \partial x_j$ and

$$a_i = D_\lambda f_i \cdot \lambda_1 + g_i .$$

Here all functions and derivatives are evaluated at $x = 0$ for $\lambda = 0$, $\rho = \rho_0$, and $\varepsilon = 0$.

Recall that $a_i = 0$ is the *adaptation condition* for the ith neuron dynamics. Suppose $a_i \neq 0$ for all i. Then the rescaling $x = \sqrt{\varepsilon} \tilde{x}$ transforms the weakly connected mapping to

$$\tilde{x}_i \mapsto \tilde{x}_i + \sqrt{\varepsilon}(a_i + p_i \tilde{x}_i^2) + \mathcal{O}(\varepsilon) ,$$

which is essentially uncoupled (although non-linear). Indeed, the ith neuron's activity is determined by the constants a_i and p_i. Depending upon the sign of their product, the dynamics can have two qualitatively different local phase portraits:

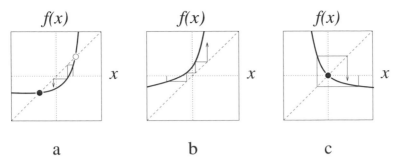

Figure 7.1. Bifurcations of a mapping. a. Saddle-node bifurcation for $a_i p_i < 0$. b. Saddle-node bifurcation for $a_i p_i > 0$. c. Flip bifurcation.

- If $a_i p_i < 0$, then there is a pair of fixed points. One of them is stable and the other one is not (see Figure 7.1a).

- If $a_i p_i > 0$, then there are no fixed points. For any initial conditions the activity eventually leaves some neighborhood of the origin (see Figure 7.1b).

The local activity of the ith neuron is not affected by the activities of the other neurons, provided that the other neuron activities are bounded. We see that the neuron activity depends only on the environment ρ_0 and internal state λ_1 (since a_i depends upon them) and does not depend on the other neuron activities.

If the adaptation condition ($a_i = 0$) is satisfied for all i, then after the rescaling $x = \varepsilon \tilde{x}/p_i$, the weakly connected system (7.3) transforms to

$$\tilde{x}_i \mapsto \tilde{x}_i + \varepsilon \left(r_i + \tilde{x}_i^2 + \sum_{j=1}^{n} \tilde{c}_{ij} \tilde{x}_j \right) + \mathcal{O}(\varepsilon^2), \tag{7.4}$$

for $\tilde{c}_{ij} = c_{ij} p_j/p_i$ and some constants $r_i \in \mathbb{R}$. System (7.4) is a canonical model for multiple saddle-node bifurcations in weakly connected maps. Thus we have proved the following result.

Theorem 7.2 (Local Canonical Model for Multiple Saddle-Node Bifurcations in Weakly Connected Mappings)

If the weakly connected mapping (7.3) is near a multiple saddle-node bifurcation point and the adaptation condition

$$D_\lambda f_i \cdot \lambda_1 + g_i = 0$$

is satisfied for all i, then there is an invertible change of variables that transforms (7.3) to a canonical model (7.4).

7.3.2 Multiple Flip Bifurcations

A weakly connected mapping (7.3) is near a flip bifurcation point $x^\star = 0$ for $\lambda^\star = 0$ if (see Section 2.3)

$$f_i = 0, \quad \frac{\partial f_i}{\partial x_i} = -1$$

as in Figure 7.1c and

$$\frac{1}{3}\frac{\partial^3 f_i}{\partial x_i^3} + \frac{1}{2}\left(\frac{\partial^2 f_i}{\partial x_i^2}\right)^2 \neq 0$$

for all i, which is a technical condition guaranteeing existence of cubic terms in the canonical model that we derive below. There is also a transversality condition to be satisfied, which we do not present here since we do not use it.

The flip bifurcation does not have an analogue for one-dimensional flows. Thus, it is natural to expect some new qualitative features. The most prominent feature of the multiple flip bifurcation is that the adaptation condition is not necessary for the local dynamics of the weakly connected mappings to exhibit "interesting" properties.

Theorem 7.3 (Local Canonical Model for Multiple Flip Bifurcations in Weakly Connected Mappings) *If the weakly connected mapping (7.3) is near a multiple flip bifurcation point, then there is an invertible change of variables that transforms (7.3) to*

$$x_i \mapsto x_i + \sqrt{\varepsilon}\left(r_i x_i \pm x_i^3 + \sum_{j=1}^{n} c_{ij} x_j\right) + \mathcal{O}(\varepsilon). \tag{7.5}$$

System (7.5) is a canonical model for multiple flip bifurcations in weakly connected mappings.

Proof. Let us denote the kth iteration of the mapping (7.3) by x^k. Thus, we have

$$x_i^{k+1} = f_i(x_i^k, \lambda) + \varepsilon g_i(x^k, \lambda, \rho, \varepsilon).$$

The initial portion of the Taylor series of the right-hand side at the origin is given by

$$-x_i^k + p_i(x_i^k)^2 + q_i(x_i^k)^3 + \varepsilon\left(a_i + \sum_{j=1}^{n} c_{ij} x_j^k\right) + \text{h.o.t.}$$

Let us find how x^{k+2} depends on x^k. For this we have to evaluate the composition

$$x_i^{k+2} = f_i(f_i(x_i^k, \lambda) + \varepsilon g_i(x^k, \lambda, \rho, \varepsilon), \lambda) + \varepsilon g_i(f(x^k, \lambda) + \varepsilon g(x^k, \lambda, \rho, \varepsilon), \lambda, \rho, \varepsilon)$$

$$= x_i^k - 2\varepsilon \left(a_i p_i x_i^k + \sum_{j=1}^{n} c_{ij} x_j^k \right) - 2(q_i + p_i^2)(x_i^k)^3 + \text{h.o.t.}$$

Now let $x_i = \sqrt{\varepsilon}\tilde{x}_i / \sqrt{2|q_i + p_i^2|}$, $\tilde{r}_i = -2a_i p_i$, and

$$c_{ij} = -\frac{1}{2}\tilde{c}_{ij} \frac{\sqrt{|q_j + p_j^2|}}{\sqrt{|q_i + p_i^2|}} .$$

Then we obtain

$$\tilde{x}_i^{k+2} = \tilde{x}_i^k + \sqrt{\varepsilon} \left(\tilde{r}_i \tilde{x}_i^k + \sigma_i (\tilde{x}_i^k)^3 + \sum_{j=1}^{n} \tilde{c}_{ij} \tilde{x}_j^k \right) + \mathcal{O}(\varepsilon) ,$$

where

$$\sigma_i = -\text{sign}(q_i + p_i^2) = \pm 1 .$$

Finally, notice that this result does not depend on the adaptation condition $a_i = 0$. $\square$

7.3.3 Multiple Secondary Andronov-Hopf Bifurcations

Let us consider the weakly connected mapping (7.1) at a fixed point corresponding to the multiple secondary Andronov-Hopf bifurcation. Recall (see Section 2.3) that the Jacobian matrix DF_i in this case has a pair of complex conjugate eigenvalues μ_i, $\bar{\mu}_i$ on the unit circle. Using complex coordinates we can rewrite the weakly connected system in the form

$$z_i \mapsto \mu_i z_i + h_i(z_i, \bar{z}_i, \lambda) + \varepsilon g_i(z, \bar{z}, \lambda, \rho, \varepsilon) .$$

As usual, we assume that $\lambda = \mathcal{O}(\varepsilon)$ and $\rho = \rho(\varepsilon)$.

No Strong Resonances

When there are no strong resonances, that is, when each μ_i satisfies $|\mu_i|^3 \neq 1$ and $|\mu_i|^4 \neq 1$, the truncated normal form is

$$z_i \mapsto \mu_i z_i + d_i z_i |z_i|^2 + \varepsilon \sum_{j=1}^{n} (s_{ij} z_j + e_{ij} \bar{z}_j)$$

for some constants $d_i, s_{ij}, e_{ij} \in \mathbb{C}$, which coincide with those from Section 5.4. It is convenient at this point to rewrite the system above as

$$z_i^{k+1} = e^{i\Omega_i} z_i^k + d_i z_i^k |z_i^k|^2 + \varepsilon \sum_{j=1}^{n} (s_{ij} z_j^k + e_{ij} \bar{z}_j^k) ,$$

where z^k is the kth iteration of the variable z_i and $e^{i\Omega_i} = \mu_i$. It is easy to see that the following change of variables,

$$\tilde{z}_i^k = \sqrt{\varepsilon} e^{ki\Omega_i} z_i^k \, ,$$

transforms the truncated normal form to

$$\tilde{z}_i^{k+1} = \tilde{z}_i^k + \varepsilon \left(\tilde{d}_i \tilde{z}_i^k |\tilde{z}_i^k|^2 + \sum_{j=1}^{n} (\tilde{s}_{ij} e^{ki(\Omega_j - \Omega_i)} \tilde{z}_j^k + \tilde{e}_{ij} e^{ki(-\Omega_j - \Omega_i)} \bar{\tilde{z}}_j^k) \right) \, ,$$

where

$$\tilde{d}_i = d_i e^{-i\Omega_i} \, , \qquad \tilde{s}_{ij} = s_{ij} e^{-i\Omega_i} \, , \quad \text{and} \quad \tilde{e}_{ij} = e_{ij} e^{-i\Omega_i} \, .$$

By averaging we can remove all terms multiplied by $e^{ki\Delta}$, $\Delta \neq 0$. Then the canonical model has the form

$$z_i \mapsto z_i + \varepsilon \left(\tilde{d}_i z_i |z_i|^2 + \sum_{j=1}^{n} c_{ij} z_j \right) \, ,$$

where $c_{ij} = 0$ if $\Omega_i \neq \Omega_j$ and $\tilde{s}_{ij}$ otherwise.

Strong Resonance Case

If there are μ_i satisfying $\mu_i^4 = 1$, then the corresponding mapping for z_i has the form

$$z_i \mapsto z_i + \varepsilon \left(w_i \bar{z}_i^3 + \tilde{d}_i z_i |z_i|^2 + \sum_{j=1}^{n} c_{ij} z_j \right)$$

for some $w_i \in \mathbb{C}$. Similarly to this case, if there are μ_i satisfying $\mu_i^3 = 1$, then the mapping for the corresponding z_i has the form

$$z_i \mapsto z_i + \sqrt{\varepsilon} w_i \bar{z}^2 + \varepsilon \left(\tilde{d}_i z_i |z_i|^2 + \sum_{j=1}^{n} c_{ij} z_j \right)$$

Notice that $c_{ij} = 0$ when the ith and the jth neurons have different frequencies Ω_i and Ω_j.

7.4   Connection with Flows

There is an intimate relationship between the canonical models for weakly connected maps and flows. The former have the form

$$x_i^{k+1} = x_i^k + \tau h_i(x^k) + \mathcal{O}(\tau^2) \, , \qquad \tau \ll 1 \, ,$$

for some functions h_i. If one considers the iteration x^k as a value of a smooth function at the moment $k\tau$, then one can rewrite the equation above as

$$x_i((k+1)\tau) = x_i(k\tau) + \tau h_i(x(k\tau)) + \mathcal{O}(\tau^2),$$

or

$$\frac{x_i(t+\tau) - x_i(t)}{\tau} = h_i(x(t)) + \mathcal{O}(\tau),$$

where $t = k\tau$. The equation above is Euler's discretization of the flow

$$\frac{dx_i}{dt} = h_i(x).$$

In this sense, we see that the canonical model for a multiple saddle-node bifurcation

$$x_i \mapsto x_i + \varepsilon \left(r_i + x_i^2 + \sum_{j=1}^{n} c_{ij}x_j \right) + \mathcal{O}(\varepsilon^2)$$

is the discretization of that for the flow

$$\frac{dx_i}{dt} = r_i + x_i^2 + \sum_{j=1}^{n} c_{ij}x_j + \mathcal{O}(\varepsilon).$$

The canonical model for a multiple flip bifurcation

$$x_i \mapsto x_i + \sqrt{\varepsilon} \left(r_i x_i \pm x_i^3 + \sum_{j=1}^{n} c_{ij}x_j \right) + \mathcal{O}(\varepsilon)$$

is the discretization of

$$\frac{dx_i}{dt} = r_i x_i \pm x_i^3 + \sum_{j=1}^{n} c_{ij}x_j + \mathcal{O}(\sqrt{\varepsilon}),$$

which is the canonical model for multiple pitchfork bifurcations. Similarly, the canonical model for secondary Andronov-Hopf bifurcations of mappings can be written in one of the forms

$$\dot{z}_i = \tilde{d}_i z_i |z_i|^2 + \sum_{j=1}^{n} c_{ij}z_j,$$

$$\dot{z}_i = w_i \bar{z}_i^3 + \tilde{d}_i z_i |z_i|^2 + \sum_{j=1}^{n} c_{ij}z_j,$$

$$\dot{z}_i = w_i \bar{z}_i^2 + \sqrt{\varepsilon} \left(\tilde{d}_i z_i |z_i|^2 + \sum_{j=1}^{n} c_{ij}z_j \right),$$

depending on the resonances.

This relationship between canonical models allows us to concentrate our efforts on studying the canonical models for multiple saddle-node, pitchfork, and Andronov-Hopf bifurcations in flows, which we undertake in subsequent chapters.

8
Saddle-Node on a Limit Cycle

In this chapter we study a weakly connected network

$$\dot{X}_i = F_i(X_i, \lambda) + \varepsilon G_i(X, \lambda, \rho, \varepsilon) , \qquad X_i \in \mathbb{R}^m, \quad i = 1, \ldots, n ,$$

of neurons each having a saddle-node bifurcation on a limit cycle. Such neurons are said to have Class 1 neural excitability. This bifurcation provides an important example of when local analysis at an equilibrium renders global information about the system behavior.

If we consider each equation in the uncoupled system

$$\dot{X}_i = F_i(X_i, \lambda) , \qquad i = 1, \ldots, n,$$

then the phase space of the ith neuron has an invariant manifold $M_i \subset \mathbb{R}^m$ homeomorphic to the unit circle $\mathbb{S}^1$. We assume that M_i is stable and normally hyperbolic so that we can use an invariant manifold reduction (Theorem 4.7), which reduces the WCNN above to the local model

$$\dot{x}_i = f_i(x_i, \lambda) + \varepsilon g_i(x, \lambda, \rho, \varepsilon) , \qquad i = 1, \ldots, n ,$$

where each $x_i \in M_i \cong \mathbb{S}^1$ and the function f_i agrees with F_i. In this chapter we simplify the local model further.

8.1   One Neuron

8.1.1   Saddle-Node Point on a Limit Cycle

Let $\mathbb{S}^1 = [-\pi, \pi]/(\{-\pi\} \equiv \{\pi\})$ be the unit circle. Suppose $x \in \mathbb{S}^1$ parametrizes the circle. Consider a dynamical system

$$\dot{x} = f(x), \tag{8.1}$$

where the function f is smooth on $\mathbb{S}^1$. Suppose (8.1) has a saddle-node point $x = 0$; i.e.,

$$f(0) = 0 \quad \text{and} \quad f'(0) = 0 \quad \text{but} \quad f''(0)/2 \neq 0 .$$

Also suppose that $f(x) > 0$ for all $x \neq 0$. This implies $p > 0$.

Since $f > 0$, the solution $x(t)$ is increasing. Obviously, if $x(0) < 0$, then $x(t)$ approaches the origin $x = 0$. If $x(0) > 0$ then $x(t)$ increases further, passes through $+\pi \equiv -\pi$, and goes toward $x = 0$; see Figure 8.1. We call such an excursion an *action potential*, or a *spike*, in analogy with a similar electrophysiological phenomenon observed in neurons.

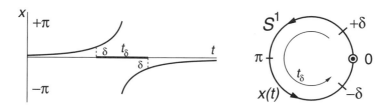

Figure 8.1. The duration of an action potential (spike) having amplitude greater than δ is t_δ given by (8.2).

Lemma 8.1 (Spike Lemma) *A spike of amplitude greater than $\delta > 0$ has time duration*

$$t_\delta = \frac{2}{p\delta} + \mathcal{O}(\ln \delta) , \tag{8.2}$$

where $p = f''(0)/2 > 0$. That is, if $x(t)$ is a solution of (8.1) with $x(0) = \delta$, then $|x(t)| > \delta$ for all $0 < t < t_\delta$ and $x(t_\delta) = -\delta$; see Figure 8.1.

Thus, $x(t)$ spends t_δ units of time outside a δ neighborhood of the origin.

Proof. Fix a small $\delta > 0$. Consider (8.1) in the form

$$\dot{x} = px^2 + qx^3 + \mathcal{O}(x^4) , \quad x(0) = \delta .$$

Its solution can be found by separation of variables. We have

$$t + c_0 = \int dt = \int \frac{dx}{px^2 + qx^3 + \mathcal{O}(x^4)}$$

$$= \int \frac{dx}{px^2(1 + qx/p + \mathcal{O}(x^2))}$$

$$= \int \frac{1}{px^2}\left(1 - qx/p + \mathcal{O}(x^2)\right) dx$$

$$= \int \left(\frac{1}{px^2} - \frac{q}{p^2 x} + \mathcal{O}(1)\right) dx$$

$$= -\frac{1}{px} - \frac{q}{p^2}\ln x + \mathcal{O}(x) .$$

Using the initial condition $x(0) = \delta$ we can determine the constant c_0:

$$c_0 = -\frac{1}{p\delta} - \frac{q}{p^2}\ln \delta + \mathcal{O}(\delta) .$$

Therefore, the solution satisfies

$$t - \frac{1}{p\delta} - \frac{q}{p^2}\ln \delta + \mathcal{O}(\delta) = -\frac{1}{px(t)} - \frac{q}{p^2}\ln x(t) + \mathcal{O}(x(t))$$

as long as it stays small. Now consider $x(t_1)$ for

$$t_1 = \frac{1}{p\delta} + \frac{q}{p^2}\ln \delta .$$

Suppose $x(t_1)$ is small. Then it satisfies

$$\mathcal{O}(\delta) = -\frac{1}{px(t_1)} - \frac{q}{p^2}\ln x(t_1) + \mathcal{O}(x(t_1)) .$$

On the other hand, it follows from the equation above that

$$x(t_1) = \mathcal{O}(1/\delta) \to \infty$$

when $\delta \to 0$. The apparent contradiction suggests that $x(t_1)$ cannot be small. Since $x(t) \in \mathbb{S}^1$, we know that $|x(t)| \leq \pi$ for all t. Therefore, $x(t_1)$ $\mathcal{O}(1)$. Thus, it takes t_1 units of time for $x(t)$ to reach a finite (nonsmall) value $x_1 = x(t_1)$. Replacing $t \to -t$ in equation (8.1) we can prove that it takes $l_2 - l_1$ units of time for $x(t)$ to reach the value $x = -\delta$ if it starts from some finite value $x_2 = \mathcal{O}(1)$; see Figure 8.2.

Since $f(x) = \mathcal{O}(1)$ for $x = \mathcal{O}(1)$, it also takes some finite time $t_3 = \mathcal{O}(1)$ for $x(t)$ to travel from x_1 to x_2. If $x_1 \succ x_2$; that is, if x_2 is not between x_1 and $-\delta$, then we take t_3 to be negative. The total time t_δ needed to travel from δ to $-\delta$ is the sum of t_1, t_3, and t_2 and is given by

$$t_\delta = \frac{2}{p\delta} + \frac{2q}{p^2}\ln \delta + \mathcal{O}(1) .$$

This completes the proof. $\square$

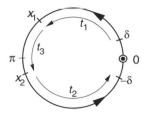

Figure 8.2. Diagram used in the proof of the spike lemma

The duration of a spike depends crucially on the constant $p = f''(0)/2$, which is determined locally. Peculiarities of the function $f(x)$ for x away from the origin make a contribution of order $\mathcal{O}(1)$ to (8.2) and therefore are negligible. Thus, the local analysis provides quantitative global information about the behavior of (8.1).

A typical example of a dynamical system having a saddle-node point on the limit cycle is a model of a voltage controlled oscillator neuron (VCON) of the form

$$\dot{\theta} = \omega - \cos\theta , \qquad \theta \in \mathbb{S}^1 , \tag{8.3}$$

for $\omega = 1$. Here the variable θ has the meaning of phase on the unit circle $\mathbb{S}^1$. The lemma below shows that VCONs are not only typical, but also canonical.

Lemma 8.2 (VCON Lemma) *The family of dynamical systems of the form (8.1) having a saddle-node point on the limit cycle has for any small δ a global canonical model*

$$\theta' = 1 - \cos\theta + R(\theta, \delta) , \qquad \theta \in \mathbb{S}^1 , \tag{8.4}$$

where $' = d/d\tau$, $\tau = \delta^2 t$ is slow time, and

$$R(\theta, \delta) = \begin{cases} 0 & \text{if } \theta = 0 , \\ \mathcal{O}(\delta) & \text{if } |\theta - \pi| > \delta , \\ \mathcal{O}(\delta \ln \delta) & \text{if } |\theta - \pi| \leq \delta . \end{cases}$$

That is, there is a continuous transformation $h : \mathbb{S}^1 \to \mathbb{S}^1$ depending on f such that if $x(t)$ solves (8.1), then $\theta(t) = h(x(t))$ is a solution of (8.4) for $t \geq 0$.

Notice that (8.1) does not depend on δ, but (8.4) does. In the limit $\delta \to 0$ equation (8.4) converges to the VCON (8.3) for $\omega = 1$. Therefore, the VCON is a canonical model for any dynamical system of the form (8.1).

Proof. The VCON lemma is a modification of Ermentrout and Kopell's (1986) Lemma 2 for parabolic bursting.

Let δ be a small fixed parameter. Let $h : \mathbb{S}^1 \to \mathbb{S}^1$ be a function defined as follows:

$$h(x) = \begin{cases} 2\operatorname{atan}(xp/\delta^2) & \text{if } x \in [-\delta/p, \ \delta/p] , \\ \hat{h}(x) & \text{if } x \in [\delta/p, \ -\delta/p] , \end{cases} \tag{8.5}$$

where $[\delta/p, \ -\delta/p] = \mathbb{S}^1 \setminus [-\delta/p, \ \delta/p]$ and the function $\hat{h}(x)$ will be constructed below.

Step I. Let us first consider the transformation

$$\theta = 2\operatorname{atan} \frac{xp}{\delta^2} , \qquad x \in [-\delta/p, \ \delta/p] . \tag{8.6}$$

Obviously, it is continuous and one-to-one for any fixed $\delta > 0$. Let us check that the image of $[-\delta/p, \ \delta/p]$ is $[\pi + 2\delta + \mathcal{O}(\delta^3), \ \pi - 2\delta + \mathcal{O}(\delta^3)]$ (thin arcs in Figure 8.3). Indeed, the image of $x = -\delta/p$ is

$$-2\operatorname{atan} \frac{1}{\delta} = -2\cot^{-1}\delta = \pi + 2\delta + \mathcal{O}(\delta^3) .$$

Similarly, the image of $x = \delta/p$ is $\theta = \pi - 2\delta + \mathcal{O}(\delta^3)$. Thus, h stretches the interval of size $2\delta/p$ in $\mathbb{S}^1$ over the interval of size $2\pi - 4\delta + \mathcal{O}(\delta^3)$ in $\mathbb{S}^1$; that is, it blows up a small neighborhood of the origin $x = 0$.

Step II. Now we want to construct $\hat{h}$ that extends h to a mapping $\mathbb{S}^1 \to \mathbb{S}^1$ so that it is smooth, one-to-one, and with a smooth inverse. Obviously, $\hat{h}$ must shrink the interval $[\delta/p, \ -\delta/p]$ of size $2\pi - 2\delta/p$ to the interval $[\pi - 2\delta, \ \pi + 2\delta] + \mathcal{O}(\delta^3)$ of size $4\delta + \mathcal{O}(\delta^3)$ (thick arcs in Figure 8.3). To define

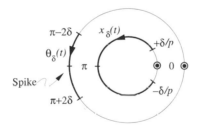

Figure 8.3. Diagram used in the proof of the VCON lemma

the mapping $\hat{h}$ we need the following: Let $x_\delta(t)$ be a solution of (8.1) such that $x_\delta(0) = \delta/p$. From the spike lemma it follows that $x_\delta(t_{\delta/p}) = -\delta/p$; that is, $x_\delta(t)$, $0 \le t \le t_{\delta/p}$, covers exactly the interval $[\delta/p, -\delta/p]$. Let $\theta_\delta(t)$ be a solution of

$$\dot{\theta} = \alpha , \qquad \theta(0) = 2\operatorname{atan} \frac{1}{\delta} ,$$

where the constant $\alpha > 0$ can be determined from the condition that $\theta_\delta(t)$, $0 \le t \le t_{\delta/p}$, covers exactly the interval

$$[2\operatorname{atan} \frac{1}{\delta}, \ -2\operatorname{atan} \frac{1}{\delta}] = [\pi - 2\delta + \mathcal{O}(\delta^3), \ \pi + 2\delta + \mathcal{O}(\delta^3)]$$

of size $4\delta + \mathcal{O}(\delta^3)$. This yields

$$\alpha = (4\delta + \mathcal{O}(\delta^3))/t_{\delta/p} = 2\delta^2 + \mathcal{O}(\delta^3 \ln \delta) \,.$$

Now we define $\hat{h}$ to be

$$\hat{h}(x_\delta(t)) = \theta_\delta(t) = 2 \operatorname{atan} \frac{1}{\delta} + \alpha t \,, \qquad t \in [0, t_{\delta/p}] \,.$$

More precisely, for any $x \in [\delta/p, -\delta/p]$ there is a unique $t \in [0, t_{\delta/p}]$ such that $x = x_\delta(t)$. Then we set $\hat{h}(x) = \theta_\delta(t)$ for such t.

Obviously, the function $h : \mathbb{S}^1 \to \mathbb{S}^1$ is well-defined, one-to-one, and continuous. It is smooth for $x \in [-\delta, \delta]$ and can be deformed to a smooth function globally by a suitable deformation of $\hat{h}$.

Step III. What is the equation for $\dot{\theta}$? According to our construction,

$$\dot{\theta} = 2\delta^2 + \mathcal{O}(\delta^3 \ln \delta) \tag{8.7}$$

for $\theta \in [h(\delta), h(-\delta)]$. To find the equation for $\theta \in [h(-\delta), h(\delta)]$, that is, when $|x| \leq \delta/p$, we plug the inverse of (8.6),

$$x = \frac{\delta^2}{p} \tan \frac{\theta}{2} \,,$$

into

$$\dot{x} = f(x) = px^2 + \mathcal{O}(\delta x^2) \,,$$

where we used the fact that $x = \mathcal{O}(\delta)$. It is easy to check that

$$\dot{x} = \frac{\delta^2}{p} \frac{\dot{\theta}}{2 \cos^2(\theta/2)}$$

and

$$f(x) = \frac{\delta^4}{p} \frac{\sin^2(\theta/2)}{\cos^2(\theta/2)} + \mathcal{O}\left(\delta^5 \frac{\sin^2(\theta/2)}{\cos^2(\theta/2)}\right) \,.$$

Therefore,

$$\dot{\theta} = 2\delta^2 \sin^2(\theta/2) + \mathcal{O}(\delta^3 \sin^2(\theta/2)) \,.$$

Taking into account that $2 \sin^2(\theta/2) = 1 - \cos \theta$, we obtain the equation

$$\dot{\theta} = \delta^2(1 - \cos \theta + \mathcal{O}(\delta \sin^2(\theta/2))) \tag{8.8}$$

for $\theta \in [h(-\delta), h(\delta)]$. Now we must glue together (8.7) and (8.8). For this, notice that (8.8) makes perfect sense for all values $\theta \in \mathbb{S}^1$. If $\theta = \pi + \mathcal{O}(\delta)$, as happens in (8.7), then $\cos(\pi + \mathcal{O}(\delta)) = -1 + \mathcal{O}(\delta^2)$, and (8.8) can be written as

$$\dot{\theta} = \delta^2(2 + \mathcal{O}(\delta)) \,.$$

Therefore, for $\theta = \pi + \mathcal{O}(\delta)$ both equations are essentially the same, up to a certain order in δ. So we take (8.8) and use slow time $\tau = \delta^2$ to obtain (8.4). The small remainder $R(\theta, \delta)$ vanishes when $\theta = 0$ because of $\sin^2(\theta/2)$ in the $\mathcal{O}$ term in (8.8). Notice also that when $|\theta| < \pi/2$, the remainder $R = \mathcal{O}(\delta^2 \tan \theta/2) = \mathcal{O}(\delta^2)$ is actually smaller than in the caption of the lemma. $\square$

8.1.2 Saddle-Node Bifurcation on a Limit Cycle

Now consider the case when dynamical system (8.1) depends on a parameter $\lambda \in \Lambda$; that is, consider a system

$$\dot{x} = f(x, \lambda) , \qquad x \in \mathbb{S}^1 ,$$

having a saddle-node bifurcation at $x = 0$ for $\lambda = 0$. As usual, we assume that $\lambda = \varepsilon \lambda_1 + \mathcal{O}(\varepsilon^2)$ for some $\lambda_1 \in \Lambda$. Then the dynamical system above can be written in the form

$$\dot{x} = f(x) + \varepsilon g(x, \varepsilon) , \qquad x \in \mathbb{S}^1 , \tag{8.9}$$

which is an ε-perturbation of (8.1).

Theorem 8.3 (Ermentrout-Kopell Theorem for Class 1 Neural Excitability)   *A family of dynamical systems of the form (8.9) having a saddle-node bifurcation on the limit cycle has a canonical model*

$$\theta' = (1 - \cos\theta) + (1 + \cos\theta)r + R(\theta, \varepsilon) , \qquad \theta \in \mathbb{S}^1 , \tag{8.10}$$

where

$$r = g(0,0) \, f''(0)/2 ,$$

$' = d/d\tau$, $\tau = \sqrt{\varepsilon} t$ *is slow time, and the remainder*

$$R(\theta, \varepsilon) = \begin{cases} \mathcal{O}(\sqrt[4]{\varepsilon}) & \text{if } |\theta - \pi| > 2\sqrt[4]{\varepsilon} , \\ \mathcal{O}(\sqrt[4]{\varepsilon} \ln \varepsilon) & \text{if } |\theta - \pi| \leq 2\sqrt[4]{\varepsilon} \end{cases}$$

is small.

Proof. We must present a continuous transformation $h : \mathbb{S}^1 \to \mathbb{S}^1$ that maps solutions of (8.9) to those of (8.10). For this consider the transformation $\theta = h(x)$ used in the VCON lemma and defined in (8.5). Let $\delta = \sqrt[4]{\varepsilon}$ be a small parameter. Then we take h in the form

$$h(x) = \begin{cases} 2 \operatorname{atan}(xp/\sqrt{\varepsilon}) & \text{if } x \in [-\sqrt[4]{\varepsilon}/p, \ \sqrt[4]{\varepsilon}/p] , \\ \hat{h}(x) & \text{if } x \in [\sqrt[4]{\varepsilon}/p, \ -\sqrt[4]{\varepsilon}/p] . \end{cases}$$

Now we apply it to (8.9).

Step I. When $|x| \le \sqrt[4]{\varepsilon}/p$, equation (8.9) can be written in the form

$$\dot{x} = \varepsilon a + px^2 + \mathcal{O}(\varepsilon^2, \varepsilon x, \sqrt[4]{\varepsilon} x^2) ,$$

where $a = g(0,0)$, $p = f''(0)/2 > 0$. Using the inverse of h for $|x| \le \sqrt[4]{\varepsilon}/p$,

$$x = \frac{\sqrt{\varepsilon}}{p} \tan \frac{\theta}{2} ,$$

we obtain

$$\sqrt{\varepsilon} \frac{\dot{\theta}}{2\cos^2(\theta/2)} = \varepsilon a p + \varepsilon \frac{\sin^2(\theta/2)}{\cos^2(\theta/2)} + \mathcal{O}\left(\varepsilon^{5/4} \frac{\sin^2(\theta/2)}{\cos^2(\theta/2)}\right) ,$$

and hence

$$\dot{\theta} = \sqrt{\varepsilon} 2 \cos^2(\theta/2) a p + \sqrt{\varepsilon} 2 \sin^2(\theta/2) + \mathcal{O}(\varepsilon^{3/4} \sin^2(\theta/2)) .$$

Now we use the trigonometric identities

$$2\sin^2(\theta/2) = 1 - \cos\theta \quad\text{and}\quad 2\cos^2(\theta/2) = 1 + \cos\theta$$

and slow time $\tau = \sqrt{\varepsilon} t$ to obtain

$$\theta' = (1 + \cos\theta)ap + (1 - \cos\theta) + \mathcal{O}(\varepsilon^{1/4}\sin^2(\theta/2)) , \qquad (8.11)$$

which coincides with (8.10).

Step II. To apply $\theta = h(x)$ when $|x| > \sqrt[4]{\varepsilon}/p$ (that is, to apply $\hat{h}$) we just notice that the term $\varepsilon g(x, \varepsilon)$ in (8.9) is negligible because $f(x) > \mathcal{O}(\sqrt{\varepsilon})$ in this case. To make this observation precise we need to estimate the size of $\hat{h}'(x)$, in particular, to prove that it is bounded. For this we use the definition of $\hat{h}$:

$$\hat{h}(x(t)) = \theta_0 + \alpha t ,$$

which we differentiate with respect to t to obtain $\hat{h}'(x)\dot{x} = \alpha$. Since $\dot{x} = f(x) > \mathcal{O}(\sqrt{\varepsilon})$ and $\alpha = \mathcal{O}(\sqrt{\varepsilon})$, we obtain $\hat{h}' = \mathcal{O}(1)$. Actually, this is what we should expect, since $\hat{h}$ shrinks an interval of size $2\pi + \mathcal{O}(\sqrt[4]{\varepsilon})$ to an interval of size $\mathcal{O}(\sqrt[4]{\varepsilon})$. Now we are ready to differentiate $\theta(t) = \hat{h}(x(t))$ with respect to t:

$$\dot{\theta} = \hat{h}'(x)\dot{x} = \hat{h}'(x)(f(x) + \varepsilon g(x, \varepsilon)) = \hat{h}'(x)f(x) + \varepsilon \hat{h}'(x)g(x, \varepsilon) .$$

From the VCON lemma it follows that

$$\hat{h}'(x)f(x) = \sqrt{\varepsilon}(1 - \cos\theta) + \mathcal{O}(\varepsilon^{3/4}\ln\varepsilon) .$$

From the fact that $\hat{h}' = \mathcal{O}(1)$ it follows that $\varepsilon \hat{h}'(x)g(x, \varepsilon) = \mathcal{O}(\varepsilon)$. Therefore, we get

$$\dot{\theta} = \sqrt{\varepsilon}(1 - \cos\theta) + \mathcal{O}(\varepsilon^{3/4}\ln\varepsilon) ,$$

which has the form

$$\theta' = (1 - \cos\theta) + \mathcal{O}(\sqrt[4]{\varepsilon}\ln\varepsilon) ,$$

in the slow time scale $\tau = \sqrt{\varepsilon}t$. Now notice that this equation coincides with (8.11) because $\theta = \pi + \mathcal{O}(\sqrt[4]{\varepsilon})$, and hence $(1 + \cos\theta)a = \mathcal{O}(\sqrt{\varepsilon})$ is negligible in this case.

It is worthwhile noting that the estimate $1 + \cos(\pi + \mathcal{O}(\sqrt[4]{\varepsilon})) = \mathcal{O}(\sqrt{\varepsilon})$, which holds for $|x| > \sqrt[4]{\varepsilon}/p$, eliminates completely the question, What is the value of $g(x, \varepsilon)$ for big x? The only fact that we need to know is that $g(x, \varepsilon)$ is bounded. Since it is multiplied by $1 + \cos(\pi + \mathcal{O}(\sqrt[4]{\varepsilon}))$, the whole term has order $\mathcal{O}(\sqrt{\varepsilon})$ and hence does not affect the dynamics of (8.10). Therefore, we can take any value for $g(x, \varepsilon)$. We take $a = g(0, 0)$ merely for the sake of convenience. $\square$

Remark. The spike lemma result holds for dynamical systems of the form (8.9) when δ is not too small, viz., $\sqrt{\varepsilon} \ll \delta \ll 1$.

In particular, the spike of an amplitude greater than $\sqrt[4]{\varepsilon}/p$ takes approximately $2/\sqrt[4]{\varepsilon}$ units of fast time t or $2\sqrt[4]{\varepsilon}$ units of slow time. Indeed, from the proof of the Ermentrout-Kopell theorem it follows that $|x| > \sqrt[4]{\varepsilon}/p$ when $|\theta - \pi| < 2\sqrt[4]{\varepsilon} + \mathcal{O}(\varepsilon^{3/4})$. Since $\theta' = 2 + \mathcal{O}(\sqrt[4]{\varepsilon}\ln\varepsilon)$, the variable $\theta(\tau)$ covers the distance $4\sqrt[4]{\varepsilon} + \mathcal{O}(\varepsilon^{3/4})$ during $2\sqrt[4]{\varepsilon} + \mathcal{O}(\sqrt[4]{\varepsilon}\ln\varepsilon)$ units of slow time τ.

When $r > 0$, the canonical model (8.10) oscillates with frequency $\mathcal{O}(1)$ (slow time). Therefore, the original system (8.9) oscillates with frequency $\mathcal{O}(\sqrt{\varepsilon})$ (fast time).

8.1.3 Analysis of the Canonical Model

Consider the canonical model (8.10) for $\varepsilon = 0$,

$$\theta' = (1 - \cos\theta) + (1 + \cos\theta)r , \qquad \theta \in \mathbb{S}^1 ,$$

which we write in the form

$$\theta' = r + 1 + (r - 1)\cos\theta . \qquad (8.12)$$

Depending on the values of the parameter r, there are many dynamical behaviors:

- Case $r > 1$. If we change the time scale $\tilde{\tau} = (r - 1)\tau$, the canonical model becomes the VCON

$$\theta' = \omega + \cos\theta ,$$

where $\omega = (r + 1)/(r - 1)$. Since $\omega > 1$, the variable $\theta \in \mathbb{S}^1$ oscillates.

- Case $r = 1$. Then the variable $\theta(\tau) \in \mathbb{S}^1$ in (8.12) oscillates with constant speed $r + 1 > 0$; that is, $\theta(\tau) = \theta(0) + (r+1)\tau$.

- Case $0 < r < 1$. The translation $\tilde{\theta} = \pi + \theta$ takes (8.12) to $\tilde{\theta}' = r + 1 + (r-1) \cos \tilde{\theta}$, which can be reduced to the VCON as shown above.

- Case $r = 0$. The dynamical system (8.12) has a nonhyperbolic equilibrium $\theta = 0$, which corresponds to the saddle-node bifurcation. Therefore, (8.12) does not reflect the dynamics of (8.10), since the latter is sensitive to small perturbations hidden in the small term $R(\theta, \varepsilon)$.

- Case $r < 0$. Equation $r + 1 + (r-1) \cos \theta = 0$ has two solutions in this case:
$$\theta^{\pm} = \pm \cos^{-1} \frac{1+r}{1-r}.$$

Therefore, (8.12) has two equilibria for any $r < 0$. One is stable (θ^-) and the other (θ^+) is not; see Figure 8.4. This case can also be reduced to the VCON model.

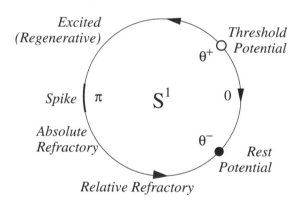

Figure 8.4. Physiological state diagram of a Class 1 neuron

Consider the dynamical system (8.12) for $r > 0$, so that $\theta(\tau) \in \mathbb{S}^1$ oscillates. What is the period of such oscillations, and how does it depend on r? To solve this problem we find the solution of (8.12) starting, say, from $\theta(0) = 0$. The reader can check (by differentiating) that

$$\theta(\tau) = 2 \operatorname{atan} \left(r \tan \sqrt{r}\tau \right)$$

is such a solution. It is periodic with period $T = \pi / \sqrt{r}$. Thus, we have proved the following result.

Proposition 8.4 *If the perturbation is inhibitory ($r < 0$), then the activity $\theta(\tau)$ of the canonical model (8.10) converges to the stable equilibrium $\theta^- + \mathcal{O}(\sqrt[4]{\varepsilon})$. If the perturbation is excitatory ($r > 0$), then $\theta(\tau)$ oscillates with period*

$$T = \frac{\pi}{\sqrt{r}} + \mathcal{O}(\sqrt[4]{\varepsilon}) \qquad \text{(slow time } \tau = \sqrt{\varepsilon}t)$$

or

$$T = \frac{\pi}{\sqrt{\varepsilon r}} + \mathcal{O}(\frac{1}{\sqrt[4]{\varepsilon}}) \qquad \text{(normal time } t),$$

where $r = g(0,0)\, f''(0)/2$.

Notice the remarkable fact that the parameter $a = g(0,0)$ affects only the period of oscillations, but not the width (time duration) of a spike (see the remark on p. 227). Another remarkable fact is that the period $T = \mathcal{O}(1)$ is a finite number in the slow time scale τ, but the spike duration $t_\delta = \mathcal{O}(\sqrt[4]{\varepsilon})$ for $\delta = \sqrt[4]{\varepsilon}$ is small. This means that the system spends most of its time in a neighborhood of the origin with short excursions, which are spikes. We use this fact below when we study weakly connected networks of neurons having saddle-node bifurcations on limit cycles.

8.1.4 Parabolic Burster

Consider a dynamical system of the form

$$\begin{cases} \dot{x} = f(x) + \varepsilon g(x,y,\varepsilon) \\ \dot{y} = \mu q(x,y,\varepsilon) \end{cases} \tag{8.13}$$

exhibiting parabolic bursting (see Section 2.9), where the parameters ε and μ are small. Moreover, we assume that $\mu = \mathcal{O}(\sqrt{\varepsilon})$, which includes $\mu = \varepsilon$. Since there could be many mechanisms of parabolic bursting, we consider the simplest one, where the fast system $\dot{x} = f(x) + \varepsilon g(x,y,\varepsilon)$ is at the saddle-node bifurcation on a limit cycle for $x = 0$ and $\varepsilon = 0$, and the slow system $\dot{y} = q(x,y,\varepsilon)$ has a normally hyperbolic limit cycle attractor γ with frequency Ω for $x = 0$ and $\varepsilon = 0$. Without loss of generality we may assume that $x \in \mathbb{S}^1$ and $y \in \mathbb{R}^m$. The following result is due to Ermentrout and Kopell (1986).

Theorem 8.5 (The Ermentrout-Kopell Theorem for a Parabolic Burster)   *A family of dynamical systems of the form (8.13) exhibiting parabolic bursting has a canonical model*

$$\begin{cases} \theta' = (1 - \cos\theta) + (1 + \cos\theta)r(\psi) \\ \psi' = \omega \end{cases} + R(\theta,\varepsilon) , \tag{8.14}$$

where $' = d/d\tau$, $\tau = \sqrt{\varepsilon}t$ is slow time, the variables θ and ψ are in $\mathbb{S}^1$, the frequency of slow oscillations $\omega = \mu\Omega/\sqrt{\varepsilon}$, the function

$$r(\psi) = g(0, \gamma(\psi), 0)\, f''(0)/2$$

is periodic, and $R(\theta, \varepsilon)$ is small as in Theorem 8.3.

Canonical models for square-wave and elliptic bursters have yet to be found.

Proof. Application of the change of variables $x = h^{-1}(\theta)$ from the VCON lemma (8.2) and introduction of slow time $\tau = \sqrt{\varepsilon}t$ transforms (8.13) to

$$\begin{cases} \theta' = (1 - \cos\theta) + (1 + \cos\theta)g(0, y, 0)\, f''(0)/2 \; + R(\theta, \varepsilon) \\ y' = \bar{\mu}q(h^{-1}(\theta), y, \varepsilon) \end{cases}$$

where $\bar{\mu} = \mu/\sqrt{\varepsilon} = \mathcal{O}(1)$ may be small. Now consider the two cases $|\theta - \pi| \geq 2\sqrt[4]{\varepsilon}$ and $|\theta - \pi| < 2\sqrt[4]{\varepsilon}$. In the first case we have $x = h^{-1}(\theta) = \mathcal{O}(\sqrt[4]{\varepsilon})$ (see the definition of h on page 223). Therefore, the slow system is governed by

$$y' = \bar{\mu}q(0, y, 0) + \mathcal{O}(\bar{\mu}\sqrt[4]{\varepsilon}) \tag{8.15}$$

for all τ for which the fast system does not generate spikes. When it does generate a spike, the variable $x = h^{-1}(\theta)$ is not small. But this happens during the time interval $\mathcal{O}(\sqrt[4]{\varepsilon})$ (see the remark on p. 227). Hence, the difference between a solutions of (8.15) and

$$y' = \bar{\mu}q(x, y, \varepsilon)$$

is of order $\mathcal{O}(\bar{\mu}\sqrt[4]{\varepsilon})$ for the slow time interval $\tau = \mathcal{O}(1)$. Therefore, it can be taken into account by the $\mathcal{O}$ term in (8.15), and the system

$$\begin{cases} \theta' = (1 - \cos\theta) + (1 + \cos\theta)g(0, y, 0)\, f''(0)/2 \; + R(\theta, \varepsilon) \\ y' = \bar{\mu}q(0, y, 0) + \mathcal{O}(\bar{\mu}\sqrt[4]{\varepsilon}) \end{cases}$$

is a canonical model for parabolic bursting. Notice that the second equation does not depend on the first one up to a certain order. Since $y' = q(0, y, 0)$ has a normally hyperbolic limit cycle attractor, the attractor persists under the disturbances. We can write the system above in the form (8.14) if we introduce the phase variable $\psi \in \mathbb{S}^1$ along the limit cycle. $\square$

Ermentrout and Kopell (1986) associated the canonical model (8.14) with the well-studied Hill's equation

$$y'' = b(\tau)y \,,$$

where $b(\tau) = -r(\omega\tau)$, by using the change of variables

$$\frac{y'}{y} = \tan\frac{\theta}{2} \,.$$

Here we dropped the small remainder for simplicity. Using Hill's equation, they deduced several interesting facts about the dynamic behavior of the parabolic burster. Some of those facts were discussed in Section 2.9 using the original canonical model (8.14), which we call the *atoll* model.

8.2   Network of Neurons

Consider a weakly connected system

$$\dot{x}_i = f_i(x_i, \lambda) + \varepsilon g_i(x, \lambda, \rho, \varepsilon) , \qquad x_i \in \mathbb{S}^1 , \quad i = 1, \dots, n , \qquad (8.16)$$

having a multiple saddle-node bifurcation on a limit cycle. Without loss of generality we may assume that the bifurcation occurs when $x_i = 0$, $i = 1, \dots, n$, for $\lambda = 0 \in \Lambda$. Thus, we have

$$f_i(0,0) = 0 , \qquad \frac{\partial f_i(0,0)}{\partial x_i} = 0 , \quad \text{and} \quad \frac{\partial^2 f_i(0,0)}{\partial x_i^2} > 0$$

for all $i = 1, \dots, n$. As usual, we assume that

$$\lambda = \lambda(\varepsilon) = \varepsilon \lambda_1 + \mathcal{O}(\varepsilon^2) \quad \text{and} \quad \rho = \rho(\varepsilon) = \rho_0 + \mathcal{O}(\varepsilon)$$

for some $\lambda_1 \in \Lambda$ and $\rho_0 \in \mathcal{R}$.

Since little is known about brain dynamics, we do not have detailed information about the connection functions g_i. In particular, we do not know the values of $g_i(0, 0, \rho_0, 0)$, even though they play a crucial role in our analysis below.

Suppose that each x_i denotes activity of a local population of neurons. Then $x_i = 0$ means that the activity of the local population is at a background steady-state (see Figure 1.8), which does not exclude the silent state. Positive (negative) $g_i(0, 0, \rho_0, 0)$ means that there is an excitatory (inhibitory) background influence from the network to the ith local population of neurons. The case $g_i(0, 0, \rho_0, 0) = 0$ might imply that there is no such background influence. This happens, for example, when $x = 0$ corresponds to a neuron's membrane potential being at rest.

Suppose that each x_i denotes activity of a single spiking neuron. Since there is no synaptic transmission when such neurons are silent, we have $g_i(x, 0, \rho_0, 0) = 0$ for x in a small, say $\sqrt[4]{\varepsilon}$-neighborhood of $x = 0$; see Figure 8.5. The assumption that there is no background excitation of inhibition when the other neurons are silent does not contradict the well-know fact

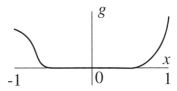

Figure 8.5. The connection function g vanishes for x in a neighborhood of the origin.

that there is a spontaneous secretion of neurotransmitter that is not caused
by firings of other neurons (Shepherd 1983; Johnston and Wu 1995). First,
such a transmission of signals is negligible in comparison with the evoked
spiking transmission. Second, the spontaneous emission of neurotransmitter
is an internal mechanism that must be taken into account in the function
f_i, not g_i. The considerations above might not be valid when there is a
graded dendro-dendritic synaptic transmission.

8.2.1 Size of PSP

Consider a pair of neurons

$$\begin{cases} \dot{x}_1 = f_1(x_1, \lambda) + \varepsilon g_1(x_1, x_2, \lambda, \rho, \varepsilon) \\ \dot{x}_2 = f_2(x_2, \lambda) + \varepsilon g_2(x_1, x_2, \lambda, \rho, \varepsilon) \end{cases} \qquad (8.17)$$

and suppose that $g_1(x_1, x_2, 0, \rho_0, 0) \equiv 0$ when $|x_2| < \sqrt[4]{\varepsilon}$; that is, there is
no synaptic transmission if x_2 is quiet. Suppose that the first neuron is at a
rest potential and the second one starts to fire a spike. Firing of x_2 perturbs
x_1. As a result, x_1 may cross the threshold value and fire a spike too. It
is frequently the case though that x_1 does not fire, but returns to the rest
potential, as in Figure 8.6. Such a perturbation is called the *postsynaptic
potential* (PSP). Below we are interested in the amplitude of the PSP.

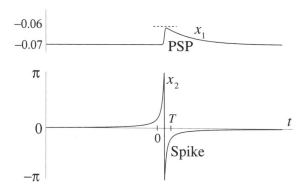

Figure 8.6. An example of a postsynaptic potential (PSP) for (8.17):
$f_1 = 1 + \lambda - \cos x_1$, $f_2 = 1 - \cos x_2$, $g_1 = (1 - \cos x_2)^4/16$, $\lambda = -\varepsilon/4$,
and $\varepsilon = 0.01$. Dashed curve is the amplitude predicted by the PSP lemma.

Lemma 8.6 (PSP Lemma) *The amplitude of a postsynaptic potential
for (8.17) is*

$$\varepsilon \int_{\mathbb{S}^1} \frac{g_1(0, x, 0, \rho, 0)}{f_2(x, 0)} \, dx + \mathcal{O}(\varepsilon^{5/4}) \,.$$

Proof. Let us write $x_1(t) = x_1(0) + \varepsilon y(t)$. Then (8.17) can be written in the form

$$\begin{cases} \dot{y} = g_1(0, x_2, 0, \rho, 0) \\ \dot{x}_2 = f_2(x_2, 0) \end{cases} + \mathcal{O}(\sqrt{\varepsilon}) .$$

Let $x_2(t)$, $t \in [0, T]$ be the solution to the second equation, such that $x_2(0) = \sqrt[4]{\varepsilon}$ and $x_2(T) = -\sqrt[4]{\varepsilon}$, where $T = \mathcal{O}(\varepsilon^{-1/4})$ is determined by the spike lemma. The solution to

$$\dot{y} = g_1(0, x_2(t), 0, \rho, 0) + \mathcal{O}(\sqrt{\varepsilon}) , \qquad y(0) = 0$$

for $t \leq T$ is given by

$$\begin{aligned} y(t) &= \int_0^T g_1(0, x_2(t), 0, \rho, 0) \, dt + \mathcal{O}(\sqrt[4]{\varepsilon}) \\ &= \int_{\sqrt[4]{\varepsilon}}^{-\sqrt[4]{\varepsilon}} \frac{g_1(0, x_2, 0, \rho, 0) \, dt}{f_2(x_2, 0)} \, dx_2 + \mathcal{O}(\sqrt[4]{\varepsilon}) . \end{aligned}$$

Since $g_1 \equiv 0$ for $|x_2| < \sqrt[4]{\varepsilon}$, we can replace the interval $[\sqrt[4]{\varepsilon}, -\sqrt[4]{\varepsilon}]$ in the last integral by the entire unit circle $\mathbb{S}^1$. This completes the proof. $\square$

Suppose system (8.17) describes the dynamics of two coupled hippocampal cells. Let us use the PSP lemma and experimental data obtained by McNaughton et al. (1981), which we described in Section 1.3.2, to estimate the value of parameter ε. In our evaluation below we take f_i and g_i as in the caption to Figure 8.6 so that (8.17) has the form

$$\begin{cases} \dot{x}_1 = 1 + \lambda - \cos x_1 + \varepsilon(1 - \cos x_2)^4/16 \\ \dot{x}_2 = 1 - \cos x_2 \end{cases} . \tag{8.18}$$

Such f_i and g_i satisfy $|f_i| < 2$ and $|g_i| < 1$.

The mean EPSP size in the hippocampal granule cells is 0.1 ± 0.03 mV. If we assume the amplitude of the action potential to be 100 mV, then the ratio of mean EPSP size to the amplitude of a spike is 0.001 ± 0.0003. Using the PSP lemma we determine the amplitude of PSP in (8.18) to be approximately ε. Since the amplitude of a spike is 2π, we obtain $0.001 \pm 0.0003 \approx \varepsilon/(2\pi)$.

Corollary 8.7 (Size of ε) *If we use (8.18) to model a pair of hippocampal cells, then we must take*

$$0.004 < \varepsilon < 0.008$$

in order to match neurophysiological data obtained by McNaughton et al. (1981).

8.2.2 Adaptation Condition Is Violated

The right-hand side of system (8.16) can be expanded locally in its Taylor series, so that the WCNN is

$$\dot{x}_i = \varepsilon a_i + p_i x_i^2 + \hat{f}_i(x_i) + \varepsilon \hat{g}_i(x, \varepsilon) \,,$$

where

$$
\begin{aligned}
a_i &= D_\lambda f_i(0,0) \cdot \lambda_1 + g_i(0,0,\rho_0,0) \,, &\qquad (8.19)\\
p_i &= \frac{1}{2} \frac{\partial^2 f_i(0,0)}{\partial x_i^2} > 0 &\qquad (8.20)
\end{aligned}
$$

as before, and the functions $\hat{f}_i$ and $\hat{g}_i$ denote higher-order terms in x and ε. In particular, $\hat{f}_i(x_i) = f_i(x_i,0) - p_i x_i^2$ and $\hat{g}_i(0,0) = 0$.

Recall that we called the condition $a_i = 0$, $i = 1,\dots,n$, the *adaptation condition*. We claimed in Section 5.3.8 that this condition is necessary for the WCNN to have interesting local dynamics. Below we substantiate this point. Moreover, we show that violation of the adaptation condition makes even global dynamics somehow less interesting.

Let us derive a canonical model for the WCNN (8.16) for the case when all a_i defined in (8.19) are nonzero.

Theorem 8.8 *Consider the weakly connected system (8.16) of neurons having saddle-node bifurcations on limit cycles. Suppose the adaptation condition is violated uniformly; that is, all $a_i \neq 0$ as defined in (8.19). Then (8.16) has a global canonical model*

$$\theta_i' = (1 - \cos\theta_i) + (1 + \cos\theta_i)c_i(\theta,\varepsilon) + R_i(\theta_i,\varepsilon) \,, \qquad \theta_i \in \mathbb{S}^1 \,, \quad (8.21)$$

where $\theta = (\theta_1,\dots,\theta_n) \in \mathbb{S}^1 \times \cdots \times \mathbb{S}^1 \cong \mathbb{T}^n$ is the variable on the n-torus, $' = d/d\tau$, and $\tau = \sqrt{\varepsilon}t$ is slow time. Each connection function $c_i : \mathbb{T}^n \times \mathbb{R} \to \mathbb{R}$ satisfies

$$
c_i(\theta,\varepsilon) = \begin{cases} a_i p_i & \text{if all } |\theta_i - \pi| > 2\sqrt[4]{\varepsilon} \,, \\ \mathcal{O}(1) & \text{if at least one } |\theta_i - \pi| \leq 2\sqrt[4]{\varepsilon} \,, \end{cases}
$$

where the parameters a_i and $p_i > 0$ are defined by (8.19) and (8.20), respectively. The remainder $R_i(\theta_i,\varepsilon)$ is small as in Theorem 8.3.

Proof. This theorem is a straightforward application of the VCON lemma. Since the proof repeats that of the Ermentrout-Kopell theorem for Class 1 neural excitability with the obvious modifications, we may leave out some tedious details and all trigonometric manipulations.

Suppose all variables are small; namely, $|x_i| \leq \sqrt[4]{\varepsilon}/p_i$. Then each function $\hat{g}_i(x,\varepsilon) = \mathcal{O}(\sqrt[4]{\varepsilon})$ is small, and the weakly connected system has the form

$$\dot{x}_i = \varepsilon a_i + p_i x_i^2 + \hat{f}_i(x_i) + \mathcal{O}(\varepsilon^{5/4}) \,, \qquad i = 1,\dots,n \,.$$

Applying the change of variables

$$x_i = \frac{\sqrt{\varepsilon}}{p_i} \tan \frac{\theta_i}{2} ,$$

introducing the slow time $\tau = \sqrt{\varepsilon} t$, and using the same trigonometric transformations as in the proof of the Ermentrout-Kopell theorem, we reduce the system above to

$$\theta_i' = (1 - \cos \theta_i) + (1 + \cos \theta_i) a_i p_i + R(\theta, \varepsilon) .$$

Now suppose at least one component satisfies $|x_i| > \sqrt[4]{\varepsilon}/p_i$, which is equivalent to $|\theta_i - \pi| < 2\sqrt[4]{\varepsilon}$. Then the assertion that each $\hat{g}_i(x, \varepsilon)$ is small cannot be guaranteed. Therefore, we assume that all functions $\hat{g}_i(x, \varepsilon)$ are of order $\mathcal{O}(1)$, which incidentally does not exclude the case of small g_i. The weakly connected system can then be written in the form

$$\dot{x}_i = \varepsilon(a_i + \hat{g}_i(x, \varepsilon)) + p_i x_i^2 + \hat{f}_i(x_i) .$$

Application of the change of variables $\theta_i = h_i(x_i)$ to each equation in the system above and introduction of slow time $\tau = \sqrt{\varepsilon} t$ yields

$$\theta_i' = (1 - \cos \theta_i) + (1 + \cos \theta_i)(a_i p_i + \hat{g}_i(h^{-1}(\theta), \varepsilon)p_i) + R(\theta, \varepsilon) .$$

Now we glue the two systems for θ_i' together by introducing the connection function $c_i(\theta, \varepsilon)$ that equals $a_i p_i$ in the first case and

$$a_i p_i + \hat{g}_i(h^{-1}(\theta), \varepsilon)p_i$$

in the second case. $\square$

The result of the theorem is valid even when there is a time-dependent input $\rho = \rho(t)$. In that case the parameters a_i are given by

$$a_i = D_\lambda f_i(0, 0) \cdot \lambda_1 + \lim_{T \to \infty} \frac{1}{T} \int_0^T g_i(0, 0, \rho_0(t), 0) \, dt ,$$

provided that the limit exists.

Notice that the connection functions $c_i(\theta, \varepsilon)$ do not depend on θ except for brief moments of order $\mathcal{O}(\sqrt[4]{\varepsilon})$ when one or more neurons generate action potentials. Thus, the transmission of a signal takes place at the moment of a spike. One expects this from a network of spiking neurons having synapses working only during a spike, but we derived this fact from a network having graded (continuous) activities x_i and nonconstant smooth synaptic functions g_i that "work" all the time, even when neurons do not generate action potentials.

A direct consequence of Theorem 8.8 is the following result.

Corollary 8.9 *Each neuron in the canonical model (8.21) is either silent or a pacemaker regardless of the activities of the other neurons; see Figure 8.7. More precisely:*

- *If $a_i < 0$, then $\theta_i(\tau)$ stays in an $\sqrt[4]{\varepsilon}$-neighborhood of an equilibrium, and the activities of the other neurons cannot make it oscillate.*

- *If $a_i > 0$, then $\theta_i(\tau)$ oscillates, and activities of the other neurons cannot silence it.*

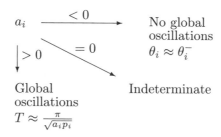

Figure 8.7. The ith neuron activity depends entirely on $a_i \neq 0$, and it is insensitive to other neuron's activities.

When $g_i(0, 0, \rho_0, 0) = 0$, the corollary follows from the PSP lemma.

Proof. Fix the index i and suppose $a_i < 0$. We claim that $\theta_i(\tau)$ cannot cross an $\sqrt[4]{\varepsilon}$-neighborhood of the origin. Indeed, suppose $\theta_i(0) = \mathcal{O}(\sqrt[4]{\varepsilon})$. Then $\cos\theta_i = 1 + \mathcal{O}(\sqrt{\varepsilon})$, and θ_i' is governed by the linear equation

$$\theta_i' = 2c_i(\theta, \varepsilon) + \mathcal{O}(\sqrt[4]{\varepsilon}\ln\varepsilon) . \tag{8.22}$$

The other neurons generate spikes and can increase the connection function $c_i(\theta, \varepsilon)$. If $c_i > 0$, then $\theta_i(\tau)$ increases. Consider a time interval of finite positive length, say $\tau = 1$. Since the total length of time of generation of the spikes is $\mathcal{O}(\sqrt[4]{\varepsilon})$ (remark on p. 227), the total increase of $\theta_i(\tau)$ during the time $\tau = 1$ is $\mathcal{O}(\sqrt[4]{\varepsilon})$. (Here we explicitly use the assumption that the number of neurons that fire simultaneously is finite and fixed. If the number were proportional to $1/\sqrt[4]{\varepsilon}$, the consideration above would not be true.) During the time interval $1 - \mathcal{O}(\sqrt[4]{\varepsilon})$ no neurons fire. The synaptic function $c_i = a_i p_i < 0$, and the total decrease of $\theta_i(\tau)$ is $2a_i p_i + \mathcal{O}(\sqrt[4]{\varepsilon})$. This moves θ_i away from the $\mathcal{O}(\sqrt[4]{\varepsilon})$-neighborhood of the origin, where the linear approximation (8.22) is valid, toward the stable equilibrium.

Consider the case $a_i > 0$. When none of the neurons generates a spike,

$$\theta_i' = (1 - \cos\theta_i) + (1 + \cos\theta_i)a_i p_i > 2\min(a_i p_i, 1) > 0 .$$

If at least one of the neurons fires, then θ_i' could become negative, but this happens only during a short time interval of order $\mathcal{O}(\sqrt[4]{\varepsilon})$. Therefore, during the finite time interval $\tau = 1$ the ith variable θ_i increases by

$2\min(a_i p_i, 1) + \mathcal{O}(\sqrt[4]{\varepsilon})$ units and decreases by $\mathcal{O}(\sqrt[4]{\varepsilon})$ units. This indicates that the activities of the other neurons cannot stop oscillations of θ_i. $\square$

Remark. It is not correct to assume that $\theta_i(\tau) \to$ const when $a_i < 0$. If other neurons generate spikes, $\theta_i(\tau)$ can wobble in a small neighborhood of an equilibrium due to the influences from the other neurons, but the oscillations have very small amplitude and again are negligible to an observer.

When $a_i > 0$, the neurons oscillate, and the canonical model is a network of pulse-coupled oscillatory neurons. It is still possible to observe many interesting phenomena such as synchronization, entrainment, and chaos. Nevertheless, caution should be used when one models these phenomena using the canonical model. Indeed, the influences from other neurons have order $\mathcal{O}(1)$ but time duration $\mathcal{O}(\sqrt[4]{\varepsilon})$ and therefore are comparable with the remainder R.

Partial Violation of Adaptation Condition

The theorem above covers the case $a_i \neq 0$ for all $i = 1, \ldots, n$. When $a_i = 0$ for some, but not all, i, it is possible to show that activity can be convergent or periodic. For example, consider a network of VCONs of the form

$$\begin{cases} \dot{\theta}_1 = 1 - \cos\theta_1 + \varepsilon c_{12}(1 \mid \cos\theta_2) \\ \dot{\theta}_2 = 1 - \cos\theta_2 + \varepsilon a_2 \end{cases} ,$$

where $a_2 > 0$, and hence θ_2 oscillates. It is easy to see that in the case $c_{12} < 0$ the variable θ_1 stays in a neighborhood of the origin, but in the case $c_{12} > 0$ it oscillates.

8.2.3 Parabolic Bursting Revisited

Now consider a dynamical system of the form

$$\begin{cases} \mu x' = f(x) + \varepsilon g(x, y, \varepsilon) \\ y' = u(y) + \varepsilon v(x, y, \varepsilon) \end{cases} \tag{8.23}$$

for the case when both the fast variable x and the slow variable y have saddle-node bifurcations on a limit cycle for $\varepsilon = 0$. The difference between this case and the one considered in Section 8.1.4 is that both subsystems are excitatory. In particular, the slow variable might not sustain periodic activity in the absence of influences from the fast variable.

First, we assume that the adaptation condition is violated for both x and y. Theorem 8.8 provides a canonical model for system (8.23); namely,

$$\begin{cases} \mu\theta' = (1 - \cos\theta) + (1 + \cos\theta)c_1(\psi, \varepsilon) + R(\theta, \varepsilon) \\ \psi' = (1 - \cos\psi) + (1 + \cos\psi)c_2(\theta, \varepsilon) + R(\psi, \varepsilon) \end{cases} . \tag{8.24}$$

The behavior of this model depends crucially on the signs of the parameters $a_1 = g(0,0,0)$, $a_2 = v(0,0,0)$, and $c_1(\psi, \varepsilon) = g(0, y, 0)$ while ψ is close to π, meaning that y fires. We summarize the basic dynamical regimes of such a system in Figure 8.8. In particular, it exhibits tonic activity (repetitive

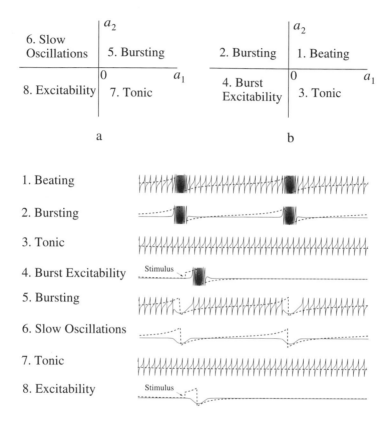

Figure 8.8. Behavior of the canonical model (8.24) depends on the signs of $a_1 = g(0,0,0)$, $a_2 = v(0,0,0)$, and $c_1(\psi, \varepsilon) = g(0, h^{-1}(\psi), 0)$ when $\psi \approx \pi$. *Case a*: $c_1(\psi, \varepsilon) < 0$. *Case b*: $c_1(\psi, \varepsilon) > 0$. The continuous curve represents activity of the fast variable, and the dashed curve represents activity of the slow variable. Simulations are performed for parameter values $|a_1| = |a_2| = 0.1$, $\mu = 0.05$, and the function $c_1(\psi) = \pm 10 \sin^{20} \psi/2$.

firing of the fast variable without quiescent phases), beating (intermittence of tonic and bursting activity) and burst excitability (short bursts of the fast variable in response to a stimulus applied to the slow variable).

Notice that repetitive bursting occurs when $a_2 = v(0,0,0) > 0$ in (8.23). Since $x = 0$ corresponds to the rest activity of the fast variable, the condition $v(0,0,0) > 0$ implies that the slow variable y is an autonomous (subcellular) oscillator. This is in accordance with the hypotheses of the

Ermentrout-Kopell Theorem for a parabolic burster. Below we present an example of a parabolic burster whose slow variable y cannot sustain periodic activity without an input from the fast variable x. The example is a modification of that of Soto-Trevino et al. (1996), and it is motivated by the work of Rinzel and Lee (1986). A necessary condition for such a behavior is that the adaptation condition, $a_2 = 0$, is satisfied for the slow variable y.

Consider the dynamical system (8.23) with the functions g and r satisfying the following conditions:

- $g(0,0,0) > 0$, but $g(0,y,0) < 0$ when $|y|$ is sufficiently large.

- $v(0,0,0) = 0$, $v(x,0,0) < 0$ when x is negative and small, and $v(x,0,0) > 0$ otherwise.

A typical choice of such g and v is depicted in Figure 8.9.

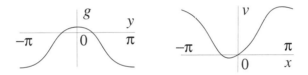

Figure 8.9. Functions $g(0,y,0)$ and $v(x,0,0)$ for parabolic burster (8.23)

The first condition implies that x fires if and only if y is quiescent; see Figure 8.10. The second condition implies that y has a stable equilibrium

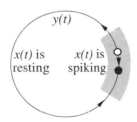

Figure 8.10. x fires only when y is quiescent.

when x is small (quiescent). We see that the slow variable y is not an autonomous subcellular oscillator. The condition $v(x,0,0) > 0$ for nonsmall $|x|$ may result in disappearance of the rest state for the slow variable y, as in the theorem below. In this case y generates spikes during which the repetitive firing of x disappears. A typical example of such bursting is depicted in Figure 8.11.

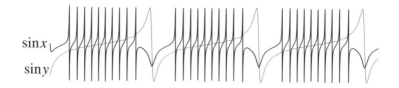

Figure 8.11. Excitable parabolic bursting in $\mu x' = 1 - \cos x + \varepsilon g(y)$, $y' = 1 - \cos y + \varepsilon v(x)$ for $\mu = 0.05$ and $\varepsilon = 0.1$. The functions $g(y) = \cos y - 0.95$ and $v(x) = \cos 0.5 - \cos(x + 0.5)$ are similar to those in Figure 8.9.

Theorem 8.10 (Excitable Parabolic Burster) *Consider the singularly perturbed dynamical system*

$$\begin{cases} \mu x' = f(x) + \varepsilon g(x, y, \varepsilon) \\ \quad y' = u(y) + \varepsilon v(x, y, \varepsilon) \end{cases} \tag{8.25}$$

where $\varepsilon \ll 1$, $\mu = o(\varepsilon^{3/2})$, and both variables $x, y \in \mathbb{S}^1$ have a saddle-node equilibrium at 0 on a limit cycle for $\varepsilon = 0$. Suppose that the functions g and v satisfy the conditions above. If

$$\int_{\mathbb{S}^1} \frac{v(x, 0, 0)}{f(x) + \varepsilon g(x, 0, 0)} \, dx > 0 \,, \tag{8.26}$$

then (8.25) is a parabolic burster.

Proof. Let us check that (8.26) implies that y leaves any $\sqrt{\varepsilon}$-neighborhood of the origin, thereby producing a spike. Suppose $y = \mathcal{O}(\sqrt{\varepsilon})$. From Proposition 8.4 it follows that the fast variable x in the equation

$$\dot{x} = f(x) + \varepsilon g(x, 0, 0) + \mathcal{O}(\varepsilon\sqrt{\varepsilon})$$

oscillates with the period

$$T = \frac{\pi}{\sqrt{\varepsilon r}} + \mathcal{O}(\frac{1}{\sqrt[4]{\varepsilon}}) \,,$$

where $r = g(0, 0, 0) \, f''(0)/2 > 0$. Let $x(t)$ be such a periodic solution, then the slow subsystem has the form

$$\dot{y} = \mu \left\{ u(y) + \varepsilon v(x(t), y, \varepsilon) \right\} \,.$$

Now we average this equation (see Section 9.3.1) to obtain

$$\dot{y} = \mu \left\{ u(y) + \varepsilon \bar{v}(y) \right\} + \mathcal{O}(\mu^2) \,, \tag{8.27}$$

where

$$\bar{v}(y) = \frac{1}{T} \int_0^T v(x(t), y, 0) \, dt = \frac{\sqrt{\varepsilon r}}{\pi} \int_{\mathbb{S}^1} \frac{v(x, 0, 0)}{f(x) + \varepsilon g(x, 0, 0)} \, dx + \mathcal{O}(\varepsilon) \,.$$

From (8.26) it follows that $\bar{v}(0) > 0$, therefore (8.27) cannot have equilibria in a small neighborhood of the origin. The slow variable y oscillates in this case, and the fast variable x has bursting pattern of activity. $\square$

It is still an open question whether similar results can be proved for other types of excitable bursters.

8.3   Adaptation Condition Is Satisfied

Now suppose $a_i = 0$ for all i in (8.19). Recall that in this case we say that the adaptation condition is satisfied. A typical example when this might happen is when $g_i(0, 0, \rho_0, 0) \equiv 0$ and $\lambda = \mathcal{O}(\varepsilon^2)$.

In Chapter 5 we showed that the local activity of

$$\dot{x}_i = f_i(x_i, \lambda) + \varepsilon g_i(x_1, \ldots, x_n, \lambda, \rho, \varepsilon) , \qquad x_i \in \mathbb{S}^1 , \tag{8.28}$$

is governed by the canonical model

$$y_i' = r_i + y_i^2 + \sum_{j=1}^{n} c_{ij} y_j , \qquad y_i \in \mathbb{R} , \tag{8.29}$$

Using phase coordinates $y_i = \tan \theta_i / 2$, we can rewrite the canonical model above in the form

$$\theta_i' = (1 - \cos \theta_i) + (1 + \cos \theta_i) \left(r_i + \sum_{j=1}^{n} c_{ij} \tan \frac{\theta_j}{2} \right) ,$$

which is valid when all $|\theta_i - \pi| > 2\sqrt{\varepsilon}$. When at least one θ_i is in the $2\sqrt{\varepsilon}$-neighborhood of π, which corresponds to firing a spike, one cannot use the canonical model to describe adequately the behavior of the WCNN (8.28).

Theorem 8.11 *Consider the weakly connected system (8.28) of neurons having saddle-node bifurcations on limit cycles. Suppose that $\lambda = \varepsilon \lambda_1 + \mathcal{O}(\varepsilon^2)$, $\rho = \rho_0 + \mathcal{O}(\varepsilon)$, and that the adaptation condition*

$$a_i \equiv D_\lambda f_i(0, 0) \cdot \lambda_1 + g_i(0, 0, \rho_0, 0) = 0$$

is satisfied for all $i = 1, \ldots, n$. Then (8.28) has the following canonical model:

$$\theta_i' = (1 - \cos \theta_i) + (1 + \cos \theta_i) c_i(\theta, \varepsilon) + R_i(\theta_i, \varepsilon) , \qquad \theta_i \in \mathbb{S}^1, \tag{8.30}$$

where $' = d/d\tau$ and $\tau = \varepsilon t$ is slow time. Each connection function c_i : $\mathbb{T}^n \times \mathbb{R} \to \mathbb{R}$ satisfies

$$c_i(\theta, \varepsilon) = \begin{cases} r_i + \sum_{j=1}^{n} c_{ij} \tan \frac{\theta_j}{2} & \text{if all } |\theta_i - \pi| > 2\sqrt{\varepsilon} , \\ \mathcal{O}(1/\varepsilon) & \text{if at least one } |\theta_i - \pi| \le 2\sqrt{\varepsilon} . \end{cases}$$

The remainder $R_i(\theta_i, \varepsilon)$ is small.

The proof repeats that of Theorem 8.8 (with $\delta = \sqrt{\varepsilon}$) and is not given here. See also the proof of Proposition 8.12 below.

8.3.1   Integrate-and-Fire Model

Consider a WCNN with the uncoupled connection function g_i,

$$\dot{x}_i = f_i(x_i, \lambda) + \varepsilon \sum_{j=1}^{n} g_{ij}(x_j, \lambda, \rho, \varepsilon) , \qquad x_i \in \mathbb{S}^1 , \quad i = 1, \ldots, n . \quad (8.31)$$

In Chapter 9 we discuss neurophysiological justifications of such g_i. As usual, we assume that $x_i = 0$ is a saddle-node point on a limit cycle for each $i = 1, \ldots, n$. Since each x_i describes the activity of a single neuron (not a local population of neurons), we assume below that there is no synaptic transmission in a neighborhood of the rest potential. This implies that $g_{ij}(x_j, \lambda, \rho_0, \varepsilon) \equiv 0$ when x_j is in a small, say $\sqrt{\varepsilon}$-neighborhood of $x_j = 0$.

A consequence of this assumption is that the adaptation condition has the form $D_\lambda f_i(0,0) \cdot \lambda_1 = 0$ in this case. Another consequence is that the connection function $c_i(\theta, \varepsilon)$ in the canonical model (8.30) is a constant unless one or more neurons generate a spike.

Proposition 8.12 (Integrate-and-Fire)   *The canonical model (8.30) for the WCNN (8.31) can be approximated by*

$$\theta_i' = (1 - \cos \theta_i) + (1 + \cos \theta_i) \left(r_i + \sum_{j=1}^{n} c_{ij}(\theta_i)\delta(\theta_j - \pi) \right) , \quad \theta_i \in \mathbb{S}^1,$$

$$(8.32)$$

where the parameter r_i is defined in the proof (see Eq. (8.37)) and δ is the Dirac delta function satisfying $\delta(y) = 0$ if $y \neq 0$, $\delta(0) = \infty$, and $\int \delta = 1$. When θ_j crosses π (fires a spike), the value of θ_i is incremented by

$$c_{ij}(\theta_i) = 2 \operatorname{atan} \left(\tan \frac{\theta_i}{2} + s_{ij} \right) - \theta_i , \quad (8.33)$$

where

$$s_{ij} = \frac{1}{2} \frac{\partial^2 f_i(0,0)}{\partial x_i^2} \int_{\mathbb{S}^1} \frac{g_{ij}(x_j, 0, \rho_0, 0)}{f_j(x_j, 0)} \, dx_j \quad (8.34)$$

is a constant.

Thus, each neuron is governed by the equation

$$\theta_i' = (1 - \cos \theta_i) + (1 + \cos \theta_i) r_i , \qquad \theta_i \in \mathbb{S}^1.$$

When $\theta_i = \pi$, the neuron fires. The neurons interact by a simple form of pulse coupling: When θ_j fires, it resets θ_i to the new value

$$\theta_i^{\text{new}} = 2 \operatorname{atan} \left(\tan \frac{\theta_i^{\text{old}}}{2} + s_{ij} \right) ,$$

which depends on the current activity θ_i^{old}. The variable θ_i integrates many such inputs from other neurons, and θ_i fires when it crosses π, hence the name *integrate-and-fire*.

Proof. Consider the initial portion of the Taylor's series of (8.31)

$$\dot{x}_i = \varepsilon^2 d_i + \varepsilon b_i x_i + p_i x_i^2 + \varepsilon \sum_{j=1}^{n} g_{ij}(x_j, 0, \rho_0, 0) , \qquad (8.35)$$

where $p_i > 0$ is defined in (8.20), and the parameters

$$d_i = D_\lambda f_i \cdot \lambda_2 + (D_\lambda^2 f_i) \cdot (\lambda_1, \lambda_1) \qquad \text{and} \qquad b_i = D_\lambda \frac{\partial f_i}{\partial x_i} \cdot \lambda_1$$

are evaluated at the origin. The translation of variables

$$x_i = y_i - \varepsilon \frac{b_i}{2 p_i}$$

transforms equation (8.35) to

$$\dot{y}_i = \varepsilon^2 \frac{r_i}{p_i} + p_i y_i^2 + \varepsilon \sum_{j=1}^{n} g_{ij}(y_j, 0, \rho_0, 0) , \qquad (8.36)$$

where

$$r_i = d_i p_i - b_i^2 / 4 . \qquad (8.37)$$

Suppose the ith neuron does not fire; that is, $|y_i| < \sqrt{\varepsilon}/p_i$. Then, using the change of variables

$$y_i = \frac{\varepsilon}{p_i} \tan \frac{\theta_i}{2}$$

we transform (8.36) to

$$\dot{\theta}_i = \varepsilon \left\{ (1 - \cos \theta_i) + (1 + \cos \theta_i) r_i \right\} + (1 + \cos \theta_i) p_i \sum_{j=1}^{n} g_{ij}(y_j, 0, \rho_0, 0) .$$

Suppose the other neurons do not fire either; i.e., all $|y_j| < \sqrt{\varepsilon}/p_i$. Since $g_{ij} = 0$ for such y_j, the equation above has the form (8.32), where $' = d/d\tau$ and $\tau = \varepsilon t$ is slow time.

Suppose the jth neuron starts to fire a spike; that is, $y_j = \sqrt{\varepsilon}/p_j$. Neglecting terms of order ε yields

$$\dot{\theta}_i = (1 + \cos \theta_i) p_i g_{ij}(y_j(t), 0, \rho_0, 0) ,$$

which must be solved with the initial condition $\theta_i(0) = \theta_i^{\text{old}}$. By separation of variables we find that

$$\int_{\theta_i^{\text{old}}}^{\theta_i(t)} \frac{d\theta_i}{1 + \cos \theta_i} = p_i \int_0^t g_{ij}(y_j(s), 0, \rho_0, 0) \, ds .$$

It is easy to see that

$$\int_{\theta_i^{\text{old}}}^{\theta_i(t)} \frac{d\theta_i}{1 + \cos\theta_i} = \tan\frac{\theta_i(t)}{2} - \tan\frac{\theta_i^{\text{old}}}{2}$$

and

$$\int_0^t g_{ij}(y_j(s), 0, \rho_0, 0)\, ds = \int_{\sqrt{\varepsilon}/p_j}^{y_j(t)} \frac{g_{ij}(y_j, 0, \rho_0, 0)}{f_j(y_j, 0)}\, dy_j \; .$$

Let t be the time when $y_j(t)$ hits the value $-\sqrt{\varepsilon}/p_j$, which corresponds to the end of the spike. Then the new value of $\theta_i^{\text{new}} = \theta_i(t)$ can be determined from

$$\tan\frac{\theta_i^{\text{new}}}{2} = \tan\frac{\theta_i^{\text{old}}}{2} + p_i \int_{\sqrt{\varepsilon}/p_j}^{y_j(t)} \frac{g_{ij}(y_j, 0, \rho_0, 0)}{f_j(y_j, 0)}\, dy_j \; ,$$

which implies

$$\theta_i^{\text{new}} = 2\operatorname{atan}\left(\tan\frac{\theta_i^{\text{old}}}{2} + s_{ij}\right) \; ,$$

where

$$s_{ij} = p_i \int_{\sqrt{\varepsilon}/p_j}^{-\sqrt{\varepsilon}/p_j} \frac{g_{ij}(y_j, 0, \rho_0, 0)}{f_j(y_j, 0)}\, dy_j \; .$$

Since we assumed that $g_{ij}(y_j, 0, \rho_0, 0) \equiv 0$ when $|y_j| < \sqrt{\varepsilon}/p_j$ we can replace the integral above by (8.34). Now notice that the jth neuron generates a spike when θ_j crosses the $2\sqrt{\varepsilon}$-neighborhood of π. This event takes $4\sqrt{\varepsilon}$ units of slow time. It looks instantaneous, and we use the Dirac delta function to approximate this. $\square$

A simulation of the WCNN (8.31) with $\varepsilon = 0.1$ is shown in Figure 8.12a and b for illustrational purposes. A magnification of a $\sqrt{\varepsilon}$-neighborhood of the spike region is shown in Figure 8.12c, where we draw the solution of the integrate-and-fire model (8.32) for comparison.

When $r_i < 0$, the ith neuron activity in (8.32) has a stable and an unstable equilibrium,

$$\theta_i^{\pm} = \pm\cos^{-1}\frac{1 + r_i}{1 - r_i} \; ,$$

(see Section 8.1.3). The unstable equilibrium θ_i^+ has the meaning of the *threshold* potential in the sense that the ith neuron does not generate spikes unless perturbations from the other neurons lead θ_i beyond θ_i^+. Such perturbations may exist when the parameters s_{ij} defined in (8.34) are large enough. Therefore, Corollary 8.9 is not applicable to the integrate-and-fire model, and we have the following result:

Corollary 8.13 *A single spike of the jth neuron in WCNNs approximated by (8.32) can make the ith neuron fire (regardless of the size of $\varepsilon \ll 1$) if*

$$s_{ij} > 2\sqrt{|r_i|} \; ,$$

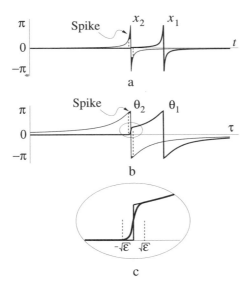

Figure 8.12. a. Simulation of a weakly connected system (8.31) with $\varepsilon = 0.1$, $f_1 = f_2 = 1 - \cos x$ and $g_{12}(x_2) = (1 - \cos x_2)^4/16$. b. The same solution written in variables $\theta_i = 2 \operatorname{atan}(2x_i/\varepsilon)$. c. Magnified neighborhood of the spike region

where s_{ij} and r_i are defined in (8.34) and (8.37), respectively.

Proof. Suppose the ith neuron is at the rest potential, i.e. $\theta_i = \theta_i^-$. To generate a spike, θ_i must cross the threshold value θ_i^+. This implies that

$$s_{ij} > \tan \frac{\theta_i^+}{2} - \tan \frac{\theta_i^-}{2} \ .$$

Simple trigonometric manipulations show that $\tan(\theta_i^\pm/2) = \pm\sqrt{|r_i|}$. $\square$

When $r_i > 0$, the neuron generates spikes repeatedly. One can study synchronization phenomena of such pulse-coupled pacemakers (Mirollo and Strogatz 1990, Kuramoto 1991, Abbott and van Vreeswijk 1993, Chawanya et al. 1993, Tsodyks 1993, Usher 1993). Unfortunately, one cannot apply the Gerstner-van-Hemmen-Cowan theorem (Gerstner et al. 1996) to (8.32) due to the fact that the increment $c_{ij}(\theta_i)$ is not a constant. It is maximal when θ_i is near the origin (rest potential). It is still an open problem to study synchronization phenomena in integrate-and-fire models with nonconstant increments of the form (8.33).

9
Weakly Connected Oscillators

In this chapter we study weakly connected networks

$$\dot{X}_i = F_i(X_i, \lambda) + \varepsilon G_i(X, \lambda, \rho, \varepsilon) , \qquad i = 1, \ldots, n , \qquad (9.1)$$

of oscillatory neurons. Our basic assumption is that there is a value of $\lambda \in \Lambda$ such that every equation in the uncoupled system ($\varepsilon = 0$)

$$\dot{X}_i = F_i(X_i, \lambda) , \qquad X_i \in \mathbb{R}^m, \qquad (9.2)$$

has a hyperbolic stable limit cycle attractor $\gamma_i \subset \mathbb{R}^m$. The activity on the limit cycle can be described in terms of its phase $\theta \in \mathbb{S}^1$ of oscillation

$$\dot{\theta}_i = \Omega_i(\lambda) ,$$

where $\Omega_i(\lambda)$ is the natural frequency of oscillations. The dynamics of the oscillatory weakly connected system (9.1) can also be described in terms of phase variables.

$$\dot{\theta}_i = \Omega_i(\lambda) + \varepsilon g_i(\theta, \lambda, \rho, \varepsilon) , \qquad i = 1, \ldots, n .$$

We prove this fact using an invariant manifold reduction, and using averaging theory we analyze the behavior of this system on the time scale $1/\varepsilon$. The behavior depends crucially on resonances among natural frequencies $\Omega_1, \ldots, \Omega_n$, namely, on the existence of relations $k_1 \Omega_1 + \cdots + k_n \Omega_n = 0$, for some integers k_i. When there are no resonances, the system behaves as if it were uncoupled; that is, its behavior is qualitatively the same whether $\varepsilon > 0$ or $\varepsilon = 0$. Resonances make the coupling important. One can observe frequency locking, synchronization, and other interesting phenomena in this case.

9.1 Introduction

In this section we define synchronization, phase locking, entrainment, etc. Unfortunately, such phenomena do not have common definitions acceptable to all scientists and engineers. Some do not distinguish between them at all, while others use definitions that are too broad or too restrictive. Our choice is inspired by and specifically adapted to weakly connected oscillatory networks.

Synchronization is a phenomenon that has been investigated for many centuries. It was observed and described in detail by Christian Huygens in the second half of seventeenth century, when he noticed that a pair of pendulum clocks synchronize when they are attached to a light-weight beam instead of a wall. The pair of clocks is among the first weakly connected systems to be studied. Astronomy and stringed musical instruments provide many other examples of frequency locking and synchronization.

Phase

Let us discuss the notion of phase. Suppose we are given an oscillator, that is, a dynamical system

$$\dot{x} = f(x) , \qquad x \in \mathbb{R}^m ,$$

having a limit cycle attractor $\gamma \subset \mathbb{R}^m$ with period T and frequency $\Omega = 2\pi/T$. There is an ambiguity in the parametrization of γ by a phase $\theta \in \mathbb{S}^1$; see Figure 9.1. For example, the parametrization depends on the point

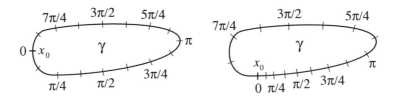

Figure 9.1. There is an ambiguity in parametrization of a limit cycle γ by a phase (angle) variable.

$x_0 \in \gamma$ that is assumed to have zero phase. In practice, the choice of x_0 is arbitrary. For example, when we consider a pacemaker neuron, we may assume the zero phase point to be the peak of the spike, its minimum, or any other point. When we consider networks of oscillators, we must fix the zero phase point in each oscillator in a consistent way. This task is trivial when the oscillators are identical or similar, but it may be tricky when the oscillators are qualitatively different, such as a regular pacemaker and a burster. Even if we fix a zero phase point, there is still an ambiguity in assigning phase values to other points on the limit cycle γ, as we illustrate in Figure 9.1.

There is, however, a natural way to assign a phase value to each point on γ. Let $x(t) \in \mathbb{R}^m$ be a T-periodic solution starting from $x(0) = x_0 \in \gamma$. Obviously, for each point $y \in \gamma$ there is a unique $t \in [0, T)$ such that $y = x(t)$. Let $\theta(t) \in \mathbb{S}^1$ be a solution to the phase equation

$$\dot{\theta} = \Omega \,,$$

such that $\theta(0) = 0$. Then the *natural phase* of each $x(t) \in \gamma$ is $\theta(t)$. While $x(t)$ makes a rotation around γ (with possibly a nonconstant speed), the natural phase θ makes a rotation around $\mathbb{S}^1$ with a constant speed $\Omega = 2\pi/T$, which is called the *natural frequency*. Obviously, this frequency coincides with the *instantaneous frequency* $\dot{\theta}$. When there are small influences from other oscillators, the instantaneous frequency $\dot{\theta}$ and the *asymptotic frequency*

$$\lim_{T \to \infty} \frac{1}{T} \int_0^T \dot{\theta}(t) \, dt \,, \tag{9.3}$$

which is also called the *rotation number*, could differ from Ω. We denote the former by $\Omega + \varepsilon\omega$, where ε is the strength of connections and $\omega \in \mathbb{R}$ is the *center frequency*, which is sometimes called frequency deviation or frequency perturbation. Notice that we cannot define the asymptotic frequency as $\lim_{T \to \infty} \theta(T)/T$, since $\theta(t) \in \mathbb{S}^1$ is bounded, and the limit vanishes. Besides, multiplication by a noninteger number is not well-defined for phases, since the result is different for θ and, say, $\theta + 2\pi$.

Locking

Next, we consider a network of two oscillators,

$$\begin{cases} \dot{\theta}_1 = \Omega_1 + \varepsilon g_1(\theta_1, \theta_2, \varepsilon) \\ \dot{\theta}_2 = \Omega_2 + \varepsilon g_2(\theta_1, \theta_2, \varepsilon) \end{cases} . \tag{9.4}$$

This network is said to be *frequency locked* when it has a stable periodic solution. Any periodic solution on the 2-torus is said to be a *torus knot*. It is said to be of type (p, q) if θ_1 makes p rotations while θ_2 makes q rotations, and p and q are relatively prime nonnegative integers (i.e., they do not have a common divisor greater than 1). Torus knots of type (p, q) produce $p : q$ frequency locking; see the illustration in Figure 9.2a and b. We say that the oscillatory network is *entrained* if there is a 1:1 frequency locking.

Suppose (9.4) is p:q frequency locked; that is there is a limit cycle attractor that is a (p, q) torus knot. We say that (9.4) is *p:q phase locked*, or just phase locked, if

$$q\theta_1(t) - p\theta_2(t) = \text{const}$$

on the limit cycle. Notice that frequency locking does not necessarily imply phase locking. Frequency locking without phase locking is called *phase trapping*. A $p : q$ phase locking occurs when a torus knot drawn in the square

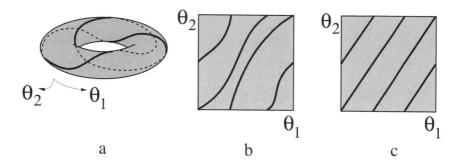

Figure 9.2. a and b. Torus knot of type (2, 3) and its representation on $[-\pi, \pi] \times [-\pi, \pi]$. Such a torus knot produces frequency locking but does not produce phase locking. c. A torus knot of type (2, 3) that produces 2:3 phase locking.

$[-\pi, \pi] \times [-\pi, \pi]$ consists of straight lines with slope q/p, as in Figure 9.2c. Existence of frequency locking does not depend on the parametrization of a limit cycle γ by a phase variable θ (see Figure 9.1), whereas existence of phase locking depends crucially on the parametrization.

A 1:1 phase locking is termed *synchronization*. Thus, synchronization means entrainment *and* phase locking. The quantity $\theta_1 - \theta_2$ is said to be *phase difference* (also known as *phase lag* or *phase lead*). When $\theta_1 - \theta_2 = 0$, the oscillators are said to be *synchronized in-phase*. When $\theta_1 - \theta_2 = \pi$, they are said to be *synchronized anti-phase*. When $\theta_1 - \theta_2$ differs from 0 or π, the synchronization is said to be *out-of-phase*; see summary in Figure 9.3. The in- and anti-phase synchronization depends on the choice of zero phase point, whereas the out-of-phase synchronization does not.

Now consider a network of n oscillators

$$\dot{\theta}_i = \Omega_i + \varepsilon g_i(\theta, \varepsilon) , \qquad i = 1, \ldots, n . \tag{9.5}$$

To be consistent with the previous definitions we say that the solution $\theta(t)$ of (9.5) is *frequency locked* when it is periodic and stable. One can define a *rotation vector* (sometimes called *winding ratios*) $p_1 : p_2 : \cdots : p_n$ to be a set of relatively prime integers p_i such that θ_1 makes p_1 rotations while θ_2 makes p_2 rotations, etc. The rotation vector coincides with the ratio of asymptotic frequencies (rotation numbers) defined by (9.3) for each $i = 1, \ldots, n$, provided the limits exist. (They may not exist for $n > 2$, which

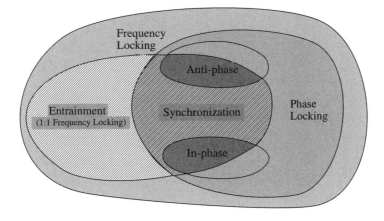

Figure 9.3. Various degrees of locking of oscillators.

corresponds to the toroidal chaos, see Baesens et al. 1991). The system is *entrained* when the rotation vector is $1 : 1 : \cdots : 1$. It is *phase locked* when there is an $(n-1) \times n$ integer matrix K having linearly independent rows such that $K\theta(t) = \text{const}$, where $\theta(t)$ is a stable periodic solution. The system is *synchronized* when it is entrained and phase locked. In this case one can take the matrix K to be

$$K = \begin{pmatrix} 1 & -1 & 0 & \cdots & 0 & 0 \\ 0 & 1 & -1 & \cdots & 0 & 0 \\ 0 & 0 & 1 & \cdots & 0 & 0 \\ \vdots & \vdots & \vdots & \ddots & \vdots & \vdots \\ 0 & 0 & 0 & \cdots & 1 & -1 \end{pmatrix} .$$

The system is synchronized *in-phase* when $K\theta = 0$ and synchronized *anti-phase* when $K\theta = \pi$ for such a matrix K. It is easy to check that the definitions of frequency locking, phase locking, entrainment and synchronization of (9.5) are equivalent to the requirement of *pairwise* frequency locking, phase locking, entrainment, and synchronization of oscillators governed by (9.5).

Phase Deviation Locking

Often it is convenient to represent θ_i as

$$\theta_i(t) = \Omega_i t + \phi_i(t) .$$

The variable $\phi_i \in \mathbb{S}^1$ is called a *phase deviation*. It describes deviations of the phase θ_i from the natural oscillation $\Omega_i t$. We say that there is *phase deviation locking* when there is an integer-valued $(n-1) \times n$ matrix K with linearly independent rows such that $K\phi(t) = \text{const}$. Phase deviation locking can differ from frequency or phase locking when the weakly connected oscillatory network is nonautonomous, e.g., of the form

$$\dot{\theta}_i = \Omega_i + \varepsilon g_i(\theta, \rho(t), \varepsilon), \qquad i = 1, \ldots, n.$$

To illustrate this issue consider two coupled oscillators of the form

$$\begin{cases} \dot{\theta}_1 = \sqrt{2} + \varepsilon \sin(\rho(t) + \theta_2 - \theta_1) \\ \dot{\theta}_2 = 1 \end{cases}. \tag{9.6}$$

Using phase deviations we can rewrite the system in the form

$$\begin{cases} \dot{\phi}_1 = \varepsilon \sin(\rho(t) + (1 - \sqrt{2})t + \phi_2 - \phi_1) \\ \dot{\phi}_2 = 0 \end{cases}.$$

It is easy to see that if $\rho(t) = (\sqrt{2} - 1)t$, then $\dot{\phi}_1 = \varepsilon \sin(\phi_2 - \phi_1)$, and the system above has an equilibrium solution $\phi_1(t) = \phi_2(t) = \phi_2(0)$. Obviously, $\phi_1(t) - \phi_2(t) = 0$; that is, there is phase deviation locking. Solutions of the original system (9.6) have the form $\theta(t) \to (\sqrt{2}t + c, t + c)$, and frequency or phase locking is impossible, since $\sqrt{2}$ and 1 are incommensurable.

Partial Phase Locking

Recall that oscillators in (9.5) are phase locked when there is an integer-valued $(n-1) \times n$ matrix K such that $K\theta(t) = \text{const}$. If this relation holds only for a smaller, say an $s \times n$ matrix K, $s < n-1$, then we say that the system is *partially phase locked*.

An example of partial phase locking occurs when two oscillators from the network, say θ_1 and θ_2 are $p : q$ phase locked and the rest are not. Then K consists of one row $(q, -p, 0, 0, \ldots, 0)$. Another example is given by the system

$$\begin{cases} \dot{\theta}_1 = \sqrt{2} + \varepsilon \sin(\theta_3 + \theta_2 - \theta_1) \\ \dot{\theta}_2 = 1 \\ \dot{\theta}_3 = \sqrt{2} - 1 \end{cases},$$

which is (9.6) with $\rho(t)$ replaced by $\theta_3(t)$. In this case

$$\theta_3 + \theta_2 - \theta_1 \to 0 \tag{9.7}$$

for almost all initial conditions. Indeed, if we write $\psi = \theta_3 + \theta_2 - \theta_1$, then $\dot{\psi} = -\varepsilon \sin \psi$, for which the origin $\psi = 0$ is a stable equilibrium. If we take $K = (-1, 1, 1)$, then $K\theta(t) = 0$ whenever $K\theta(0) = 0$. It is easy to see that the oscillators are not phase locked. Nevertheless, the solution $\theta \in \mathbb{T}^3$ does not fill $\mathbb{T}^3$ but approaches a proper submanifold of $\mathbb{T}^3$ that is homeomorphic to the 2-torus $\mathbb{T}^2$. Obviously, there is some kind of coherence in this system, since the relationship (9.7) holds among phases.

It is often quite restrictive to require that $K\theta(t) \equiv$ const for all t. In practice we require only that $K\theta(t) =$ const $+ o(1)$ on a certain time scale, say $1/\varepsilon$. (Here $o(1)$ is the Landau "little oh" function satisfying $o(1) \to 0$ as $\varepsilon \to 0$.)

9.2 Phase Equations

We show here how a general oscillatory weakly connected system can be reduced to a phase model

$$\dot{\theta}_i = \Omega_i + \varepsilon g_i(\theta_1, \ldots, \theta_n, \varepsilon) , \qquad \theta_i \in \mathbb{S}^1, \quad i = 1, \ldots, n .$$

We present two theorems below: The first one deals with weakly connected oscillatory systems having arbitrary natural frequencies $\Omega_1, \ldots, \Omega_n$. It is based on the invariant manifold reduction theorem (Theorem 4.7) for weakly connected systems that we proved in Chapter 4. Unfortunately, it does not provide an explicit formula for the connection functions g_i. The second theorem, which is due to Malkin, provides such an explicit formula, but it does require that the vector $\Omega \in \mathbb{R}^n$ of natural frequencies be proportional to a vector of integers.

Theorem 9.1 (Phase Equations for Oscillatory Neural Networks)
Consider a family of weakly connected systems

$$\dot{X}_i = F_i(X_i, \lambda) + \varepsilon G_i(X, \lambda, \rho, \varepsilon) , \qquad i = 1, \ldots, n , \qquad (9.8)$$

such that each equation in the uncoupled system ($\varepsilon = 0$)

$$\dot{X}_i = F_i(X_i, \lambda) , \qquad X_i \in \mathbb{R}^m, \qquad i = 1, \ldots, n ,$$

has an exponentially orbitally stable limit cycle attractor $\gamma_i \subset \mathbb{R}^m$ having natural frequency $\Omega_i(\lambda) \neq 0$. Then, the dynamical system

$$\dot{\theta}_i = \Omega_i(\lambda) + \varepsilon y_i(\theta_1, \ldots, \theta_n, \lambda, \rho, \varepsilon) , \qquad \theta_i \in \mathbb{S}^1, \quad i = 1, \ldots, n , \qquad (9.9)$$

defined on the n-torus $\mathbb{T}^n = \mathbb{S}^1 \times \cdots \times \mathbb{S}^1$ is a local model for (9.8). That is, there is an open neighborhood W of $M = \gamma_1 \times \cdots \times \gamma_n \subset \mathbb{R}^{mn}$ and a continuous function $h : W \to \mathbb{T}^n$ that maps solutions of (9.8) to those of (9.9).

Proof. First, we use Lemma 4.5 to establish that the direct product of hyperbolic limit cycles, $M = \gamma_1 \times \cdots \times \gamma_n$, is a normally hyperbolic invariant manifold. Then, we use invariant manifold reduction (Theorem 4.7) for the weakly connected system (9.8) to obtain an open neighborhood W of M and a function $\tilde{h}_\varepsilon : W \to M$ that maps solutions of (9.8) to those of the system

$$\dot{x}_i = F_i(x_i, \lambda) + \varepsilon \tilde{G}_i(x, \lambda, \rho, \varepsilon) , \qquad x_i \in \gamma_i . \qquad (9.10)$$

Since γ_i is homeomorphic to the unit circle $\mathbb{S}^1$, we can parametrize it using the natural phase variable $\theta_i \in \mathbb{S}^1$. Let $x_i^0 \in \gamma_i$ be a zero phase point. Consider $x_i(t)$, the periodic solution to $\dot{x}_i = F_i(x_i, \lambda)$ such that $x_i(0) = x_i^0$. Let $\theta_i(t)$ be a phase variable that rotates with constant speed Ω_i; that is, $\theta_i(t) = \Omega_i t$. The natural parametrization is given by the mapping

$$\Gamma_i : \mathbb{S}^1 \to \gamma_i , \qquad \Gamma_i(\theta_i(t)) = x_i(t) \in \gamma_i , \quad t \in [0, 2\pi/\Omega_i] ;$$

see Figure 9.4. Since $x_i(t) = \Gamma_i(\theta_i(t))$ is a periodic solution to $\dot{x}_i =$

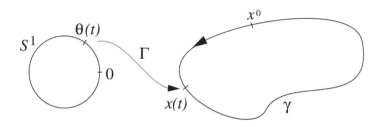

Figure 9.4. Parametrization Γ of a limit cycle γ using a phase variable $\theta \in \mathbb{S}^1$.

$F_i(x_i, \lambda)$, we have

$$\dot{x}_i = \frac{d\Gamma_i(\theta_i(t))}{dt} = \Gamma_i'(\theta_i)\dot{\theta}_i = \Gamma_i'(\theta_i)\Omega_i = F_i(\Gamma_i(\theta_i(t)), \lambda)$$

for all t. Therefore,

$$\Gamma_i'(\theta_i) = F_i(\Gamma_i(\theta_i), \lambda)/\Omega_i \qquad (9.11)$$

for all $\theta_i \in \mathbb{S}^1$. Now we substitute $x_i(t) = \Gamma_i(\theta_i(t))$ into (9.10) to obtain

$$\Gamma_i'(\theta_i)\dot{\theta}_i = F_i(\Gamma_i(\theta_i), \lambda) + \varepsilon\tilde{G}_i(\Gamma(\theta), \lambda, \rho, \varepsilon) .$$

Multiplying both sides by

$$\alpha_i(\theta_i) = \frac{\Omega_i F_i(\Gamma_i(\theta_i), \lambda)^\top}{|F_i(\Gamma_i(\theta_i), \lambda)|^2}$$

and using (9.11), we obtain

$$\dot{\theta}_i = \Omega_i + \varepsilon\alpha_i(\theta_i)\tilde{G}_i(\Gamma(\theta), \lambda, \rho, \varepsilon) ,$$

which coincides with (9.9) if we set

$$g_i(\theta_1, \ldots, \theta_n, \lambda, \rho, \varepsilon) = \alpha_i(\theta_i)\tilde{G}_i(\Gamma(\theta), \lambda, \rho, \varepsilon) .$$

Finally, notice that Γ_i is a one-to-one mapping of $\mathbb{S}^1$ onto $\gamma_i \subset \mathbb{R}^m$. Let $\Gamma_i^{-1} : \gamma_i \rightarrow \mathbb{S}^1$ be its inverse. Then the function $h : W \rightarrow \mathbb{T}^n$ that maps solutions of WCNN (9.8) to those of phase model (9.9) is a composition of $\tilde{h}_\varepsilon : W \rightarrow M$ and $\Gamma^{-1} = (\Gamma_1^{-1}, \ldots, \Gamma_n^{-1}) : M \rightarrow \mathbb{T}^n$. $\square$

It follows from the proof above and from Theorem 4.7 that

$$g_i = \frac{\Omega_i F_i^\top}{|F|^2}(G_i + \mathrm{ad}_F \, P_i) + \mathcal{O}(\varepsilon) \,, \tag{9.12}$$

where

$$\mathrm{ad}_F \, P = [F, P] = DF \, P - DP \, F$$

is the Poisson bracket of the vector-function $F = (F_1, \ldots, F_n)^\top$ and some vector-function $P = (P_1, \ldots, P_n)^\top$ that denotes the rescaled perturbations of the invariant manifold M due to the weak connections. Each function P_i can be determined from the condition

$$G_i + \mathrm{ad}_F \, P_i \in \mathrm{span} \, F_i \qquad \text{for all } x_i \in \gamma_i \,.$$

It could be difficult to find a function P_i for which this condition holds even though such a function always exists (the existence follows from the persistence of normally hyperbolic compact manifolds; see Section 4.3). A way to avoid this problem is to assume that $P \equiv 0$ in (9.12), which could be interpreted as an "infinite attraction" to the invariant manifold M (Ermentrout and Kopell 1990, 1991). In this case $g_i = \Omega_i F_i^\top G_i / |F|^2$ is easily computed.

9.2.1 Malkin's Theorem for Weakly Connected Oscillators

Malkin's theorem has an advantage over Theorem 9.1 in that it provides an explicit way to determine the connection functions g_i without the assumption of "infinite attraction". The disadvantage is that it requires that the vector of natural frequencies $\Omega = (\Omega_1, \ldots, \Omega_n)^\top \in \mathbb{R}^n$ be proportional to a vector of integers, which is equivalent to requiring that the natural frequencies Ω_i, $i = 1, \ldots, n$, be pairwise commensurable. Without loss of generality (by rescaling the time variable t if necessary) we may assume that each Ω_i is an integer. The ith oscillator in this case has a least period $2\pi/\Omega_i$. Obviously, 2π is a period too, though not the least one. If we do not require that the periods under consideration be least periods, then we may assume that each $\Omega_i = 1$. The general case $\Omega_i \in \mathbb{Z}$ is discussed later.

For the sake of clarity of notation we omit the external input $\rho \in \mathcal{R}$ and the bifurcation parameter $\lambda \in \Lambda$ in our analysis below. That is, we study the weakly connected system

$$\dot{X}_i = F_i(X_i) + \varepsilon G_i(X) \,, \qquad i = 1, \ldots, n \,, \tag{9.13}$$

assuming implicitly that $F_i = F_i(X_i, \lambda)$ and $G = G_i(X, \lambda, \rho, \varepsilon)$.

The theorem below is due to Malkin (1949, 1956, see also Blechman 1971, p.187), who proved it for general (not necessarily weakly connected) nonautonomous systems having parametrized families of solutions. We follow Blechman (1971), Ermentrout (1981) and Ermentrout and Kopell (1991) and state the theorem in terms of weakly connected oscillators below. General statement and its proof are provided in Appendix (Section 9.6).

Theorem 9.2   (Malkin Theorem for Weakly Connected Oscillators)   *Consider a weakly connected system of the form (9.13) such that each equation in the uncoupled system*

$$\dot{X}_i = F_i(X_i) , \qquad X_i \in \mathbb{R}^m , \qquad i = 1, \ldots, n ,$$

has an exponentially orbitally stable 2π-periodic solution $\gamma_i \subset \mathbb{R}^m$. Let $\tau = \varepsilon t$ be slow time and let $\phi_i(\tau) \in \mathbb{S}^1$ be the phase deviation from the natural oscillation $\gamma_i(t), t \geq 0$. Then, the vector of phase deviations $\phi = (\phi_1, \ldots, \phi_n)^\top \in \mathbb{T}^n$ is a solution to

$$\phi_i' = H_i(\phi - \phi_i, \varepsilon) , \qquad i = 1, \ldots, n , \tag{9.14}$$

where $' = d/d\tau$, the vector $\phi - \phi_i = (\phi_1 - \phi_i, \ldots, \phi_n - \phi_i)^\top \in \mathbb{T}^n$, and the function

$$H_i(\phi - \phi_i, 0) = \frac{1}{2\pi} \int_0^{2\pi} Q_i(t)^\top G_i(\gamma(t + \phi - \phi_i)) \, dt , \tag{9.15}$$

where $Q_i(t) \in \mathbb{R}^m$ is the unique nontrivial 2π-periodic solution to the linear system

$$\dot{Q}_i = -\{DF_i(\gamma_i(t))\}^\top Q_i \tag{9.16}$$

satisfying the normalization condition

$$Q_i(0)^\top F_i(\gamma_i(0)) = 1 . \tag{9.17}$$

Proof. Consider each limit cycle $\gamma_i \subset \mathbb{R}^m$ as a periodic solution $X_i(t) = \gamma_i(t), \ t \in [0, 2\pi]$ of the system $\dot{X}_i = F_i(X_i)$ starting from the initial point $X_i(0) = \gamma_i(0)$. Since each limit cycle γ_i has period 2π, there is a natural parametrization of γ_i by $\theta_i \in \mathbb{S}^1$; namely, the phase of $\gamma_i(t), \ t \in [0, 2\pi]$ is just t. Now consider the weakly connected system (9.13) for $\varepsilon > 0$ and suppose that $\tau = \varepsilon t$ is slow time. Let

$$x_i(t) = \gamma_i(t + \phi_i(\tau)) + \varepsilon P_i(t + \phi_i(\tau), \varepsilon) ,$$

where $\phi_i(\tau) \in \mathbb{S}^1, \ i = 1, \ldots, n$, are slow phase deviations that account for the dynamical changes due to the weak connections, and the smooth vector-functions εP_i account for the ε-perturbation of the invariant manifold $M =$

$\gamma_1 \times \cdots \times \gamma_n \subset \mathbb{R}^{mn}$. We differentiate the equation above with respect to t to obtain

$$
\begin{aligned}
\dot{x}_i &= \gamma_i'(t + \phi_i)\left(1 + \varepsilon \frac{d\phi_i}{d\tau}\right) + \varepsilon \frac{dP_i(t + \phi, \varepsilon)}{dt} + \mathcal{O}(\varepsilon^2) \\
&= F_i(\gamma_i(t + \phi_i)) + \varepsilon G_i(\gamma(t + \phi)) \\
&\quad + \varepsilon DF_i(\gamma_i(t + \phi_i))\, P_i(t + \phi, \varepsilon) + \mathcal{O}(\varepsilon^2) \, .
\end{aligned}
$$

Equating terms of order ε and taking into account that

$$
\gamma_i'(t + \phi_i) = F_i(\gamma_i(t + \phi_i)) \, ,
$$

we obtain

$$
F_i(\gamma_i(t + \phi_i)) \frac{d\phi_i}{d\tau} + \frac{dP_i(t + \phi, 0)}{dt}
$$
$$
= G_i(\gamma(t + \phi)) + DF_i(\gamma_i(t + \phi_i))\, P_i(t + \phi, 0) \, .
$$

It is convenient to rewrite the equation above in the form

$$
\frac{dy_i(t, \phi)}{dt} = A_i(t, \phi)\, y_i(t, \phi) + b_i(t, \phi) \, , \tag{9.18}
$$

where $y_i(t, \phi) = P_i(t + \phi, 0)$ is an unknown vector variable ($\phi \in \mathbb{T}^n$ is treated as a parameter) and the matrix

$$
A_i(t, \phi) = DF_i(\gamma_i(t + \phi_i))
$$

and the vector

$$
b_i(t, \phi) = G_i(\gamma(t + \phi)) - F_i(\gamma_i(t + \phi_i)) \frac{d\phi_i}{d\tau} \tag{9.19}
$$

are 2π-periodic in t. The theory of linear inhomogeneous equations of the form (9.18) is well developed (Hale 1969, Farkas 1994). To study existence and uniqueness of solutions to (9.18) one must consider the *adjoint* linear homogeneous system

$$
\frac{dq_i(t, \phi_i)}{dt} = -A_i(t, \phi_i)^\top q_i(t, \phi_i) \tag{9.20}
$$

with a normalization condition, which we take in the form

$$
\frac{1}{2\pi} \int_0^{2\pi} q_i(t, \phi_i)^\top F_i(\gamma_i(t + \phi_i))\, dt = 1 \, . \tag{9.21}
$$

Each limit cycle γ_i is exponentially orbitally stable. Hence, a homogeneous ($b \equiv 0$) linear system of the form (9.18) and the adjoint system (9.20) have 1 as a simple Floquet multiplier, and the other multipliers are not on the unit circle. This implies, in particular, that the adjoint system (9.20) has a unique nontrivial periodic solution, say $q_i(t, \phi_i)$, which can easily be found using standard numerical methods. Now we use the Fredholm alternative to conclude that the linear system (9.18) has a unique periodic solution $y_i(t, \phi)$ if and only if the *orthogonality condition*

$$\langle q, b \rangle = \frac{1}{2\pi} \int_0^{2\pi} q_i(t, \phi_i)^\top b_i(t, \phi) \, dt = 0$$

holds. Due to the normalization condition (9.21) and expression (9.19) for b_i, this is equivalent to

$$\frac{d\phi_i}{d\tau} = \frac{1}{2\pi} \int_0^{2\pi} q_i(t, \phi_i)^\top G_i(\gamma(t + \phi)) \, dt \ .$$

Due to the special form of the matrix $A_i(t, \phi_i)$, it suffices to find a solution $q_i(t, \phi_i)$ to the adjoint system (9.20) for $\phi_i = 0$, and any other solution $q_i(t, \phi_i)$ has the form $q_i(t, \phi_i) = q_i(t + \phi_i, 0)$. Now we rewrite the equation above in the form

$$\begin{aligned}
\frac{d\phi_i}{d\tau} &= \frac{1}{2\pi} \int_0^{2\pi} q_i(t + \phi_i, 0)^\top G_i(\gamma(t + \phi)) \, dt \\
&= \frac{1}{2\pi} \int_0^{2\pi} q_i(s, 0)^\top G_i(\gamma(s + \phi - \phi_i)) \, ds \ , \qquad s = t + \phi_i \ .
\end{aligned}$$

Finally, notice that $x_i(t) = F_i(\gamma_i(t))$ is a solution to the homogeneous linear problem $\dot{x}_i = A_i(t, 0)x_i$, since

$$\frac{d}{dt} F_i(\gamma_i(t)) = DF_i(\gamma_i(t))\gamma_i'(t) = DF_i(\gamma_i(t))F_i(\gamma_i(t)) \ .$$

The vector-function $Q(t) = q_i(t, 0)$ is a solution to the adjoint linear problem $\dot{Q}_i = -A_i(t, 0)^\top Q_i$. Any such solutions satisfy

$$Q_i(t)^\top x_i(t) = \text{const} \qquad \text{for any } t.$$

Indeed,

$$\begin{aligned}
\frac{d}{dt}(Q_i(t)^\top x_i(t)) &= \left(-A_i(t, 0)^\top Q_i(t)\right)^\top x_i(t) + Q_i(t)^\top A_i(t, 0)x_i(t) \\
&= -Q_i(t)^\top A_i(t, 0)x_i(t) + Q_i(t)^\top A_i(t, 0)x_i(t) \\
&= 0 \ .
\end{aligned}$$

From the normalization condition (9.21) we conclude that

$$Q_i(t)^\top x_i(t) = Q_i(t)^\top F_i(\gamma_i(t)) = 1 \qquad \text{for any } t,$$

which results in (9.17) if we take $t = 0$. $\square$

Malkin's theorem can be generalized for WCNNs having delayed functions G_i.

Proposition 9.3 *Consider an oscillatory neural network that has an explicit transmission delay; that is, it is described by the system*

$$\dot{X}_i = F_i(X_i(t)) + \varepsilon G_i(X(t - \eta_i)) , \qquad i = 1, \ldots, n ,$$

where $\eta_i = (\eta_{i1}, \ldots, \eta_{in})^\top \in \mathbb{R}^n$ is a vector of time delay constants and $X(t - \eta_i) = (X_1(t - \eta_{i1}), \ldots, X_n(t - \eta_{in}))^\top \in \mathbb{T}^n$. Then the phase model has the form

$$\phi_i' = H_i(\phi + \eta_i - \phi_i, \varepsilon) , \qquad i = 1, \ldots, n ,$$

where the function H_i is defined by (9.15).

Thus, an explicit synaptic delay creates a simple phase shift in the coupling function H_i. This was pointed out by Ermentrout (1994) when he considered weakly connected oscillatory networks with nontrivial temporal synaptic dynamics.

Now suppose that each Ω_i may differ from 1; that is, 2π is not the least period of a periodic solution $\gamma_i \subset \mathbb{R}^m$ of the dynamical system

$$\dot{X}_i = F_i(X_i) .$$

Let $2\pi/\Omega_i$ be the least period. Let $\Gamma_i : \mathbb{S}^1 \to \gamma_i$ be the parametrization of γ_i by the phase variable $\theta_i \in \mathbb{S}^1$ as in the proof of Theorem 9.1, so that

$$\gamma_i(t) = \Gamma_i(\Omega_i t) .$$

Then, proceeding as in the proof above we obtain

$$\frac{d\phi_i}{d\tau} = \frac{1}{2\pi} \int_0^{2\pi} Q_i(\Omega_i t + \phi_i)^\top G_i(\Gamma(\Omega t + \phi)) \, dt ,$$

where $\phi_i(\tau)$ is the phase deviation from the natural oscillation $\Omega_i t$. Later in this chapter we show (Theorem 9.6) that the integral on the right-hand side can be written in the form

$$H_i(\Omega_i \phi_1 - \Omega_1 \phi_i, \; \Omega_i \phi_2 - \Omega_2 \phi_i, \; \ldots, \; \Omega_i \phi_n - \Omega_n \phi_i)$$

for some quasi-periodic function H_i (i.e., H_i is 2π-periodic in each argument).

Finally, we note the relationship between phase models (9.9) and (9.14): we can rewrite the former using phase deviations $\phi = \theta - \Omega t$ as

$$\dot{\phi}_i = \varepsilon g_i(\Omega_1 t + \phi_1, \; \ldots, \; \Omega_n t + \phi_n, \varepsilon)$$

and then average it to obtain the latter. Ermentrout and Kopell (1991) prove that both procedures yield the same result.

9.2.2 Example: Coupled Wilson-Cowan Neural Oscillators

Studying phase locking of two identical oscillators is a classical problem in mathematical neuroscience. Weakly coupled Wilson-Cowan oscillators have been considered, for example, by Ermentrout and Kopell (1991) and by Borisyuk et al. (1995).

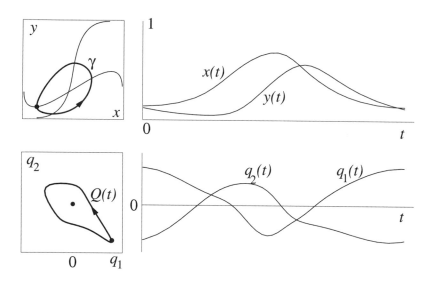

Figure 9.5. Top: Limit cycle attractor $\gamma \subset \mathbb{R}^2$ in the Wilson-Cowan neural oscillator model. Parameters are $a = b = c = 10$, $d = -2$, $\rho_x = -2$, $\rho_y = -6$. Bottom: Periodic solution $Q(t) = (q_1(t), q_2(t))^\top \in \mathbb{R}^2$ of the adjoint problem (9.20) for such γ. Filled dots denote the zero phase point.

Consider two coupled identical oscillators described by the Wilson-Cowan model

$$
\left\{
\begin{array}{rcl}
\dot{x}_1 & = & -x_1 + S(\rho_x + ax_1 - by_1 + s_1 x_2 - s_2 y_2) \\
\dot{y}_1 & = & -y_1 + S(\rho_y + cx_1 - dy_1 + s_3 x_2 - s_4 y_2) \\
\dot{x}_2 & = & -x_2 + S(\rho_x + ax_2 - by_2 + s_1 x_1 - s_2 y_1) \\
\dot{y}_2 & = & -y_2 + S(\rho_y + cx_2 - dy_2 + s_3 x_1 - s_4 y_1)
\end{array}
\right.
$$

Suppose that the synaptic coefficients s_i, $i = 1, \ldots, 4$, are small, say of order $\varepsilon \ll 1$. If we rescale them as $s_i = \varepsilon c_i$, then the system above can be written in the form

$$
\left\{
\begin{array}{rcl}
\dot{x}_1 & = & -x_1 + S(\rho_x + ax_1 - by_1) + \varepsilon G_x(x_1, y_1, x_2, y_2, \varepsilon) \\
\dot{y}_1 & = & -y_1 + S(\rho_y + cx_1 - dy_1) + \varepsilon G_y(x_1, y_1, x_2, y_2, \varepsilon) \\
\dot{x}_2 & = & -x_2 + S(\rho_x + ax_2 - by_2) + \varepsilon G_x(x_2, y_2, x_1, y_1, \varepsilon) \\
\dot{y}_2 & = & -y_2 + S(\rho_y + cx_2 - dy_2) + \varepsilon G_y(x_2, y_2, x_1, y_1, \varepsilon)
\end{array}
\right.
$$

where

$$
\begin{aligned}
G_x(x_1,y_1,x_2,y_2,0) &= S'(\rho_x + ax_1 - by_1)(c_1x_2 - c_2y_2) \\
G_y(x_1,y_1,x_2,y_2,0) &= S'(\rho_y + cx_1 - dy_1)(c_3x_2 - c_4y_2) \\
G_x(x_2,y_2,x_1,y_1,0) &= S'(\rho_x + ax_2 - by_2)(c_1x_1 - c_2y_1) \\
G_y(x_2,y_2,x_1,y_1,0) &= S'(\rho_y + cx_2 - dy_2)(c_3x_1 - c_4y_1) \,.
\end{aligned}
$$

We can use Figure 2.12 to select the parameters $\rho_x, \rho_y, a, b, c, d$ so that each oscillator has an exponentially stable limit cycle attractor, such as the one depicted in Figure 9.5. Then, we can use Theorem 9.2 to derive phase equations for the coupled Wilson-Cowan neural oscillators. The Jacobian matrix at the limit cycle attractor can easily be computed, and the adjoint problem (9.16) can be solved using standard numerical methods, e.g., those described by Cymbalyuk et al. (1994) or Ermentrout (1996). A typical solution $Q(t)$ is depicted in Figure 9.5.

Since the connection function $G = (G_x, G_y)^\top$ is linear in c_1, c_2, c_3, and c_4, it suffices to compute H defined by (9.15) for each c_i separately. Such functions are depicted in Figure 9.6, where the case $c_1 > 0$ corresponds

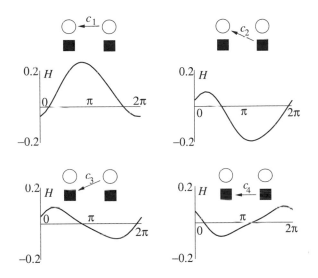

Figure 9.6. Functions $H(\chi,0)$, $\chi = \phi_2 - \phi_1$, for four different synaptic organizations.

to excitatory $\rightarrow$ excitatory synaptic organization, $c_2 < 0$ corresponds to inhibitory $\rightarrow$ excitatory synaptic organization, etc. From Theorem 9.2 the phase equations have the form

$$
\left\{
\begin{aligned}
\phi_1' &= H(\phi_2 - \phi_1, \varepsilon) \\
\phi_2' &= H(\phi_1 - \phi_2, \varepsilon)
\end{aligned}
\right. ,
$$

where H is a linear combination of the functions depicted in Figure 9.6 with coefficients $c_1, |c_2|, c_3, |c_4|$. Williams (1992) suggests a method for determining such H experimentally, which we discuss later.

To study phase locking we use the new variable $\chi = \phi_2 - \phi_1$, so that

$$\chi' = H(-\chi, \varepsilon) - H(\chi, \varepsilon) \ .$$

Notice that the in-phase and anti-phase solutions always exist, since both $\chi = 0$ and $\chi = \pi$ are equilibria. They are stable if $H'(0, \varepsilon) > 0$ (respectively $H'(\pi, \varepsilon) > 0$). The in-phase solution is stable in excitatory $\rightarrow$ inhibitory or inhibitory $\rightarrow$ excitatory synaptic organizations, while the anti-phase solution is stable in excitatory $\rightarrow$ excitatory or inhibitory $\rightarrow$ inhibitory synaptic organizations. Similar results were obtained by Ermentrout and Kopell (1991) and Borisyuk et al. (1995). We return to the issue of dynamics vs. synaptic organizations in Chapter 13, where we show that some of the results above are universal, i.e. they are model-independent.

9.3   Averaging the Phase Equations

In the rest of this chapter we study the phase model

$$\dot{\theta}_i = \Omega_i(\lambda) + \varepsilon g_i(\theta_1, \dots, \theta_n, \lambda, \rho, \varepsilon) \ , \quad \theta_i \in \mathbb{S}^1, \quad i = 1, \dots, n \ .$$

For the sake of clarity of notation we omit the external input $\rho \in \mathcal{R}$ and the bifurcation parameter $\lambda \in \Lambda$ in our analysis below. That is, we study the dynamical system

$$\dot{\theta} = \Omega + \varepsilon g(\theta, \varepsilon) \ , \quad \theta \in \mathbb{T}^n,$$

and implicitly assume that $\Omega = \Omega(\lambda)$ and $g = g(\theta, \lambda, \rho, \varepsilon)$.

The phase model can be simplified further. First of all, we write $\theta(t)$ in the form

$$\theta = \Omega t + \phi \ ,$$

where $\phi = (\phi_1, \dots, \phi_n)^\top \in \mathbb{T}^n$ is a vector of phase deviations. Substituting this into the system

$$\dot{\theta} = \Omega + \varepsilon g(\theta, \varepsilon) \ , \quad \theta \in \mathbb{T}^n,$$

gives

$$\dot{\phi} = \varepsilon g(\Omega t + \phi, \varepsilon) \ . \tag{9.22}$$

We see that the phase deviations ϕ change relatively slowly in comparison with the fast oscillatory term Ωt. To stress this fact we can rewrite (9.22) using slow time $\tau = \varepsilon t$ as $\phi' = g(\frac{\Omega}{\varepsilon}\tau + \phi, \varepsilon)$, where $' = d/d\tau$.

The function $g(\theta, \varepsilon)$ is 2π-periodic in each component θ_i of the vector $\theta = (\theta_1, \dots, \theta_n)$. Therefore, $g(\Omega t + \varphi, \varepsilon)$ is a *quasi-periodic* function of t. It is not periodic unless the vector of frequencies Ω is proportional to a vector of integers. The averaging theory is well developed for such quasi-periodic functions. It provides a time-dependent near-identity change of variables

$$\phi = \varphi + \varepsilon h(\varphi, t)$$

that transforms (9.22) to

$$\dot{\varphi} = \varepsilon \bar{g}(\varphi) + \mathcal{O}(\varepsilon^2) \, ,$$

where

$$\bar{g}(\varphi) = \lim_{T \to \infty} \frac{1}{T} \int_0^T g(\Omega t + \varphi, 0) \, dt$$

is the "average" of g. Since g is a quasi-periodic function, this limit exists, and the function $\bar{g}$ is well defined. Its form depends crucially on the vector of frequencies Ω. It should be stressed that the full system (9.22) and the averaged system have distinct solutions (see Figure 9.7 for an illustration), which may diverge from each other even when they have identical initial

Figure 9.7. Numerical simulation of $\dot{\phi} = \varepsilon(\cos t - \phi)$ and the averaged system $\dot{\varphi} = -\varepsilon\varphi$ for $\varepsilon = 0.05$).

states. Averaging theory guarantees their closeness on time scale $1/\varepsilon$ and even on the infinite time scale if additional requirements are imposed.

9.3.1 Averaging Theory

In this section we give a short exposition of the averaging theory that we use here. Even though we assume that $\phi \in \mathbb{T}^n$, the theory is valid if we substitute $\mathbb{T}^n$ by any compact subset of $\mathbb{R}^n$.

There is an intimate relationship between averaging and normal form theories (Arnold 1982, Sanders and Verhulst 1985). Both theories employ a change of variables that simplifies the right-hand side of a dynamical system. The normal form theory provides a change of variables that removes nonresonant monomials. Averaging theory provides a change of variables that removes time-dependent terms, which are in some sense nonresonant too.

Theorem 9.4 (Formal Averaging) *Consider a dynamical system*

$$\dot{\phi} = \varepsilon g(\phi, t, \varepsilon) , \qquad \phi \in \mathbb{T}^n , \tag{9.23}$$

and suppose the average of g, given by

$$\bar{g}(\varphi) = \lim_{T \to \infty} \frac{1}{T} \int_0^T g(\varphi, t, 0) \, dt , \tag{9.24}$$

exists as a smooth function for $\varphi \in \mathbb{T}^n$. Then

$$\dot{\varphi} = \varepsilon \bar{g}(\varphi) + \mathcal{O}(\varepsilon^2) \tag{9.25}$$

is a model *of (9.23) in the sense that there is a change of variables*

$$\phi = \varphi + \varepsilon h(\varphi, t) , \tag{9.26}$$

where

$$h(\varphi, t) = \int_0^t g(\varphi, s, 0) - \bar{g}(\varphi) \, ds , \tag{9.27}$$

that maps solutions of (9.23) to those of (9.25).

Proof. Let us substitute $\phi = \varphi + \varepsilon h(\varphi, t)$ into (9.23). The result is

$$\dot{\phi} = (I + \varepsilon D_\varphi h(\varphi, t))\dot{\varphi} + \varepsilon \frac{\partial h(\varphi, t)}{\partial t} = \varepsilon g(\varphi, t, 0) + \mathcal{O}(\varepsilon^2) ,$$

where I is the identity matrix. Since $(I + \varepsilon D_\varphi h(\varphi, t))^{-1} = I - \varepsilon D_\varphi h(\varphi, t) + \mathcal{O}(\varepsilon^2)$ we have that

$$\dot{\varphi} = \varepsilon g(\varphi, t, 0) - \varepsilon \frac{\partial h(\varphi, t)}{\partial t} + \mathcal{O}(\varepsilon^2) . \tag{9.28}$$

If we define h as in (9.27), then

$$\frac{\partial h(\varphi, t)}{\partial t} = g(\varphi, t, 0) - \bar{g}(\varphi) ,$$

and (9.28) reduces to (9.25). $\square$

Since both systems (9.23) and (9.25) are defined on the same space $\mathbb{T}^n$, one can ask whether or not $|\phi(t) - \varphi(t)|$ is small. That is, whether the change of variables (9.26) is nearly the identity. The answer is NO, due to the fact that the function $\varepsilon h(\varphi, t)$ is not small when $t \to \infty$. This is why we call it *formal averaging*. Classical averaging theory studies this issue (Bogoliubov and Mitropolsky 1961, Sanders and Verhulst 1985, Hoppensteadt 1993). One way to guarantee smallness of $\varepsilon h(\varphi, t)$ is to restrict the time scale.

Lemma 9.5 (Smallness of εh) *Suppose the conditions of Theorem 9.4 hold. Then the function $\varepsilon h(\varphi, t)$ defined in (9.27) is small over a large time interval. More precisely, for any $\tau_0 > 0$ and any small $\delta > 0$ there is $\varepsilon_0 > 0$ such that*

$$\sup_{\substack{\varphi \in \mathbb{T}^n \\ t \in [0, \frac{\tau_0}{\varepsilon}]}} |\varepsilon h(\varphi, t)| < \delta$$

for all $0 < \varepsilon \leq \varepsilon_0$.

Proof. Since the limit (9.24) exists for all $\varphi \in \mathbb{T}^n$ and $\mathbb{T}^n$ is a compact manifold, there is a continuous monotone decreasing function $a(T)$, $T \geq 0$, tending to zero when $T \to \infty$, such that

$$\left| \frac{1}{T} \int_0^T g(\varphi, t, 0) - \bar{g}(\varphi) \, dt \right| \leq a(T) .$$

Let $\tau = \varepsilon t$ be slow time. It is easy to see that the function

$$b(\varepsilon) = \max_{\tau \leq \tau_0} \tau a\left(\frac{\tau}{\varepsilon}\right)$$

vanishes when $\varepsilon \to 0$. The proof of the lemma follows from the observation that

$$|\varepsilon h(\varphi, t)| \leq \varepsilon t a(t) = \tau a(\tau/\varepsilon) \leq b(\varepsilon) .$$

$\square$

If the original system (9.23) is periodic in t and the averaged system (9.25) has a hyperbolic equilibrium, then the function $\varepsilon h(\varphi, t)$ is uniformly small on the semi-infinite interval $t \in [0, \infty)$ (Guckenheimer and Holmes 1983, Sanders and Verhulst 1985, Hoppensteadt 1993). We do not use this result in this book, since estimates to t of order $\mathcal{O}(1/\varepsilon)$ suffice for our analysis of weakly connected systems.

9.3.2 Resonant Relations

A vector of frequencies Ω is said to have *resonant relations* if

$$k \cdot \Omega = k\Omega = k_1 \Omega_1 + \cdots + k_n \Omega_n = 0$$

for some nonzero integer row-vector $k = (k_1, \ldots, k_n) \in \mathbb{Z}^n$. We call such a vector Ω *resonant*. There might be many resonant relations

$$\begin{cases} k_{11}\Omega_1 + \cdots + k_{1n}\Omega_n = 0 \\ k_{21}\Omega_1 + \cdots + k_{2n}\Omega_n = 0 \\ \vdots \quad \vdots \quad \vdots\vdots\vdots \quad \vdots \quad \vdots \quad \vdots \end{cases} .$$

Each such relation is written as $k\Omega = 0$ for some $k \in \mathbb{Z}^n$. Moreover, the set of all such k form an additive Abelian group, say $G \subset \mathbb{Z}^n$, which we call the *resonant group*.

We can consider $k\Omega$ as a scalar product of two vectors. Then any $k \in G$ lies in the orthogonal complement of span Ω. Since the dimension of the orthogonal complement is $(n-1)$, the rank of G, which we denote by s, is at most $n-1$. Simple considerations show that $s = n-1$ if and only if Ω is proportional to a vector of integers. Let $k_1, \ldots k_s \in G$ denote a set of generators of G; that is, any element of G can be represented as a linear integer combination of $k_1, \ldots, k_s$. The matrix K is said to be a *resonant matrix* for Ω if its rows are generators of the resonant group G. That is, any vector $k \in \mathbb{Z}^n$ for which $k\Omega = 0$ can be represented as $k = aK$ for some row-vector $a \in \mathbb{Z}^s$; i.e., k is an integer linear combination of rows of K.

Some properties of resonant matrix K are as follows:

- $K\Omega = 0$.

- Each row of K consists of relatively prime integers.

- Rows of K are linearly independent.

- Any other resonant matrix K_1 can be written in the form $K_1 = AK$ for some nonsingular $s \times s$ matrix A having integer entries.

Examples

An example of a nonresonant vector Ω is $(1, \sqrt{2})^\top \in \mathbb{R}^2$. If there were a resonance relation $k\Omega = 0$, this would mean that $\sqrt{2}$ is a rational number. Still, $|k\Omega|$ can be arbitrarily small for some k.

An example of a resonant vector Ω is $(1, 1)^\top \in \mathbb{R}^2$. In this case the resonant group $G = \{(m, -m) \subset \mathbb{Z}^2 | \ m \in \mathbb{Z}\}$ has rank 1. The vector $\Omega = (1, \sqrt{2}, \sqrt{2} - 1)^\top \in \mathbb{R}^3$ is also resonant, since there is a resonance relation corresponding to $k = (1, -1, 1) \in \mathbb{Z}^3$ even though the elements of Ω are pairwise incommensurable. In this case the resonant group $G = \{(m, -m, m) \subset \mathbb{Z}^3 | \ m \in \mathbb{Z}\}$ has rank 1.

The vector $\Omega = (1, 2, \sqrt{2}, 2\sqrt{2})^\top \in \mathbb{R}^4$ has a resonant matrix

$$K = \begin{pmatrix} -2 & 1 & 0 & 0 \\ 0 & 0 & -2 & 1 \end{pmatrix}.$$

The resonant group here is $G = \{(-2m, m, -2k, k) \subset \mathbb{Z}^4 | \ m, k \in \mathbb{Z}\}$. It has rank 2 in this case. Notice that the rows of K generate the full resonant group G. The matrix

$$\begin{pmatrix} -2 & 1 & 0 & 0 \\ -2 & 1 & -4 & 2 \end{pmatrix}$$

has linearly independent rows of relatively prime integers. But it is not a resonant matrix, since the rows are not generators of G (none of their linear combinations with integer coefficients produce the vector $(0, 0, -2, 1)$).

9.3.3 Generalized Fourier Series

A powerful method of computing integrals of the form

$$\bar{g}(\varphi) = \lim_{T \to \infty} \frac{1}{T} \int_0^T g(\Omega t + \varphi) \, dt$$

uses Fourier series.

We consider smooth functions $g(\theta)$ that can be represented as absolutely and uniformly convergent *generalized Fourier series*

$$g(\theta) = \sum_{k=-\infty}^{+\infty} a_k e^{ik\theta} , \qquad (9.29)$$

where

$$a_k = \frac{1}{(2\pi)^n} \int_0^{2\pi} \cdots \int_0^{2\pi} g(\theta) e^{-ik\theta} \, d\theta \qquad (9.30)$$

are the Fourier coefficients, the integer row-vector $k \in \mathbb{Z}^n$ is a multi-index, and $k\theta = k_1\theta_1 + \cdots + k_n\theta_n$. The coefficient a_0 corresponds to the constant term, and the other coefficients correspond to oscillatory terms. Absolute convergence of this series implies that

$$\sum_{k=-\infty}^{+\infty} |a_k| \qquad (9.31)$$

converges. Many results presented in this chapter are based on the following Theorem.

Theorem 9.6 *Let g be a smooth function defined on $\mathbb{T}^n$. Let Q be defined by*

$$Q(\varphi) = \lim_{T \to \infty} \frac{1}{T} \int_0^T g(\Omega t + \varphi) \, dt ,$$

where $\Omega \subset \mathbb{R}^n$ is a vector of frequencies.

- *If Ω is nonresonant (that is, $k\Omega \neq 0$ for all nonzero $k \in \mathbb{Z}^n$), then $Q(\varphi) = a_0$, where a_0 is the constant term in the generalized Fourier series (9.29).*

- *If Ω is resonant and K is the corresponding resonant matrix, then there is a smooth function P defined on a lower-dimensional torus $\mathbb{T}^s = K\mathbb{T}^n$ such that $Q(\varphi) = P(K\varphi)$.*

Proof. Since the Fourier series (9.29) converges absolutely and uniformly, we can exchange integration and summation to obtain

$$\frac{1}{T} \int_0^T \sum_{k=-\infty}^{+\infty} a_k e^{ik(\Omega t + \varphi)} = \sum_{k=-\infty}^{+\infty} a_k e^{ik\varphi} \frac{1}{T} \int_0^T e^{ik\Omega t} .$$

It is easy to check that

$$\frac{1}{T} \int_0^T e^{ik\Omega t}\,dt = \begin{cases} \frac{e^{ik\Omega T}-1}{ik\Omega T} & \text{if } k\Omega \neq 0\,, \\ 1 & \text{if } k\Omega = 0\,. \end{cases} \tag{9.32}$$

Therefore,

$$Q(\varphi) = \sum_{\substack{k=-\infty \\ k\Omega=0}}^{+\infty} a_k e^{ik\varphi} + \lim_{T\to\infty} \sum_{\substack{k=-\infty \\ k\Omega\neq0}}^{+\infty} a_k e^{ik\varphi} \frac{e^{ik\Omega T}-1}{ik\Omega T}\,.$$

First, we claim that

$$\lim_{T\to\infty} \sum_{\substack{k=-\infty \\ k\Omega\neq0}}^{+\infty} a_k e^{ik\varphi} \frac{e^{ik\Omega T}-1}{ik\Omega T} = 0\,. \tag{9.33}$$

The quantity $|k\Omega|$ in the denominator could become arbitrarily small as $|k| \to \infty$. (For example, the set $\{k_1 + k_2\sqrt{2} \mid k_1, k_2 \in \mathbb{Z}\}$ is dense in $\mathbb{R}$.) This obstacle is known as the *small divisor* problem. Fortunately, when $|k\Omega| \ll 1$, the numerator vanishes too, and

$$\left| \frac{e^{ik\Omega T}-1}{ik\Omega T} \right| \leq 1$$

for any choice of $k\Omega T \neq 0$, so

$$\sum_{\substack{k=-\infty \\ k\Omega\neq0}}^{+\infty} \left| a_k e^{ik\varphi} \frac{e^{ik\Omega T}-1}{ik\Omega T} \right| \leq \sum_{\substack{k=-\infty \\ k\Omega\neq0}}^{+\infty} |a_k|\,.$$

Since series on the right-hand side converges (it is a subseries of (9.31)), the series in (9.33) converges absolutely. Therefore, the exchange of summation and limit $T \to \infty$ is justified in (9.33). Since for each fixed k

$$\lim_{T\to\infty} \frac{e^{ik\Omega T}-1}{ik\Omega T} = 0\,,$$

claim (9.33) is proved, and

$$Q(\varphi) = \sum_{\substack{k=-\infty \\ k\Omega=0}}^{+\infty} a_k e^{ik\varphi}\,. \tag{9.34}$$

Suppose the frequency vector Ω is nonresonant. Then the only vector k for which $k\Omega = 0$ is $k = 0$, and hence $Q(\varphi) = a_0$. This completes proof of the first part of the theorem.

Suppose Ω is resonant. Then any vector $k \in \mathbb{Z}^n$ for which $k\Omega = 0$ is a linear combination of rows of the resonant matrix K, say $k = mK$ for some unique $m \in \mathbb{Z}^s$. Moreover, any $m \in \mathbb{Z}^s$ defines a vector $k = mK$ satisfying $k\Omega = 0$. Therefore, there is a one-to-one correspondence between the resonant group $G \subset \mathbb{Z}^n$ and $\mathbb{Z}^s$. If we set $\chi = K\varphi$, then $k\varphi = mK\varphi = m\chi$, and (9.34) can be rewritten as

$$P(\chi) = \sum_{m=-\infty}^{+\infty} b_m e^{\mathrm{i}m\chi} , \qquad \chi = K\varphi ,$$

where $b_m = a_{mK}$. $\square$

The first part of this theorem follows from ergodic theory: When the vector of frequencies Ω does not have resonant relations, the flow on the n-torus is ergodic (Hoppensteadt 1993). It follows from the weak ergodic theorem that the time average equals the space average in the sense that

$$Q(\varphi) = \lim_{T\to\infty} \frac{1}{T} \int_0^T g(\Omega t + \varphi)\, dt = \frac{1}{(2\pi)^n} \int_0^{2\pi} \cdots \int_0^{2\pi} g(\theta)\, d\theta .$$

Since the space average does not depend on the phase deviations φ, nor does the function Q. Therefore, it is a constant. According to (9.30) its value is a_0.

9.3.4   Nonresonant Ω

A direct application of Theorem 9.6 is the following result.

Theorem 9.7 *Consider a weakly connected oscillatory system*

$$\dot{\theta} = \Omega + \varepsilon g(\theta, \varepsilon) , \qquad \theta \in \mathbb{T}^n, \tag{9.35}$$

where $\Omega \in \mathbb{R}^n$ is a nonresonant vector of frequencies; that is, $k\Omega \neq 0$ for any nonzero $k \in \mathbb{Z}^n$. Such system has a canonical model

$$\dot{\varphi} = 0 + \mathcal{O}(\varepsilon^2) . \tag{9.36}$$

More precisely, there is a change of variables

$$\theta(t) = (\Omega + \varepsilon a_0)t + \varphi + \varepsilon h(\varphi, t) \tag{9.37}$$

mapping solutions of (9.35) to those of (9.36).

Proof. First, the weakly connected oscillatory system is transformed to $\dot{\phi} = \varepsilon g(\Omega t + \phi, e)$ by the change of variables $\theta = \Omega t + \phi$, then to $\dot{\varphi} = \varepsilon \bar{g}(\tilde{\varphi}) + \mathcal{O}(\varepsilon^2)$ by the transformation $\phi = \varphi + \varepsilon h(\varphi, t)$ defined in Theorem

9.4. From Theorem 9.6 it follows that $\bar{g} = a_0$; that is, the averaged system has the form

$$\dot{\tilde{\varphi}} = \varepsilon a_0 + \mathcal{O}(\varepsilon^2) .$$

Now we use the transformation $\tilde{\varphi} = \varepsilon a_0 t + \varphi$ to obtain (9.36). □

Consider (9.35) and its model (9.36) on the time scale $1/\varepsilon$. In this case the function $\varepsilon h(\varphi, t)$ is small (Lemma 9.5), and $\theta(t) \approx (\Omega + \varepsilon a_0)t + \varphi(t)$. From (9.36) it follows that $\theta(t) \approx (\Omega + \varepsilon a_0)t + \varphi(0)$; that is, the oscillators (neurons or local populations of neurons) do not interact on the time scale $1/\varepsilon$. The behavior of the network is simple (uncoupled), and any information processing is impossible on this time scale.

Even when the vector $\Omega + \varepsilon a_0$ is nonresonant and the behavior of $\theta(t)$ appears to be ergodic on the time scale $1/\varepsilon$, we cannot confirm genuine ergodicity of (9.35), since that is an asymptotic notion requiring an infinite time interval.

To study (9.35) on an infinite time interval, we use the approach employed by Bogoliubov, Kolmogorov, Arnold, Moser, and others: We consider (9.35) in the form

$$\dot{\theta} = \Omega + \Delta + \varepsilon g(\theta, \varepsilon) , \qquad \theta \in \mathbb{T}^n, \qquad (9.38)$$

where $\Delta \in \mathbb{R}^n$ is an additional small parameter that is determined in such a manner that this system can be transformed to the uncoupled form

$$\dot{\varphi} = 0 , \qquad \varphi \in \mathbb{T}^n , \qquad (9.39)$$

by a suitable change of variables

$$\theta(t) = \Omega t + \varphi + \varepsilon h(\Omega t + \varphi, \varepsilon) . \qquad (9.40)$$

Such a change of variables exists when the nonresonant vector of natural frequencies $\Omega \in \mathbb{R}^n$ satisfies the *Liouville condition* (sometimes called the *Diophantine condition*): There exist constants $l > 0$ and $r > n - 1$ such that

$$|k\Omega| \geq l|k|^{-r} \qquad (9.41)$$

for all nonzero $k \in \mathbb{Z}^n$. The set of nonresonant Ω for which Liouville's condition is not satisfied has Lebesgue measure zero.

The following theorem was proved by Arnold (1961) for analytic g and extended by Moser (1966) for differentiable functions. The theorem is a part of the KAM (Kolmogorov-Arnold-Moser) theory.

Theorem 9.8 (Arnold-Moser)   *There exists a function $\Delta = \Delta(\varepsilon)$, $\Delta(0) = 0$, such that the weakly connected oscillatory system (9.38) with the nonresonant Ω satisfying the Liouville condition (9.41) can be transformed by a coordinate change (9.40) into the uncoupled system (9.39). Therefore, (9.39) is a canonical model for such a system (9.38).*

The proof can be found elsewhere. The only oscillatory regimes that we pursue in the following are the resonant ones, since nonresonant networks cannot exhibit coupling phenomena.

9.3.5 Resonant Ω

Theorem 9.9 *Consider a weakly connected oscillatory system*

$$\dot\theta = \Omega + \varepsilon g(\theta,\varepsilon)\,, \qquad \theta \in \mathbb{T}^n, \tag{9.42}$$

having a resonant vector of frequencies Ω. Let K be its resonant matrix. Then, there is a transformation

$$\chi = K\theta + \varepsilon h(\theta,t,\varepsilon) \tag{9.43}$$

that maps solutions of (9.42) to those of the lower-dimensional system

$$\dot\chi = \varepsilon X(\chi) + \mathcal{O}(\varepsilon^2)\,, \qquad \chi \in \mathbb{T}^s\,. \tag{9.44}$$

Therefore, (9.44) is a model of (9.42).

Proof. We use standard averaging technique. First, we transform (9.42) to $\dot\phi = \varepsilon g(\Omega t + \phi, t, \varepsilon)$ using $\theta = \Omega t + \phi$. Then we average this system to obtain $\dot\varphi = \varepsilon \bar{g}(\varphi) + \mathcal{O}(\varepsilon^2)$, where $\phi = \varphi + \varepsilon \bar{h}(\varphi, t)$. From Theorem 9.6 it follows that $\bar{g}(\varphi) = P(K\varphi)$ for some function P. Thus, the averaged system has the form

$$\dot\varphi = \varepsilon P(K\varphi) + \mathcal{O}(\varepsilon^2)\,, \qquad \varphi \in \mathbb{T}^n.$$

Let $\chi = K\varphi$. Then

$$\dot\chi = \varepsilon K P(\chi) + \mathcal{O}(\varepsilon^2)\,, \qquad \chi \in \mathbb{T}^s,$$

which coincides with (9.44) if we set $X(\chi) = KP(\chi)$. $\square$

Let $k_1, \ldots, k_s \subset \mathbb{Z}^n$ be the rows of K. Then $\chi = (\chi_1, \ldots, \chi_s) \in \mathbb{T}^s$ is given by $\chi_i = k_i\theta_i + \mathcal{O}(\varepsilon)$; that is, each $\chi_i \in \mathbb{S}^1$ is associated with the ith generator of the resonant group $G \subset \mathbb{Z}^n$. Physicists might say that each χ_i represents the ith *oscillatory mode* in the weakly connected system, and the function $X(\chi)$ describes how the oscillatory modes interact. The number of different modes equals the number of independent resonances s and is not affected by the size of the system n. The importance of oscillatory mode interactions governed by (9.44) for neuroscience applications has yet to be understood.

Corollary 9.10 *When (9.44) has a stable hyperbolic equilibrium, system (9.42) is partially phase locked on the time scale $1/\varepsilon$. If additionally $s = n-1$, which happens exactly when Ω is proportional to a vector of integers, system (9.42) is phase locked on an infinite time scale.*

Further Analysis

Next, we discuss how solutions of (9.44) are related to those of (9.42). For this we scrutinize transformation (9.43). We use ideas similar to those used in the rotation vector method (Hoppensteadt 1993).

Without loss of generality (by renaming variables if needed) we may assume that the first s columns of the $s \times n$ matrix K are linearly independent, so that we can represent it as

$$K = (A, B) ,$$

where A is a nonsingular $s \times s$ matrix and B is an $(n-s) \times s$ matrix that complements A to K. Let $\theta_a = (\theta_1, \ldots, \theta_s)^\top \in \mathbb{T}^s$ and $\theta_b = (\theta_{s+1}, \ldots, \theta_n)^\top \in \mathbb{T}^{n-s}$ so that

$$\theta = \begin{pmatrix} \theta_a \\ \theta_b \end{pmatrix} .$$

Similarly, we define Ω_a, Ω_b, g_a, and g_b. In these new variables, system (9.42) has the form

$$\begin{cases} \dot{\theta}_a = \Omega_a + \varepsilon g_a(\theta_a, \theta_b, \varepsilon) , & \theta_a \in \mathbb{T}^s , \\ \dot{\theta}_b = \Omega_b + \varepsilon g_b(\theta_a, \theta_b, \varepsilon) , & \theta_b \in \mathbb{T}^{n-s} . \end{cases} \tag{9.45}$$

Since A is nonsingular, we can rewrite the transformation

$$\chi = K\theta + \mathcal{O}(\varepsilon) = A\theta_a + B\theta_b + \mathcal{O}(\varepsilon)$$

as

$$\theta_a = A^{-1}(\chi - B\theta_b) + \mathcal{O}(\varepsilon) . \tag{9.46}$$

Now we substitute this for θ_a in the second equation in (9.45) and obtain

$$\dot{\theta}_b = \Omega_b + \varepsilon g_b(A^{-1}(\chi - B\theta_b), \theta_b, 0) + \mathcal{O}(\varepsilon^2) , \qquad \theta_b \in \mathbb{T}^{n-s} ,$$

which we rewrite concisely as

$$\dot{\theta}_b = \Omega_b + \varepsilon Y(\chi, \theta_b) + \mathcal{O}(\varepsilon^2) , \qquad \theta_b \in \mathbb{T}^{n-s} . \tag{9.47}$$

We consider this equation together with (9.44). Since χ does not depend on θ_b, we find $\chi(t)$ and then use it in (9.47) to find $\theta_b(t)$. The full solution $\theta(t)$ is then defined using (9.46):

$$\theta(t) = \begin{pmatrix} A^{-1}(\chi(t) - B\theta_b(t)) \\ \theta_b(t) \end{pmatrix} + \mathcal{O}(\varepsilon) .$$

When system (9.44) has an attractor, say $\chi^\star$, system (9.42) has an attractor too that lies in an invariant set $M \subset \mathbb{T}^n$ homeomorphic to $\chi^\star \times \mathbb{T}^{n-s}$. In particular, its dimension is less than or equal to the dimension of M, which is that of $\chi^\star$ plus $n - s$. Below, we consider the most trivial case, when $\chi^\star$ is a point. In the analysis of (9.47) the following result is useful.

Lemma 9.11 *The frequency vector $\Omega_b \in \mathbb{R}^{n-s}$ in (9.47) does not have resonant relations.*

Proof. Indeed, suppose there is a nonzero $k_b \in \mathbb{Z}^{n-s}$ such that $k_b \Omega_b = 0$. Then $k = (0, k_b) \in \mathbb{Z}^n$ gives a resonant relation, $k\Omega = 0$, but k is a linear combination of rows of matrix K. In particular, k_b is a linear combination of rows of matrix B, which is fine, but $0 \in \mathbb{R}^s$ is a linear combination of rows of matrix A, which is impossible, since A is nonsingular. $\square$

This lemma allows us to use Theorem 9.7 and the Arnold-Moser theorem (Theorem 9.8) when χ converges to an equilibrium χ^*. In this case, the behavior of $\theta_b(t)$ appears to be ergodic on a time scale $1/\varepsilon$ or larger.

9.3.6 A Neighborhood of Resonant Ω

Let us consider a weakly connected oscillatory system

$$\dot{\theta} = \Omega + \Delta + \varepsilon g(\theta, \varepsilon)$$

having frequency vector $\Omega + \Delta$, where Ω is resonant and $\Delta \ll 1$. Such a system is said to be close to a resonant one.

Proceeding as in the proof of Theorem 9.9 we introduce the new variable $\chi = K\varphi$, where $\varphi = \theta - \Omega t + \mathcal{O}(\varepsilon)$. Then the averaged system has the form

$$\dot{\chi} = K\Delta + \varepsilon X(\chi) + \mathcal{O}(\varepsilon^2), \qquad \Delta \ll 1,$$

where $X(\chi)$ is as defined earlier. Whether such a system has a stable equilibrium or not depends on the relative sizes of $K\Delta$ and εX. This affects the occurrence of frequency or phase locking. When $\varepsilon \ll |K\Delta|$, one could average the system again to lower its dimension.

9.4 Conventional Synaptic Connections

In this section we continue to study weakly connected oscillatory networks of the form

$$\dot{\theta} = \Omega + \varepsilon g(\theta, \varepsilon), \qquad \theta \in \mathbb{T}^n,$$

with the assumption that the connection function $g = (g_1, \ldots, g_n)$ has a special form; namely, we suppose that

$$g_i(\theta, \varepsilon) = \sum_{j=1}^{n} g_{ij}(\theta_i, \theta_j) + \mathcal{O}(\varepsilon) \tag{9.48}$$

for all $i = 1, \ldots, n$. It should be stressed that (9.48) is just an assumption. It is *not* a consequence of weakness of connections, as some tend to believe.

However, it *does* follow from weakness of connections when the limit cycle amplitudes are small, as in the case of multiple Andronov-Hopf bifurcations; see Section 5.4.

First, let us discuss the neurophysiological basis of such an assumption.

9.4.1   Synaptic Glomeruli

A synapse is a contact between two neurons. A reasonable way to classify synapses is to determine the contacting parts. For example, a contact from an axon to a dendrite is termed an *axo-dendritic* synapse, whereas that onto a cell body is termed *axo-somatic*. Those are conventional (classical) synapses in the sense that they represent our preconception of a neuron as a polarized device receiving input via dendrites and cell body and sending output via the axon.

There are many other types of synapses, which are called *unconventional* or even *nonusual* (Shepherd 1983). Among them the most prominent are *dendro-dendritic* and *axo-axonic* synapses, which are frequently found inside tightly grouped clusters of terminals, called *synaptic glomeruli*; see Figure 9.8.

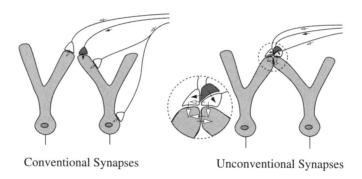

Conventional Synapses   Unconventional Synapses

Figure 9.8. Axo-dendritic and axo-somatic synapses are conventional. Axo-axonic and dendro-dendritic synapses are unconventional. They can frequently be found inside synaptic glomeruli.

The anatomical distinction between conventional and unconventional synapses is obvious. Let us discuss the functional difference. Synaptic transmission via conventional synapses depends on activities of pre- and postsynaptic neurons and is relatively independent of activities of other neurons. This leads to (9.48). In contrast, synaptic transmission inside a glomerulus is more complicated. It depends not only on pre- and postsynaptic neurons, but also on activities of other neurons sending terminals to the glomerulus. For example, the inhibitory axo-dendritic synapse in Figure 9.9 can be shut down by another inhibitory axo-axonic synapse. Both synapses may be a part of a sophisticated glomerulus with many intricate reciprocal interconnections. Obviously, we cannot assume that synaptic transmissions are relatively independent and can be described by (9.48) in this case.

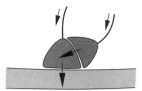

Figure 9.9. Transmission of the conventional inhibitory synapse can be shut down by an unconventional axo-axonic inhibitory synapse.

It should be stressed that the assumption that the connection function g has form (9.48) when synaptic interactions are conventional is neither a fact nor a theorem, but a reasonable principle used in mathematical neuroscience, which still requires rigorous mathematical justification.

9.4.2 Phase Equations

The special form of the connection function g leads to many interesting results. In particular, to exhibit nontrivial behavior, it is not enough for the weakly connected system to have a resonant frequency vector $\Omega = (\Omega_1, \ldots, \Omega_n)^\top \in \mathbb{R}^n$. Some of the frequencies must be commensurable, which we discuss next.

Nonzero real numbers a and b are said to be *commensurable*, denoted by $a \sim b$, when there are positive integers k and m such that $ka - mb = 0$; that is, the vector $(a, b)^\top \in \mathbb{R}^2$ is resonant. To be commensurable is an *equivalence relation*, since it satisfies the following conditions, which are easy to verify.

- *Reflexivity.* $a \sim a$ for any $a \in \mathbb{R}$.

- *Symmetry.* If $a \sim b$, then $b \sim a$.

- *Transitivity.* If $a \sim b$, and $b \sim c$, then $a \sim c$.

As a consequence, any set of real numbers, e.g., the set $\{\Omega_1, \ldots, \Omega_n\}$, can be partitioned into disjoint equivalence classes. Accordingly, a weakly connected oscillatory network having natural frequencies $\Omega_1, \ldots, \Omega_n$ can also be partitioned into noninteracting subnetworks, which follows from the theorem below.

Theorem 9.12 *Consider a weakly connected oscillatory system*

$$\dot{\theta}_i = \Omega_i + \varepsilon\omega_i + \varepsilon\sum_{j=1}^{n} g_{ij}(\theta_i, \theta_j) + \mathcal{O}(\varepsilon^2)\,, \qquad \theta_i \in \mathbb{S}^1\,, \tag{9.49}$$

where the functions g_{ij} have zero average (the average is taken into account by ω_i). There is a change of variables of the form

$$\theta = \Omega t + \varphi + \varepsilon h(\varphi, t)$$

that transforms (9.49) to

$$\dot{\varphi}_i = \varepsilon \omega_i + \varepsilon \sum_{j=1}^{n} q_{ij}(\varphi_i, \varphi_j) + \mathcal{O}(\varepsilon^2), \qquad \varphi_i \in \mathbb{S}^1, \qquad (9.50)$$

where each function q_{ij} satisfies

$$q_{ij}(\varphi_i, \varphi_j) = \begin{cases} 0 & \text{if } \Omega_i \not\sim \Omega_j, \\ H_{ij}(k_{ij}\varphi_j - m_{ij}\varphi_i) & \text{if } \Omega_i \sim \Omega_j, \end{cases}$$

where k_{ij} and m_{ij} are positive relatively prime integers for which $k_{ij}\Omega_j - m_{ij}\Omega_i = 0$, and the function H_{ij} has zero average; i.e.,

$$\int_0^{2\pi} H_{ij}(\chi)\,d\chi = 0. \qquad (9.51)$$

In particular, if $\Omega_j \not\sim \Omega_i$, then φ_j and φ_i do not depend on each other (up to order ε^2).

Proof. The averaging technique that we used in Theorems 9.7 and 9.9 above (namely, changing variables to $\phi = \theta - \Omega t$ in (9.49) and averaging of $\dot{\phi} = \varepsilon g(\Omega t + \phi, \varepsilon)$) leads to the averaged system

$$\dot{\varphi}_i = \varepsilon \sum_{j=1}^{n} \bar{g}_{ij}(\varphi_i, \varphi_j) + \mathcal{O}(\varepsilon^2), \qquad i = 1, \ldots, n,$$

where

$$\bar{g}_{ij}(\varphi_i, \varphi_j) = \lim_{T \to \infty} \frac{1}{T} \int_0^T g_{ij}(\Omega_i t + \varphi_i, \Omega_j t + \varphi_j)\,dt.$$

Now we use Theorem 9.6 to compute $\bar{g}_{ij}$:

$$\bar{g}_{ij}(\varphi_i, \varphi_j) = \begin{cases} a_{ij0} & \text{if } \Omega_i \not\sim \Omega_j, \\ H_{ij}(k_{ij}\varphi_j - m_{ij}\varphi_i) & \text{if } \Omega_j \sim \Omega_i, \end{cases}$$

where

$$a_{ij0} = \frac{1}{(2\pi)^2} \int_0^{2\pi} \int_0^{2\pi} g_{ij}(\theta_i, \theta_j)\,d\theta_i\,d\theta_j.$$

Since each g_{ij} has zero average, the average of H_{ij} is zero, and each $a_{ij0} = 0$. $\square$

When the frequencies $\Omega_1, \ldots, \Omega_n$ are pairwise incommensurable, the system (9.50) reduces to

$$\dot{\varphi}_i = \varepsilon \omega_i + \mathcal{O}(\varepsilon^2),$$

which is uncoupled on the time scale $1/\varepsilon$ even when there *are* resonant relations among the frequencies $\Omega_1, \ldots, \Omega_n$. One can still study such a system by performing another averaging, paying attention to resonances among $\omega_1, \ldots, \omega_n$. The averaged system might exhibit interesting dynamical features, but they could be observable only on the longer time scale $1/\varepsilon^2$.

9.4.3 Commensurable Frequencies and Interactions

A direct consequence of Theorem 9.12 is the following corollary, which is a generalization of Corollary 5.9 that we proved for WCNNs at Andronov-Hopf bifurcations.

Corollary 9.13 *All oscillators can be divided into pools, or ensembles, according to their frequencies Ω_i. Oscillators from different pools have incommensurable frequencies, and interactions between them are negligible.*

Notice that the corollary contains two statements: (1) The network is partitioned into pools of oscillators, and (2) Interactions between oscillators from different pools are negligible. The first statement agrees with the well-known fact (Kuramoto 1984, Gerstner and van Hemmen 1994, Daido 1996) from statistical mechanics that a network of oscillators can be decomposed into coherent pools if the strength of connections is strong enough in comparison with the distribution of frequencies. The second statement is novel. It claims that oscillators having incommensurable frequencies work independently from each other even when they have synaptic contacts between them. Although physiologically present and active, those synaptic connections average to zero. Therefore, they are functionally insignificant and do not play any role in the network's dynamics. We have already discussed such a phenomenon in Section 5.4.2.

Barely Commensurable Frequencies

Let us discuss interactions between oscillators from the same pool. For the sake of illustration, consider a network of three oscillators

$$\begin{cases} \dot{\theta}_1 = \Omega_1 + \varepsilon g(\theta_1, \theta_2) + \varepsilon y(\theta_1, \theta_3) \\ \dot{\theta}_2 = \Omega_2 + \varepsilon g(\theta_2, \theta_1) + \varepsilon g(\theta_2, \theta_3) \\ \dot{\theta}_3 = \Omega_3 + \varepsilon g(\theta_3, \theta_1) + \varepsilon g(\theta_3, \theta_2) \end{cases}$$

having identical connection functions and pairwise commensurable frequencies $\Omega_1 = \Omega_2 = 1$ and $\Omega_3 = 1.1$. Therefore, the oscillators belong to the same pool. Notice that if $k\Omega_3 - m\Omega_1 = 0$, then $k + m \geq 21$. Whenever $k + m$ is a large number, we say that the frequencies are *barely commensurable*.

Are interactions between θ_1 and θ_2 the same as those between θ_1 and θ_3? In particular, is there any qualitative difference between the averaged connection functions $H_{12}(\varphi_2 - \varphi_1)$ and $H_{13}(10\varphi_3 - 11\varphi_1)$ in the model

$$
\begin{cases}
\dot\varphi_1 = & \varepsilon H_{12}(\varphi_2 - \varphi_1) & + & \varepsilon H_{13}(10\varphi_3 - 11\varphi_1) & + & \mathcal{O}(\varepsilon^2) \\
\dot\varphi_2 = & \varepsilon H_{21}(\varphi_1 - \varphi_2) & + & \varepsilon H_{23}(10\varphi_3 - 11\varphi_2) & + & \mathcal{O}(\varepsilon^2) \\
\dot\varphi_3 = & \varepsilon H_{31}(11\varphi_1 - 10\varphi_3) & + & \varepsilon H_{32}(11\varphi_2 - 10\varphi_3) & + & \mathcal{O}(\varepsilon^2)
\end{cases}
$$
(9.52)

To study this issue we recall (see the proof of Theorem 9.6) that

$$
H_{12} = \sum_{j=-\infty}^{+\infty} a_{j(-1,1)} e^{\mathbf{i} j(\varphi_2 - \varphi_1)} \quad\text{and}\quad H_{13} = \sum_{j=-\infty}^{+\infty} a_{j(-11,10)} e^{\mathbf{i} j(10\varphi_2 - 11\varphi_1)} .
$$

Since the series $\sum |a_k|$ converges, each Fourier coefficient $|a_k| \to 0$ as $|k| \to \infty$. Therefore, we expect that

$$
|a_{j(-11,10)}| \ll |a_{j(-1,1)}|
$$

for most j, meaning that $|H_{13}| \ll |H_{12}|$. Thus, the interactions between θ_1 and θ_2 are much stronger than those between θ_1 and θ_3. This result is not valid for all functions g. For example, it is not valid for $g(x,y) = \sin(10y - 11x)$. Such g has zero Fourier coefficients except $a_{(-11,10)}$ and $a_{(11,-10)}$, and therefore $H_{12} \equiv 0$, but $H_{13} = g \not\equiv 0$. Nevertheless, if we fix the connection function g_{ij} and consider the natural frequencies Ω_i and Ω_j as parameters, the following result, which is a direct consequence of the Riemann-Lebesgue lemma (Ermentrout 1981), holds.

Proposition 9.14 *The strength of interactions between a pair of oscillators having barely commensurable frequencies Ω_i and Ω_j ($k\Omega_i - m\Omega_j = 0$ for relatively prime large integers k and m) vanishes when either k or m increases.*

Thus, interactions among oscillators from the same ensemble could be negligible when the frequencies are barely commensurable ($k + m$ is sufficiently large). Whether or not $k + m$ is "bad" enough depends on peculiarities of the connection function g and the strength of connections $\varepsilon \ll 1$.

The example above can be studied using methods from Section 9.3.6. We can assume that the system has frequency vector $\Omega = (1, 1, 1)^\top + \Delta$, where $\Delta = (0, 0, 0.1)^\top$ is a small perturbation. The averaged system in this case has the form

$$
\begin{cases}
\dot\varphi_1 = & \varepsilon H(\varphi_2 - \varphi_1) + \varepsilon H(\varphi_3 - \varphi_1) + \mathcal{O}(\varepsilon^2) \\
\dot\varphi_2 = & \varepsilon H(\varphi_1 - \varphi_2) + \varepsilon H(\varphi_3 - \varphi_2) + \mathcal{O}(\varepsilon^2) \\
\dot\varphi_3 = 0.1 + & \varepsilon H(\varphi_1 - \varphi_3) + \varepsilon H(\varphi_2 - \varphi_3) + \mathcal{O}(\varepsilon^2)
\end{cases}
$$

for some function H. Whether such a system has an equilibrium depends on the relative sizes of 0.1 and ε. If $\varepsilon \ll 0.1$, one might have to average it again. This results in (9.52).

9.4.4 Example: A Pair of Oscillators

Consider a pair of weakly coupled oscillators having a nearly resonant vector of natural frequencies. We can apply Theorem 9.1 to find the phase model for such oscillators

$$\begin{cases} \dot\theta_1 = \Omega_1 + \varepsilon\omega_1 + \varepsilon g_1(\theta_1,\theta_2,\varepsilon) \\ \dot\theta_2 = \Omega_2 + \varepsilon\omega_2 + \varepsilon g_2(\theta_1,\theta_2,\varepsilon) \end{cases} , \qquad (9.53)$$

where $\theta_i \in \mathbb{S}^1$ are phase variables, $\omega_i \in \mathbb{R}$ are the frequency deviations, and the average of each g_i is zero. Let k and m be positive relatively prime integers such that

$$k\Omega_2 - m\Omega_1 = 0 \ .$$

Application of Theorem 9.12 results in the system

$$\begin{cases} \dot\varphi_1 = \varepsilon\omega_1 + \varepsilon H_1(k\varphi_2 - m\varphi_1) \\ \dot\varphi_2 = \varepsilon\omega_2 + \varepsilon H_2(m\varphi_1 - k\varphi_2) \end{cases} ,$$

where we omit $\mathcal{O}(\varepsilon^2)$ terms for the sake of simplicity. To study phase locking we introduce a new variable

$$\chi = k\varphi_2 - m\varphi_1 \ .$$

It follows that

$$\chi' = \omega + H(\chi) , \qquad \chi \in \mathbb{S}^1 , \qquad (9.54)$$

where $' = d/d\tau$, $\tau = \varepsilon t$ is slow time, $\omega = k\omega_2 - m\omega_1$, and

$$H(\chi) = kH_2(-\chi) - mH_1(\chi) \ .$$

The stable equilibria of (9.54) are in one-to-one correspondence with various $k : m$ phase locked periodic solutions of (9.53). Geometrically, the equilibria are the intersections of the graph of H and the horizontal line $-\omega$ (see Figure 9.10), and they are stable if the slope of the graph is negative at the intersection.

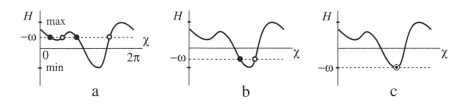

Figure 9.10. Equilibria of (9.54) for various ω.

Let $\chi^\star$ be a stable equilibrium of (9.54). Then, the phases θ_1 and θ_2 satisfy the relationship

$$k\theta_2 - m\theta_1 = \chi^\star \ .$$

On the other hand, if we have control over $\omega = k\omega_2 - m\omega_1$ and can measure the difference $\chi^\star$ experimentally, then we can use the equation

$$\omega + H(\chi^\star) = 0$$

to reconstruct the part of the function H that has negative slope, which is relevant to the phase locking; see Figure 9.11. This procedure, which was

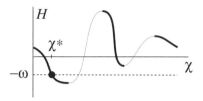

Figure 9.11. Knowing the phase difference $\chi^\star$ and the frequency difference ω, one can reconstruct the part of H having negative slope, which is responsible for phase locking.

suggested by Williams (1992), allows one to find H experimentally when the equations that govern the oscillators are unknown.

Let us determine the frequencies $\Omega_i + \varepsilon\omega_i$, $i = 1, 2$, that ensure $k : m$ phase locking in (9.53): We use

$$\frac{\Omega_1 + \varepsilon\omega_1}{\Omega_2 + \varepsilon\omega_2} = \frac{k}{m} + \varepsilon\frac{\omega}{mk} + \mathcal{O}(\varepsilon^2)$$

and $\omega \in (-\max H, \ -\min H)$ to plot the sets corresponding to various types of locking in Figure 9.12. Triangular bands, which are sometimes called *Arnold tongues*, emanate from the rational numbers k/m and denote the regions of $k : m$ phase locking. The tongues in the figure are linear, since we have neglected small terms in ε in (9.54) and in the equation above. These terms can distort the tongues (see the dotted curve in Figure 9.12) and endow them with quite complicated geometry. A continuation of the Arnold tongues for moderate ε shows that they can overlap, bifurcate, form mushroom-like objects, etc. Some interesting pictures obtained in the study of forced biological oscillators can be found in Glass and Belair (1985) and Flaherty and Hoppensteadt (1978).

How does the $k : m$ phase locking disappear? To study this question we consider (9.54) when the parameter ω passes either $-\min H$ or $-\max H$; see, e.g., Figure 9.10c. A saddle-node bifurcation occurs in this case. This corresponds to the annihilation of a stable and an unstable limit cycle

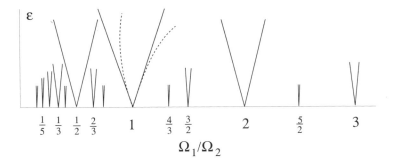

$$\Omega_1/\Omega_2$$

Figure 9.12. Illustration of Arnold tongues.

of (9.53), and the new behavior is called *drifting*, which means that the activity of (9.53) is either quasi-periodic or a high-order locking.

When the drift is a stable periodic solution, which is a high-order torus knot, the activity of (9.53) is frequency locked. Since $\chi \in \mathbb{S}^1$, the saddle-node bifurcation is on a limit cycle, so that we can use the theory developed in Chapter 8 to conclude that χ oscillates with an arbitrarily small frequency that depends on how far ω is from the bifurcation value. The spike lemma suggests that χ spends most of its time in a neighborhood of χ^*, where χ^* is the saddle-node equilibrium. In this case oscillators (9.53) are nearly $k : m$ locked with a phase difference near χ^*. When χ generates a spike (an excursion from χ^*), the oscillators unlock for a brief moment. In Figure 9.13 we illustrate 1:1 phase locking and drifting that is at a 10:9 frequency lock.

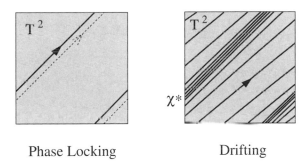

Phase Locking   Drifting

Figure 9.13. *Left:* Stable limit cycle (bold line) corresponds to phase locking. *Right:* Drifting (high order frequency locking) occurs when the stable limit cycle coalesces with the unstable one (dashed line) and disappears.

Finally note that the flow on the 2-torus $\mathbb{T}^2$ defined by (9.53) can be reduced to a mapping of the unit circle $\mathbb{S}^1$ to itself. Indeed, such a flow has a cross-section, say $\theta_1 = 0$, and the Poincaré (first return) map captures all qualitative and most of the important quantitative features of the flow. Similarly, the flow on the 3-torus $\mathbb{T}^3$, describing dynamics of a network of three oscillators, can be reduced to a Poincaré mapping of $\mathbb{T}^2$ to itself, whose studying reveals many interesting dynamical features, such as locking, contractible, annular, and toroidal chaos (Baesens et al. 1991).

9.4.5 Example: A Chain of Oscillators

Chains of oscillators can describe many processes in nature, e.g., the undulatory locomotion of some fish, such as lamprey or dogfish, peristalsis in vascular and intestinal smooth muscle, communication of fireflies, synchronization of emerging oscillations in sensory processing in the cortex, etc., see reviews by Kopell (1995) and Rand et al. (1987).

A typical traveling wave in a chain is a synchronized solution in which oscillators oscillate with a common frequency but with various phase deviations that increase (or decrease) monotonically along the chain. There are many mechanisms leading to such synchronized solutions even in chains with a nearest-neighbor pattern of connections. Some of them are summarized in Figure 9.14 and discussed briefly below.

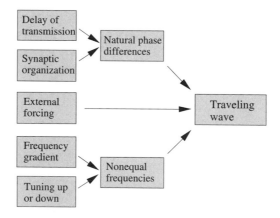

Figure 9.14. Various mechanisms leading to traveling waves in chains of oscillators.

Let $\dot{X}_i = F_i(X_i, \lambda)$, $X_i \in \mathbb{R}^m$, describe the dynamics of the ith oscillator having a hyperbolic limit cycle attractor with natural frequency Ω_i for $\lambda = 0$. Consider a chain of n such oscillators weakly connected in *only one direction*, e.g., from X_{i-1} to X_i; see Figure 9.15.

$$\begin{cases} \dot{X}_1 &= F_1(X_1, \lambda)\,, \\ \dot{X}_i &= F_i(X_i, \lambda) + \varepsilon G_i(X_{i-1}, X_i)\,, \qquad i = 2, \ldots, n\,. \end{cases}$$

Figure 9.15. A chain of $n = 7$ oscillators connected in one direction.

We assume that $\Omega_1 = \cdots = \Omega_n$ and that $\lambda = \mathcal{O}(\varepsilon)$ so that the oscillators have ε-close natural frequencies. In this case we can use Malkin's theorem (Theorem 9.2) to obtain the canonical model

$$\begin{cases} \phi_1' = \omega_1 \,, \\ \phi_i' = \omega_i + H_i(\phi_{i-1} - \phi_i) \,, \qquad i = 2, \ldots, n \,, \end{cases}$$

where $' = d/d\tau$ and $\tau = \varepsilon t$ is slow time, each $\phi_i \in \mathbb{S}^1$ denotes the phase deviation from the natural oscillation with natural frequency Ω_i, each function H_i is 2π-periodic, and we have omitted $\mathcal{O}(\varepsilon)$ terms for the sake of simplicity. The parameters ω_i account for frequency deviations due to $\lambda = \mathcal{O}(\varepsilon)$. It is convenient to use new variables $\phi_i = \omega_1 \tau + \varphi_i$ and to rewrite this system in the form

$$\begin{cases} \varphi_1' = 0 \,, \\ \varphi_i' = \omega_i - \omega_1 + H_i(\varphi_{i-1} - \varphi_i) \,, \qquad i = 2, \ldots, n \,. \end{cases} \qquad (9.55)$$

Stable equilibria of this system correspond to synchronization of the chain's dynamics and possibly to traveling waves.

If connections between oscillators are diffusive (e.g., gap junctions), then $G_i(X_{i-1}, X_i) = c_i(X_{i-1} - X_i)$ for some $c_i > 0$, and from (9.15) it follows that $H_i(0) = 0$ and $H_i'(0) > 0$. If $\omega_1 = \cdots = \omega_n$, then the chain's activity is synchronized in-phase; that is, $\varphi_1 = \cdots = \varphi_n$. There is a traveling wave when all $\omega_i < \omega_1$ or all $\omega_i > \omega_1$, $i = 2, \ldots, n$, provided that $|\omega_i - \omega_1|$ are small enough. The first case corresponds to ω_1 being tuned up. The wave moves from left to right, since the solution to $\omega_i - \omega_1 + H_i = 0$ is positive with the result that $\varphi_{i-1} > \varphi_i$ for all i. In the second case, when ω_1 is tuned down, the wave moves backwards (from right to left), since $\varphi_{i-1} < \varphi_i$ for all $i = 2, \ldots, n$.

The assumption that connections between oscillators are diffusive may be plausible in some cases, but it is not plausible for modeling the spinal cord of lamprey or information processing in the cortex. What happens if we drop this assumption? First of all, the condition $H_i(0) = 0$ may still hold when connections between oscillators are not diffusive. Quite often however, $H_i(0) \neq 0$. In this case a crucial role is played by the values $\chi_i \in \mathbb{S}^1$ such that $H_i(\chi_i) = 0$ and $H_i'(\chi_i) > 0$. Such χ_i are called *natural phase differences*. They depend on many factors among which the most important are transmission delays (see Proposition 9.3) and the type of connections between oscillators. The latter is closely related to the issue of synaptic organization of the brain that we discuss in Chapter 13. A given function H_i may have many natural phase differences or none. We call them *natural* because they arise naturally between two identical oscillators connected in one direction (System (9.55) for $n = 2$ and $\omega_1 = \omega_2$). A nonzero natural phase difference is sometimes called an *edge effect*. It allows traveling waves in chains of identical ($\omega_1 = \cdots = \omega_n$) oscillators.

Now consider a chain of oscillators connected in *both* directions; see Figure 9.16. For the sake of simplicity we consider nearly identical oscillators

Figure 9.16. A chain of $n = 7$ nearly identical oscillators connected in both directions.

$$
\begin{cases}
\dot{X}_1 &= F_1(X_1, \lambda) + \varepsilon G^+(X_1, X_2) \\
\dot{X}_i &= F_i(X_i, \lambda) + \varepsilon G^-(X_{i-1}, X_i) + \varepsilon G^+(X_i, X_{i+1}) \\
\dot{X}_n &= F_n(X_n, \lambda) + \varepsilon G^-(X_{n-1}, X_n)
\end{cases} ,
$$

where $i = 2, \ldots, n-1$ and $F_i = F + \mathcal{O}(\varepsilon)$ for some function F. Proceeding as above, we obtain a canonical model

$$
\begin{cases}
\phi'_1 &= \omega_1 + H^+(\phi_2 - \phi_1) \\
\phi'_i &= \omega_i + H^-(\phi_{i-1} - \phi_i) + H^+(\phi_{i+1} - \phi_i) \\
\phi'_n &= \omega_n + H^-(\phi_{n-1} - \phi_n)
\end{cases} ,
$$

where ω_i accounts for small differences between F_i and F. Any phase-locked solution of this system has the form $\phi_i(\tau) = \omega_0 \tau + \varphi_i$, $i = 1, \ldots, n$, where ω_0 is a common frequency and $\varphi_1, \ldots, \varphi_n$ are some constants. It is easy to see that these satisfy the n conditions

$$
\begin{cases}
\omega_0 &= \omega_1 + H^+(\varphi_2 - \varphi_1) \\
\omega_0 &= \omega_i + H^-(\varphi_{i-1} - \varphi_i) + H^+(\varphi_{i+1} - \varphi_i) \\
\omega_0 &= \omega_n + H^-(\varphi_{n-1} - \varphi_n)
\end{cases} .
$$

Suppose that the connections between oscillators are diffusive, which leads to $H^\pm(0) = 0$ and $H^{\pm'}(0) > 0$. If the oscillators are identical ($\omega_1 = \cdots = \omega_n$), then in-phase synchronization always exists. A traveling wave may occur when the frequency deviations ω_i are not all equal. It is easy (Cohen et al. 1982) to find conditions on ω_i that guarantee the existence of a traveling wave with a constant phase difference $\chi = \phi_i - \phi_{i-1}$ along the chain. Simple arithmetic shows that $\omega_2 = \cdots = \omega_{n-1}$,

$$
H^-(-\chi) = \omega_1 - \omega_2 , \qquad \text{and} \qquad H^+(\chi) = \omega_n - \omega_2 ,
$$

and $\omega_0 = \omega_1 + \omega_n - 2\omega_2$ in this case. In particular, if $\omega_1 > \omega_2 = \cdots = \omega_{n-1} > \omega_n$, which corresponds to the first oscillator being tuned up and the last oscillator being tuned down, then the traveling wave moves from left to right.

When there is a gradient of frequencies ($\omega_1 > \omega_2 > \cdots > \omega_n$ or vice versa), one may still observe a traveling wave but with a non-constant phase difference along the chain. When the gradient is large enough, the synchronized solution corresponding to a single traveling wave disappears, and frequency plateaus may appear (Ermentrout and Kopell 1984). That is, solutions occur in which the first $k < n$ oscillators are phase locked and the last $n - k$ oscillators are phase locked, but the two pools oscillate with different frequencies. There may be many frequency plateaus.

If the number n of oscillators in the chain is sufficiently large and the frequency deviations ω_i are sufficiently close between adjacent oscillators, then a continuum limit approximation can be justified (Kopell 1986, Kopell and Ermentrout 1990). In this case phase locked behavior can be described by a singularly perturbed partial differential equation, which we do not present here.

9.4.6 Oscillatory Associative Memory

From Corollary 9.13 it follows that a weakly connected network of limit-cycle oscillators can be partitioned into pools of oscillators according to their natural frequencies Ω_i. Since interactions between oscillators from different pools are negligible, we study only interactions between oscillators from the same pool here. Without loss of generality we may assume that the entire network is one such pool.

For the sake of simplicity we assume that these oscillators have equal natural frequencies $\Omega_1 = \cdots = \Omega_n$. If we use slow time $\tau = \varepsilon t$ and discard high-order terms in ε, then the canonical model (9.50) can be written in the form (9.56). The following theorem shows that under some conditions, oscillatory neural network (9.56) can have interesting neurocomputational properties. In particular, it can act as a multiple attractor type neural network, where attractors are limit cycles.

Theorem 9.15   (Convergence Theorem for Oscillatory Neural Networks) *Consider the oscillatory neural network*

$$\varphi_i' = \omega_i + \sum_{j=1}^{n} H_{ij}(\varphi_j - \varphi_i) , \qquad \varphi_i \in \mathbb{S}^1 , \qquad (9.56)$$

and suppose that $\omega_1 = \cdots = \omega_n = \omega$ and

$$H_{ij}(-\chi) = -H_{ji}(\chi) , \qquad \chi \in \mathbb{S}^1 , \qquad (9.57)$$

for all $i, j = 1, \ldots, n$. Then the network dynamics converge to a limit cycle attractor. On the limit cycle, all neurons oscillate with equal frequencies and constant phase deviations. This corresponds to synchronization of the network activity.

The proof is a modification of the one given by Cohen and Grossberg (1983) to prove absolute stability of a general network of nonlinear nonoscillatory neurons; see page 112.

Proof. Let $\varphi = \omega t + \phi$. Then

$$\phi_i' = \sum_{j=1}^{n} H_{ij}(\phi_j - \phi_i) , \qquad \phi_i \in \mathbb{S}^1 . \tag{9.58}$$

If this system always converges to an equilibrium, say $\phi^\star$, then (9.56) converges to a limit cycle $\varphi(\tau) = \omega\tau + \phi^\star$, and the original weakly connected system converges to a limit cycle $\theta(t) = (\Omega + \varepsilon\omega)t + \phi^\star$. Obviously, phase deviations (vector $\phi^\star \in \mathbb{T}^n$) are constant on the limit cycle.

Let

$$R_{ij}(\chi) = \int_0^\chi H_{ij}(t)\, dt$$

be the antiderivative of $H_{ij}(\chi)$; that is, $dR_{ij}/d\chi = H_{ij}$. Due to (9.51) $R_{ij}(2\pi) = 0$, and therefore $R_{ij}(\chi)$ is continuous. Let us check that the function $U : \mathbb{T}^n \to \mathbb{R}$ given by

$$U(\phi) = \frac{1}{2} \sum_{i=1}^{n} \sum_{j=1}^{n} R_{ij}(\phi_j - \phi_i)$$

is a global Liapunov function for (9.58). Indeed, since it is continuous and $\mathbb{T}^n$ is compact, U is bounded below. Moreover, it satisfies

$$\frac{\partial U}{\partial \phi_k} = \frac{1}{2}\left(\sum_{i=1}^{n} H_{ik}(\phi_k - \phi_j) - \sum_{j=1}^{n} H_{kj}(\phi_j - \phi_k)\right)$$

$$= -\sum_{j=1}^{n} H_{kj}(\phi_j - \phi_k) = -\phi_k' ,$$

and hence

$$\frac{dU(\phi)}{d\tau} = \sum_{k=1}^{n} \frac{\partial U}{\partial \phi_k} \phi_k' = -\sum_{k=1}^{n} |\phi_k'|^2 \leq 0 .$$

Notice that $dU/d\tau = 0$ precisely when $\phi_1' = \cdots = \phi_n' = 0$, i.e., at an equilibrium point of (9.58). $\square$

There could be many equilibria of (9.56) depending on connection functions H_{ij}. Therefore, the network can remember and reproduce many previously memorized oscillatory patterns. It is still an open problem how to "teach" connection functions H_{ij} so that the network memorizes a given set of patterns. Ermentrout and Kopell (1994) suggest some mechanisms. A similar problem for an oscillatory network with symmetry is studied by Woodward (1990). We study the problem of teaching the connection functions H_{ij} below, where we consider specific H_{ij}, namely $H_{ij}(\varphi_j - \varphi_i) = s_{ij}\sin(\varphi_j + \psi_{ij} - \varphi_i)$ for some constants s_{ij} and ψ_{ij}.

9.4.7 Kuramoto's Model

When all connection functions in the neural network are equal, namely, $H_{ij}(\chi) = H_{ji}(\chi) = H(\chi)$ for some function $H(\chi)$, then condition (9.57) implies that $H(\chi)$ is an odd function. Any odd function on $\mathbb{S}^1$ can be represented as a Fourier series

$$H(\chi) = \sum_{k=1}^{+\infty} a_k \sin k\chi \ .$$

Kuramoto (1984) suggested that the first term, $a_1 \sin \chi$, is dominant, and he disregarded the rest of the series. This assumption can be motivated by Figure 9.6. Indeed, the connection function H obtained for weakly connected Wilson-Cowan neural oscillators does resemble $\sin \chi$ for inhibitory $\to$ excitatory and excitatory $\to$ inhibitory synaptic organizations. Kuramoto's phase model is

$$\varphi_i' = \omega_i + \sum_{j=1}^{n} s_{ij} \sin(\varphi_j - \varphi_i) \ , \qquad \varphi_i \in \mathbb{S}^1 \ ,$$

where $s_{ij} \in \mathbb{R}$ are synaptic coefficients. The convergence theorem above is applicable to Kuramoto's model when $\omega_1 = \cdots = \omega_n$ and the synaptic matrix $S = (s_{ij})$ is symmetric; that is, $s_{ij} = s_{ji}$ for all i and j.

Notice that $\sin \chi$ is a bad approximation for the excitatory $\to$ excitatory and inhibitory $\to$ inhibitory synaptic organizations in Figure 9.6. The former can be better approximated by $-\cos \chi$ and the latter by $\cos \chi$. To take this into account one can adjust Kuramoto's model to include the *natural phase differences* ψ_{ij}, so that the model is

$$\varphi_i' = \omega_i + \sum_{j=1}^{n} s_{ij} \sin(\varphi_j + \psi_{ij} - \varphi_i) \ , \qquad \varphi_i \in \mathbb{S}^1 \ , \qquad (9.59)$$

where each connection is encoded by a pair of numbers: the strength of synapse s_{ij} and its phase ψ_{ij}. It is convenient to represent such a synapse by a complex number $c_{ij} = s_{ij} e^{i\psi_{ij}} \in \mathbb{C}$. Theorem 9.15 imposes the additional requirement that $\psi_{ij} = -\psi_{ji}$, which means that the complex-valued synaptic matrix $C = (c_{ij})$, is self-adjoint; i.e., $c_{ij} = \bar{c}_{ji}$, where $\bar{c}$ is the complex conjugate of c. We encounter this requirement when we prove an analogue of Theorem 9.15 for a WCNN near a multiple Andronov-Hopf bifurcation (Theorem 10.5).

Learning Rule

There is an intimate relation between the phase model of the form (9.59) and the canonical model for multiple Andronov-Hopf bifurcations in WC-NNs, which is derived in Section 5.4 and analyzed in Chapter 10. The former can be obtained from the latter if we discard the amplitudes of oscillators and consider only their phases. In this case it is convenient to represent the activity of the ith oscillator φ_i as a complex number

$$z_i = e^{\mathrm{i}\varphi_i} \in \mathbb{C} .$$

In Hoppensteadt and Izhikevich (1996b, see also Section 13.4.1) we use Hebb's assumptions to derive a synaptic modification rule

$$c'_{ij} = -\gamma c_{ij} + k z_i \bar{z}_j ,$$

where $\gamma > 0$ denotes the rate of forgetting, $k > 0$ is a parameter, and $\bar{z}_j$ is the complex conjugate of z_j. We can use the representation $c_{ij} = s_{ij}e^{\mathrm{i}\psi_{ij}}$ to rewrite the modification rule in the form

$$\begin{aligned}
s'_{ij} &= -\gamma s_{ij} + k \cos(\varphi_i - \varphi_j - \psi_{ij}) \\
\psi'_{ij} &= \frac{k}{s_{ij}} \sin(\varphi_i - \varphi_j - \psi_{ij}) .
\end{aligned}$$

If, during the learning period, the ith and jth oscillators are synchronized, then c_{ij} memorizes the phase difference. That is, if $\varphi_i(\tau) - \varphi_j(\tau) = \chi^\star$, then $\psi_{ij}(\tau) \to \chi^\star$ (memorization of phase information) and $s_{ij}(\tau) \to k/\gamma$ (increase of synaptic strength). On the other hand, if the oscillators are incoherent, then $\sin(\varphi_i(\tau) - \varphi_j(\tau) - \psi_{ij})$ and $\cos(\varphi_i(\tau) - \varphi_j(\tau) - \psi_{ij})$ may average to zero. In this case, ψ_{ij} is relatively unaffected (persistence of phase memory), but $s_{ij} \to 0$ (synaptic fading). The learning rule establishes the relation $c_{ij} = \bar{c}_{ji}$ so that the convergence theorem (Theorem 9.15) applies.

Finally, note that it is still an open question whether or not the adjusted Kuramoto model (9.59) with the learning rule above are canonical models for weakly connected limit-cycle oscillators having nearly identical natural frequencies.

9.5   Time-Dependent External Input ρ

In this section we study the weakly connected, weakly forced oscillatory system

$$\dot{\theta} = \Omega(\lambda) + \varepsilon g(\theta, \lambda, \rho(t), \varepsilon) , \qquad (9.60)$$

where the external input $\rho(t)$ is no longer assumed to be a constant. It is still an open problem how to obtain such a system from the WCNN

$$\dot{X} = F(X, \lambda) + \varepsilon G(X, \lambda, \rho(t), \varepsilon)$$

for general $\rho(t)$, although periodic or quasi-periodic $\rho(t)$ poses no difficulties. Indeed, suppose $\rho(t) = r(\omega_1 t, \dots, \omega_m t)$ for some function r periodic in each argument. Then we introduce dummy variables $\eta_i \in \mathbb{S}^1$, $i = 1, \ldots, m$, and rewrite the WCNN in the autonomous form

$$\begin{cases} \dot{X}_i = F_i(X_i, \lambda) + \varepsilon G_i(X, \lambda, r(\eta_1, \ldots, \eta_m), \varepsilon) , & i = 1, \ldots, n \\ \dot{\eta}_i = \omega_i , & i = 1, \ldots, m \end{cases}$$

and we proceed as in the general case to reduce this system to (9.60).

The case of chaotic input does not pose a problem either, provided that the signal $\rho(t)$ is generated by a dynamical system, say

$$\dot{\rho} = R(\rho) , \qquad \rho \in \mathcal{R} .$$

Then the WCNN can be written in the autonomous form

$$\begin{cases} \dot{X}_i = F_i(X_i, \lambda) + \varepsilon G_i(X, \lambda, \rho, \varepsilon) , & i = 1, \ldots, n \\ \dot{\rho} = R(\rho) \end{cases}$$

and analyzed using the invariant manifold reduction, where the invariant manifold has the form $M \times \mathcal{R}$, $M \cong \mathbb{T}^n$.

To apply averaging theory to (9.60) we rewrite the system in new coordinates $\theta = \Omega t + \phi$ as

$$\dot{\phi} = \varepsilon g(\Omega t + \phi, \lambda, \rho(t), \varepsilon) .$$

The averaged system has the form

$$\dot{\varphi} = \varepsilon \bar{g}(\varphi, \lambda) + \mathcal{O}(\varepsilon^2) ,$$

where

$$\bar{g}(\varphi, \lambda) = \lim_{T \to \infty} \frac{1}{T} \int_0^T g(\Omega(\lambda)t + \varphi, \lambda, \rho(t), 0) \, dt ,$$

provided that the limit exists. The integral depends not only on resonances among $\Omega_1, \ldots, \Omega_n$, but also on resonances between Ω and time-dependent input $\rho(t)$. That is, it could be nonconstant even when Ω is nonresonant (contrast with Theorem 9.7) or vanish for resonant Ω (contrast with Theorem 9.9).

We do not study here this issue for general $\rho(t)$. The case of quasiperiodic ρ poses no problem, since it is reducible to the $(n+m)$-dimensional oscillatory system

$$\begin{cases} \dot{\theta} = \Omega(\lambda) + \varepsilon g(\theta, \lambda, r(\eta), \varepsilon) \\ \dot{\eta} = \omega \end{cases} ,$$

where $\eta \in \mathbb{T}^m$ is a dummy variable. The results developed earlier are applicable to this system, provided that the resonances are searched for the extended vector of frequencies $\tilde{\Omega} = (\Omega_1, \ldots, \Omega_n, \omega_1, \ldots, \omega_m)^\top \in \mathbb{R}^{n+m}$.

9.5.1 Conventional Synapses

In this section we consider the oscillatory neural network

$$\dot\theta_i = \Omega_i + \varepsilon \sum_{j=1}^{n} g_{ij}(\theta_i, \theta_j, \rho(t)) + \mathcal{O}(\varepsilon^2) . \tag{9.61}$$

Notice that such a connection architecture violates the assumption that the synapses are conventional; see Section 9.4. We interpret this as follows: The synaptic contacts between neurons from the network are conventional, but the synapses from the other brain structures that carry the signal $\rho(t)$ may be unconventional, as in Figure 9.17. Such synapses modulate interactions between neurons.

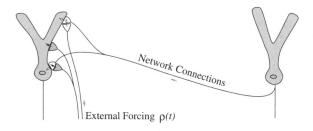

Figure 9.17. System (9.61) describes a network of oscillating neurons having conventional synapses. External forcing may modulate the network's activity via unconventional synapses.

Dynamical features of (9.61) when the external input ρ is constant are known: Oscillators having incommensurable frequencies do not interact, and those having barely commensurable frequencies interact ineffectively. Can time-dependent external forcing $\rho(t)$ change the qualitative picture? The answer is YES.

For the sake of illustration consider the case of periodic external input $\rho(t) = \Omega_0 t + \phi_0$. Proceeding as in the proof of Theorem 9.12 we find that interactions between the ith and the jth oscillators are governed by the function

$$\bar g_{ij}(\varphi_i, \varphi_j, \varphi_0) = \lim_{T \to \infty} \frac{1}{T} \int_0^T g_{ij}(\Omega_i t + \varphi_i, \ \Omega_j t + \varphi_j, \ \Omega_0 t + \varphi_0) \, dt , \tag{9.62}$$

which depends crucially on the resonances among Ω_i, Ω_j, and Ω_0.

Incommensurable Frequencies

Suppose frequencies Ω_i and Ω_j are incommensurable, so that the oscillators would not interact without external input $\rho(t) = \Omega_0 t + \varphi_0$. The input can link them if Ω_0 is chosen such that the vector $(\Omega_i, \Omega_j, \Omega_0)^\top \in \mathbb{R}^3$ is resonant. There may be various cases:

- Ω_0 is commensurable with Ω_i; that is, there are relatively prime k and m such that $k\Omega_0 - m\Omega_i = 0$. Then, according to Theorem 9.6 the function $\bar{g}_{ij}$ defined by (9.62) has the form

$$\bar{g}_{ij}(\varphi_i, \varphi_j, \varphi_0) = H(k\varphi_0 - m\varphi_i)$$

for some H. We see that the ith oscillator is affected by the external input, but it is not affected by the jth oscillator.

- Ω_0 is commensurable with Ω_j. Suppose $k\Omega_0 - m\Omega_j = 0$. Then

$$\bar{g}_{ij}(\varphi_i, \varphi_j, \varphi_0) = H(k\varphi_0 - m\varphi_j) .$$

Thus, the ith oscillator is affected by both the jth oscillator and the external input, but the efficacy does not depend on its own activity φ_i. It is still an open question how to interpret this result and what might be its use for brain dynamics.

- Ω_0 is commensurable with some linear combination of Ω_i and Ω_j. Hence $k_i\Omega_i + k_j\Omega_j + k_0\Omega_0 = 0$ for some integers, and

$$\bar{g}_{ij}(\varphi_i, \varphi_j, \varphi_0) = H(k_i\varphi_i + k_j\varphi_j + k_0\varphi_0) \tag{9.63}$$

for some H. We see that external input can link the oscillators having incommensurable frequencies.

Notice that there are many values of Ω_0 that can link the oscillators. To make the interactions as strong as possible, the sum $|k_i| + |k_j| + |k_0|$ must be as small as possible. Its smallest value is 3, which is achieved when $\Omega_0 = \Omega_i + \Omega_j$ or $\Omega_0 = \pm\Omega_i \mp \Omega_j$.

Barely Commensurable Frequencies

Periodic external input can amplify connections between oscillators having barely commensurable frequencies. Indeed, suppose $k\Omega_j - m\Omega_i = 0$ only for very large $k \mid m$. Without external input this leads to an averaged function of the form $H(k\varphi_j - m\varphi_i)$ that is small (Proposition 9.14). If the frequency of the external input is tuned so that $\Omega_0 = \Omega_i + \Omega_j$, then the resonance $\Omega_i + \Omega_j - \Omega_0 = 0$ has small order (order 3), which can increase substantially the strength of interactions.

Commensurable Frequencies

When the frequencies Ω_i and Ω_j are commensurable, say $k\Omega_j - m\Omega_i = 0$ for some small integers k and m, the external periodic input cannot remove the resonance $k\Omega_j - m\Omega_i + 0\Omega_0 = 0$, but it could add a new one. This makes the vector $(\Omega_i, \Omega_j, \Omega_0)^\top \in \mathbb{R}^3$ proportional to a vector of integers. Since the resonance group has rank 2 in this case, there is an ambiguity in the choice of its generators. The choice affects the form of connection function. For example, it can be

$$\bar{g}_{ij}(\varphi_i, \varphi_j, \varphi_0) = H(k\varphi_j - m\varphi_i, \, k_0\varphi_0 - k_i\varphi_i)$$

or

$$\bar{g}_{ij}(\varphi_i, \varphi_j, \varphi_0) = H(k\varphi_j - m\varphi_i, \, k_j\varphi_j - k_0\varphi_0) \,.$$

Any such function can be transformed to another one by an invertible change of variables.

Finally, notice that when it is necessary to link many oscillators with pairwise incommensurable frequencies, periodic input cannot cope with this task. The input must be quasi-periodic or chaotic in this case.

9.6   Appendix: The Malkin Theorem

Below we provide a general statement of Malkin's theorem (Malkin 1949, 1956). The theorem can be generalized and applied to discontinuous systems (Kolovskii 1960, see also Blechman 1971 p. 215).

Theorem 9.16 (Malkin) *Consider a T-periodic dynamical system of the form*

$$\dot{X} = F(X, t) + \varepsilon G(X, t) \,, \qquad X \in \mathbb{R}^m \,, \tag{9.64}$$

and suppose that the unperturbed system, $\dot{X} = F(X, t)$, has a k-parameter family of T-periodic solutions

$$X(t) = U(t, \alpha) \,,$$

where $\alpha = (\alpha_1, \ldots, \alpha_k)^\top \in \mathbb{R}^k$ is a vector of independent parameters, which implies that the rank of the $n \times k$ matrix $D_\alpha U$ is k. Suppose the adjoint linear problem

$$\dot{Q}_i = - \{DF(U(t, \alpha))\}^\top Q_i$$

has exactly k independent T-periodic solutions $Q_1(t, \alpha), \ldots, Q_k(t, \alpha) \in \mathbb{R}^m$. Let Q be the matrix whose columns are these solutions such that

$$Q^\top D_\alpha U = I \,, \tag{9.65}$$

where I is the identity $k \times k$ matrix. Then the perturbed system (9.64) has a solution of the form

$$X(t) = U(t, \alpha(\varepsilon t)) + \mathcal{O}(\varepsilon) \,,$$

where

$$\frac{d\alpha}{d\tau} = \frac{1}{T} \int_0^T Q(t, \alpha)^\top G(U(t, \alpha), t) \, dt \,, \tag{9.66}$$

where $\tau = \varepsilon t$ is slow time. If (9.66) has a stable equilibrium, then system (9.64) has a T-periodic solution.

Proof. Let us substitute

$$X(t) = U(t, \alpha(\varepsilon t)) + \varepsilon P(t) + \mathcal{O}(\varepsilon^2)$$

into (9.64) to obtain

$$\dot{U} + \varepsilon D_\alpha U \frac{d\alpha}{d\tau} + \varepsilon \dot{P} + \mathcal{O}(\varepsilon^2) = F(U, t) + \varepsilon D_X F(U, t) P + \varepsilon G(U, t) + \mathcal{O}(\varepsilon^2) .$$

Since $\dot{U} = F(U, t)$, we have

$$\dot{P} = D_X F(U, t) P + G(U, t) - D_\alpha U \frac{d\alpha}{d\tau} ,$$

which we rewrite in the form

$$\dot{P} = A(t, \alpha) P + b(t, \alpha) , \tag{9.67}$$

where

$$A(t, \alpha) = D_X F(U(t, \alpha), t) \quad \text{and} \quad b(t, \alpha) = G(U(t, \alpha), t) - D_\alpha U(t, \alpha) \frac{d\alpha}{d\tau} .$$

Using the Fredholm alternative we conclude that the linear T-periodic equation (9.67) has a unique solution if and only if the following orthogonality condition

$$\langle q, b \rangle = \int_0^T q(t, \alpha)^\top b(t, \alpha) \, dt = 0$$

is satisfied for any solution $q(t, \alpha) \in \mathbb{R}^m$ of the linear homogeneous adjoint system

$$\dot{q} = -A^\top(t, \alpha) q .$$

According to the conditions of the theorem this system has exactly k independent solutions $Q_1, \ldots, Q_k \in \mathbb{R}^m$, which satisfy (9.65). We rewrite the orthogonality condition in matrix form as

$$
\begin{aligned}
\langle Q, b \rangle &= \int_0^T Q(t, \alpha)^\top b(t, \alpha) \, dt \\
&= \int_0^T Q(t, \alpha)^\top \left(G(U(t, \alpha), t) - D_\alpha U(t, \alpha) \frac{d\alpha}{d\tau} \right) dt \\
&= 0 .
\end{aligned}
$$

Now notice that $Y(t, \alpha) = D_\alpha U(t, \alpha)$ is a matrix solution to the homogeneous linear problem $\dot{Y} = A(t, \alpha) Y$. One can easily check (by differentiating, as we did in the proof of Theorem 9.2) that

$$Q^\top(t, \alpha) Y(t, \alpha) \equiv Q^\top(0, \alpha) Y(0, \alpha) = I .$$

Therefore, the orthogonality condition above can be written in the form (9.66). $\square$

It is easy to see that Theorem 9.2 is a direct consequence of Malkin's Theorem 9.16. Indeed, one takes $F(X, t) = (F_1(X_1), \ldots, F_n(X_n))$ in this case.

Part III

Analysis of Canonical Models

10
Multiple Andronov-Hopf Bifurcation

A WCNN of the form

$$\dot{X}_i = F_i(X_i, \lambda) + \varepsilon G_i(X, \lambda, \rho, \varepsilon)$$

near a multiple Andronov-Hopf bifurcation point was shown (Theorem 5.8) to have a canonical model of the form

$$z_i' = b_i z_i + d_i z_i |z_i|^2 + \sum_{j \neq i}^{n} c_{ij} z_j, \qquad (10.1)$$

where $' = d/d\tau$, τ is slow time, and $b_i, c_{ij}, d_i, z_i \in \mathbb{C}$. In this chapter we study general properties of this canonical model. In particular, we are interested in the stability of the origin $z_1 = \cdots = z_n = 0$ and in the possibility of in-phase and anti-phase locking.

10.1 Preliminary Analysis

It is sometimes convenient to rewrite (10.1) in polar coordinates: Let

$$z_i = r_i e^{\mathrm{i}\varphi_i}, \quad b_i = \alpha_i + \mathrm{i}\omega_i, \quad d_i = \sigma_i + \mathrm{i}\gamma_i, \quad c_{ij} = |c_{ij}| e^{\mathrm{i}\psi_{ij}}.$$

Then

$$
\begin{aligned}
z_i' &= r_i' e^{\mathrm{i}\varphi_i} + \mathrm{i}\varphi_i' r_i e^{\mathrm{i}\varphi_i} \\
&= (\alpha_i + \mathrm{i}\omega_i) r_i e^{\mathrm{i}\varphi_i} + (\sigma_i + \mathrm{i}\gamma_i) r_i^3 e^{\mathrm{i}\varphi_i} + \sum_{j \neq i}^{n} |c_{ij}| r_j e^{\mathrm{i}(\varphi_j + \psi_{ij})}.
\end{aligned}
$$

Dividing everything by $e^{\mathrm{i}\varphi_i}$ gives

$$r_i' + \mathrm{i}\varphi_i' r_i = (\alpha_i + \mathrm{i}\omega_i)r_i + (\sigma_i + \mathrm{i}\gamma_i)r_i^3 + \sum_{j \neq i}^n |c_{ij}|r_j e^{\mathrm{i}(\varphi_j + \psi_{ij} - \varphi_i)} .$$

The real and imaginary parts of this equation can be written separately as

$$\begin{cases} r_i' = \alpha_i r_i + \sigma_i r_i^3 + \sum |c_{ij}|r_j \cos(\varphi_j + \psi_{ij} - \varphi_i) \\ \varphi_i' = \omega_i + \gamma_i r_i^2 + \frac{1}{r_i} \sum |c_{ij}|r_j \sin(\varphi_j + \psi_{ij} - \varphi_i) . \end{cases}$$

This representation of the canonical model is useful, since it explains the meanings of the parameters.

10.1.1 A Single Oscillator

Let us discuss the meaning of each parameter and variable in this system. For this let us consider the (uncoupled) equation describing the dynamics of a single oscillator

$$\begin{cases} r' = \alpha r + \sigma r^3 \\ \varphi' = \omega + \gamma r^2 . \end{cases} \tag{10.2}$$

We see that activity of the variable r does not depend on that of the variable φ.

Parameters α and σ

The equation

$$r' = \alpha r + \sigma r^3$$

is the canonical model for a pitchfork bifurcation (see Section 2.2.3) that occurs at $\alpha = 0$. The parameter σ affects the criticality of the bifurcation; see Figure 10.1.

A supercritical ($\sigma < 0$) pitchfork bifurcation in the system above corresponds to a supercritical Andronov-Hopf bifurcation in (10.2). The origin is stable, and there are no oscillations for $\alpha < 0$. The origin is unstable, and there are stable oscillations of radius $\sqrt{-\alpha/\sigma}$ for positive α. Increasing α increases the radius of stable oscillations.

A subcritical ($\sigma > 0$) pitchfork bifurcation in the system above corresponds to a subcritical Andronov-Hopf bifurcation in (10.2). The origin is stable, and there are unstable oscillations of radius $\sqrt{-\alpha/\sigma}$ for negative α. Increasing α shrinks the radius of unstable oscillations. The unstable limit cycle coalesces with the origin and disappears when $\alpha = 0$. The origin is unstable for $\alpha > 0$.

Since the radius r must be a nonnegative number, it is customary not to draw the branch $r < 0$ in the bifurcation diagrams; see bottom of Figure 10.1.

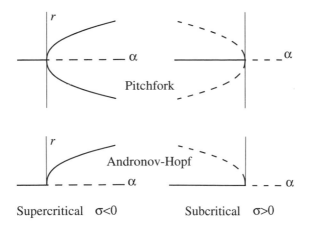

Figure 10.1. Two types of criticality of the pitchfork and Andronov-Hopf bifurcations. The radius r is always positive in the Andronov-Hopf bifurcation.

Parameters ω and γ

The parameter ω, which is called the *center frequency*, affects the frequency of oscillations of the canonical model

$$\begin{cases} r' = \alpha r + \sigma r^3 \\ \varphi' = \omega + \gamma r^2 \end{cases},$$

and the parameter γ, which is called the *shear* parameter, determines how the frequency depends on the amplitude.

A peculiar phenomenon occurs when $\omega\gamma < 0$: The oscillations described by the model above can change their directions (see Figure 10.2). Indeed,

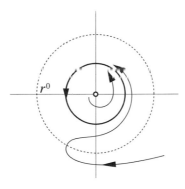

Figure 10.2. The oscillations can change direction when $\omega\gamma < 0$ in the canonical model.

there is $r^0 = \sqrt{|\omega/\gamma|}$ such that φ' has different signs for $r(\tau) > r^0$ and for $r(\tau) < r^0$. In particular, if $r^0 = \sqrt{|\alpha/\sigma|}$, then $\varphi(\tau) = \text{const}$; that is, the oscillator has a nonzero amplitude, but does not oscillate.

This phenomenon is a property of the model above, but not necessarily of the original system $\dot{X} = F(X, \lambda)$. The phase φ of z measures the deviation of phase of the oscillations of $X(t)$ from the natural oscillations Ωt. Therefore, the phenomenon above (phase $\varphi(\tau) = \text{const}$) corresponds to $X(t)$ oscillating precisely with the natural frequency Ω.

10.1.2 Complex Synaptic Coefficients c_{ij}

Let us return to the canonical model (10.1) written in the form

$$\begin{cases} r_i' = \alpha_i r_i + \sigma_i r_i^3 + \sum |c_{ij}| r_j \cos(\varphi_j + \psi_{ij} - \varphi_i) \\ \varphi_i' = \omega_i + \gamma_i r_i^2 + \frac{1}{r_i} \sum |c_{ij}| r_j \sin(\varphi_j + \psi_{ij} - \varphi_i) \end{cases}.$$

The complex coefficient $c_{ij} = |c_{ij}| e^{i\psi_{ij}}$ describes the amplitude and polarity of the connection from the jth to the ith oscillator. $|c_{ij}|$ gives the synaptic strength, and $\psi_{ij} = \text{Arg}\, c_{ij}$ encodes phase information about the synaptic connection, which we call the polarity or *natural phase difference*.

Proposition 10.1 (Natural Phase Difference) *The natural phase difference $\psi_{12} = Arg\, c_{12}$ occurs between two oscillators connected in one direction, say from z_2 to z_1, and having equal center frequencies $\omega_1 = \omega_2 = \omega$ and no shear, viz., in the system*

$$\begin{cases} r_1' = \alpha_1 r_1 + \sigma_1 r_1^3 + |c_{12}| r_2 \cos(\varphi_2 + \psi_{12} - \varphi_1) \\ r_2' = \alpha_2 r_2 + \sigma_2 r_2^3 \\ \varphi_1' = \omega + \frac{1}{r_1} |c_{12}| r_2 \sin(\varphi_2 + \psi_{12} - \varphi_1) \\ \varphi_2' = \omega \end{cases}.$$

Proof. Let $\chi = \varphi_1 - \varphi_2$. From the last two equations we obtain

$$\chi' = \frac{1}{r_1} |c_{12}| r_2 \sin(\psi_{12} - \chi)$$

from which we see that $\chi(\tau) \to \psi_{12}$. $\square$

This proposition motivates our definition of the natural phase difference, since it occurs "naturally" between a pair of oscillators connected in one direction. We cannot relax the assumption that the oscillators have equal center frequencies. Indeed, if $\omega_1 \neq \omega_2$, then we obtain

$$\chi' = \omega_1 - \omega_2 + a \sin(\psi_{12} - \chi),$$

where $a = r_1^{-1} |c_{12}| r_2 > 0$. When the center frequencies are not too far apart (namely, $|\omega_1 - \omega_2| \leq a$), the equation above has a stable equilibrium

$$\chi^\star = \psi_{12} + \sin^{-1}(\omega_1 - \omega_2)/a,$$

and the oscillators synchronize with a specific phase difference $\chi^\star$ that is in a $\pi/2$-neighborhood of the natural phase difference ψ_{12}. When the center frequencies are different enough ($|\omega_1 - \omega_2| > a$), the equation above does not have equilibria. The oscillators do not synchronize; they *drift* relative to each other. Additional analysis shows that they may phase lock, but on different frequencies. We call such phase locking $p : q$ phase locking, where p and q are relatively prime positive integers, and the rational number p/q is the ratio of frequencies on which they lock.

Remark. When $\psi_{12} < 0$, an observer sees that one of the oscillators (in this case z_1) oscillates with some delay. Obviously, it would be incorrect to ascribe this to spike propagation or synaptic transmission delays. As we will see later, the coefficient ψ_{12} may take on many values, depending on the synaptic organization of the network, but not the speed of transmission of spikes through axons or dendrites.

If the neurons are connected in both directions (i.e., $c_{12} \neq 0$ and $c_{21} \neq 0$), then the phase difference between them generically differs from $\operatorname{Arg} c_{12}$ or $\operatorname{Arg} c_{21}$ even when they have equal center frequencies (we study this case later in this chapter). Nevertheless, for a network of n interconnected neurons we can prove the following result:

Lemma 10.2 *If $c_{ij} \neq 0$, then there are values of the parameters $\alpha_1, \ldots, \alpha_n$ such that*

$$\varphi_i(\tau) \to \varphi_j(\tau) + \psi_{ij} ; \qquad (10.3)$$

i.e., the ith and jth oscillators have constant phase differences ψ_{ij}.

Notice that we do not require the equality of center frequencies ω_i in this case.

Proof. Fix j and let $\alpha_j = 1, \alpha_i = -\frac{1}{\delta}$ for $i \neq j$, where $\delta \ll 1$. After the rescaling ($z_i \to \delta z_i$ for $i \neq j$), the system (10.1) transforms to

$$\begin{cases} z_j{}' = (1 + \mathrm{i}\omega_j)z_j + d_j z_j |z_j|^2 + \mathcal{O}(\delta) \\ z_i{}' = \frac{1}{\delta}(c_{ij}z_j - z_i) + \mathcal{O}(1), \qquad i \neq j \end{cases} .$$

We see that the jth oscillator forces the others. Applying singular perturbation methods (Hoppensteadt 1993) to the ith equation, we see that

$$z_i(\tau) \to c_{ij}z_j(\tau) + \mathcal{O}(\delta) .$$

After a short initial transient period this equation gives (10.3) for $\delta \to 0$. $\square$

It follows from this proof that the choice of parameters corresponds to the case when the amplitude of the jth oscillator is much bigger than that of the others. One can consider the jth oscillator as the leading one that synchronizes the whole network. A similar phenomenon was studied by Kazanovich and Borisyuk (1994).

It is easy to see that the impact of one oscillator on the amplitude of another is maximal when their phases are locked with the natural phase difference. Indeed,

$$\cos(\varphi_j + \psi_{ij} - \varphi_i) = \cos(\psi_{ij} - \psi_{ij}) = 1$$

reaches its maximal value then. If the two oscillators are completely out of phase, i.e. if

$$\varphi_i(\tau) - \varphi_j(\tau) = \psi_{ij} \pm \frac{\pi}{2} \,,$$

then $\cos(\varphi_j + \psi_{ij} - \varphi_i) = 0$, and the influence of jth oscillator on the amplitude r_i of ith one is negligible even when $|c_{ij}|$ is very large! Such "phase gaps" play crucial role in Dynamic Link Architecture theory (Von der Malsburg 1995).

There is another interesting observation: The larger an oscillator's amplitude r_j, the larger the oscillator's impact on the other oscillators. Conversely, if the ith oscillator has very small amplitude r_i, then it is susceptible to influences from the other oscillators because of the term

$$\frac{1}{r_i} \sum_{\substack{j \neq i}}^{n} |c_{ij}| r_j \sin(\varphi_j + \psi_{ij} - \varphi_i) \,,$$

which can grow as $r_i \to 0$.

We next study some local properties of (10.1), in particular, the stability of the origin $z_1 = \cdots = z_n = 0$.

10.2   Oscillator Death and Self-Ignition

Note that the canonical model (10.1) always has an equilibrium point $z_1 = \cdots = z_n = 0$ for any choice of parameters. If all α_i are negative numbers with sufficiently large absolute values, then the equilibrium point is stable.

In this section we study how the equilibrium can lose its stability. Using the canonical model, we also illustrate two well-known phenomena: oscillator death (or quenching, or the Bar-Eli effect) and coupling-induced spontaneous activity (or self-ignition). We consider (10.1) for the most interesting case, when $\operatorname{Re} d_i = \sigma_i < 0$ for all i. This corresponds to a supercritical Andronov-Hopf bifurcation; i.e., to the birth of a stable limit cycle.

We start from the observation that in the canonical model each oscillator is governed by a dynamical system of the form

$$z' = (\alpha + i\omega)z + (\sigma + i\gamma)z|z|^2 \,. \tag{10.4}$$

Obviously, if $\alpha \leq 0$, then the equilibrium $z = 0$ is stable. As α increases through $\alpha = 0$, equation (10.4) undergoes Andronov-Hopf bifurcation. As a result, the equilibrium loses its stability, and (10.4) has a stable limit cycle of radius $\mathcal{O}(\sqrt{\alpha})$ for $\alpha > 0$.

We can characterize the qualitative differences in dynamic behavior of (10.4) for $\alpha \leq 0$ and $\alpha > 0$ as follows:

- When $\alpha \leq 0$, the dynamical system (10.4) describes an *intrinsically passive* element incapable of sustaining periodic activity.

- When $\alpha > 0$, the dynamical system (10.4) describes an *intrinsically active* oscillator, or pacemaker.

In the uncoupled network ($C = 0$)

$$z_i' = (\alpha + \mathrm{i}\omega_i)z_i + (\sigma_i + \mathrm{i}\gamma_i)z_i|z_i|^2 , \quad i = 1,\ldots,n ,$$

of such oscillators, the equilibrium point $z_1 = \cdots = z_n = 0$ is stable for $\alpha \leq 0$ and unstable for $\alpha > 0$.

We ask what happens when we consider the canonical model (10.1) with nonzero matrix $C = (c_{ij})$. A partial answer is given in the following result.

Lemma 10.3 *Let b denote the largest real part of all the eigenvalues of the connection matrix $C = (c_{ij})$. Consider the network of identical oscillators governed by*

$$z_i' = (\alpha + \mathrm{i}\omega)z_i + (\sigma_i + \mathrm{i}\gamma_i)z_i|z_i|^2 + \sum_{j=1}^{n} c_{ij}z_j , \quad \sigma < 0 , \quad i = 1,\ldots,n.$$

The equilibrium point $z_1 = \cdots = z_n = 0$ is stable if $\alpha < -b$ and unstable if $\alpha > -b$.

Proof. The full system has the form

$$\begin{cases} z_i' &= (\alpha + \mathrm{i}\omega)z_i + (\sigma_i + \mathrm{i}\gamma_i)z_i|z_i|^2 + \sum_{j=1}^{n} c_{ij}z_j \\ \bar{z}_i' &= (\alpha - \mathrm{i}\omega)\bar{z}_i + (\sigma_i - \mathrm{i}\gamma_i)\bar{z}_i|z_i|^2 + \sum_{j=1}^{n} \bar{c}_{ij}\bar{z}_j \end{cases} , \quad i = 1,\ldots,n .$$

$$(10.5)$$

It is easy to see that the origin $z_1 = \bar{z}_1 = \cdots = z_n = \bar{z}_n = 0$ is always an equilibrium point of (10.5). The $(2n) \times (2n)$ Jacobian matrix J at the origin has the form

$$J = \begin{pmatrix} (\alpha + \mathrm{i}\omega)I + C & 0 \\ 0 & (\alpha - \mathrm{i}\omega)I + \bar{C} \end{pmatrix} ,$$

where 0 and I are the $n \times n$ zero and identity matrices, respectively.

Suppose $\lambda_1,\ldots,\lambda_n$ are eigenvalues of C counted with their multiplicity, and suppose that $v_1,\ldots,v_n$ are corresponding (generalized) eigenvectors. Direct computation shows that J has $2n$ eigenvalues

$$\alpha + \mathrm{i}\omega + \lambda_1, \quad \alpha - \mathrm{i}\omega + \bar{\lambda}_1, \quad \ldots, \quad \alpha + \mathrm{i}\omega + \lambda_n, \quad \alpha - \mathrm{i}\omega + \bar{\lambda}_n$$

and $2n$ corresponding eigenvectors

$$\begin{pmatrix} v_1 \\ 0 \end{pmatrix}, \begin{pmatrix} 0 \\ \bar{v}_1 \end{pmatrix}, \dots, \begin{pmatrix} v_n \\ 0 \end{pmatrix}, \begin{pmatrix} 0 \\ \bar{v}_n \end{pmatrix},$$

where 0 denotes a vector of zeros. Stability of (10.5) is determined by the eigenvalues of J with maximal real part. These eigenvalues are of the form

$$\alpha + \mathrm{i}\omega + \lambda, \quad \alpha - \mathrm{i}\omega + \bar{\lambda},$$

where λ are eigenvalues of C. Let $b = \operatorname{Re}\lambda$; then the origin is stable if

$$\alpha + b < 0$$

and unstable if

$$\alpha + b > 0.$$

□

In the rest of this section we use Lemma 10.3 to illustrate two interesting effects.

- If $b < 0$, then the neural network is stable even when

$$0 < \alpha < -b.$$

  That is, even though each oscillator is a pacemaker, the coupled system may approach $z = 0$. This effect, which can be called *oscillator death*, was studied numerically by Bar-Eli (1985) and analytically for general systems by Aronson et al. (1990).

- If $b > 0$, then the neural network can exhibit spontaneous activity even when

$$-b < \alpha < 0;$$

  i.e., when each oscillator is intrinsically passive, coupling can induce synchronous activity in the network. This effect, which can be called *self-ignition* (Rapaport 1952), is discussed in details by Kowalski et al. (1992). Smale (1974) interpreted this phenomena in terms of live and dead cells: Two non-oscillating (dead) cells start to oscillate (become alive) when they are coupled together.

Remark. If C has only one eigenvalue with maximal real part, then the canonical model undergoes an Andronov-Hopf bifurcation as α increases through $\alpha = -b$. In this case the coordinates of the limit cycle depend upon the matrix C, more precisely, upon the eigenvector that corresponds to the "leading" eigenvalue of C. Thus, to understand the dynamics of the canonical model, one should understand possible structures of the connection matrix $C = (c_{ij})$. We will do this in Chapter 13.

10.3   Synchronization of Two Identical Oscillators

The study of synchronization of n oscillators of the form (10.1) is a difficult problem, even when $n = 2$ (see Aronson et al. 1990). It is a feasible task if we restrict ourselves to the case of two identical oscillators. In this section we study such a pair; in particular, we are interested when in-phase or anti-phase synchronization is possible.

Theorem 10.4 (In- and Anti-phase Locking of Two Identical Oscillators)   *Consider a pair of weakly connected identical oscillators at a multiple supercritical Andronov-Hopf bifurcation, which is governed by the canonical model*

$$\begin{cases} z_1' = (\alpha + i\omega)z_i + (\sigma + i\gamma)z_1|z_1|^2 + cz_2 \\ z_2' = (\alpha + i\omega)z_2 + (\sigma + i\gamma)z_2|z_2|^2 + cz_1 \ . \end{cases} \tag{10.6}$$

for $\sigma < 0$. Then

- *The oscillators are in-phase synchronized ($z_1 = z_2$) if*

$$\alpha + \operatorname{Re} c \ > \ 0 \tag{10.7}$$
$$\alpha + 3\operatorname{Re} c \ > \ 0 \tag{10.8}$$
$$(\alpha + \operatorname{Re} c)(\gamma/\sigma \operatorname{Im} c + \operatorname{Re} c) + |c|^2 \ > \ 0 \ . \tag{10.9}$$

- *The oscillators are anti phase synchronized ($z_1 = -z_2$) if*

$$\alpha - \operatorname{Re} c \ > \ 0 \tag{10.10}$$
$$\alpha - 3\operatorname{Re} c \ > \ 0 \tag{10.11}$$
$$(\alpha - \operatorname{Re} c)(-\gamma/\sigma \operatorname{Im} c - \operatorname{Re} c) + |c|^2 \ > \ 0 \ . \tag{10.12}$$

Proof. First, we rewrite the system (10.6) in polar coordinates:

$$\begin{cases} r_1' = \alpha r_1 + \sigma r_1^3 + |c|r_2 \cos(\varphi_2 + \psi - \varphi_1) \\ r_2' = \alpha r_2 + \sigma r_2^3 + |c|r_1 \cos(\varphi_1 + \psi - \varphi_2) \\ \varphi_1' = \omega + \gamma r_1^2 + r_1^{-1}|c|r_2 \sin(\varphi_2 + \psi - \varphi_1) \\ \varphi_2' = \omega + \gamma r_2^2 + r_2^{-1}|c|r_1 \sin(\varphi_1 + \psi - \varphi_2) \end{cases} \ .$$

Let $\chi = \varphi_1 - \varphi_2$ be the phase difference between the oscillators. Then, the system above can be rewritten in the form

$$\begin{cases} r_1' = \alpha r_1 + \sigma r_1^3 + |c|r_2 \cos(\psi - \chi) \\ r_2' = \alpha r_2 + \sigma r_2^3 + |c|r_1 \cos(\psi + \chi) \\ \chi' = \gamma(r_1^2 - r_2^2) + |c| \left(r_1^{-1}r_2 \sin(\psi - \chi) - r_2^{-1}r_1 \sin(\psi + \chi) \right) \end{cases} \ . \tag{10.13}$$

The in-phase locking corresponds to $\chi = 0$. The anti-phase locking corresponds to $\chi = \pi$. If we exclude the degenerate case when $\sigma + i\gamma$ is proportional to c, then the in- and anti-phase solutions satisfy

$$\begin{aligned} r^2 \equiv r_1^2 = r_2^2 = -\sigma^{-1}(\alpha + \operatorname{Re} c) & \quad \text{(in-phase locking } \chi = 0), \\ r^2 \equiv r_1^2 = r_2^2 = -\sigma^{-1}(\alpha - \operatorname{Re} c) & \quad \text{(anti-phase locking } \chi = \pi), \end{aligned}$$

where $\operatorname{Re} c = |c| \cos \psi$ is used for simplicity of notation. To determine the stability of the in-phase locking, we linearize (10.13) at the equilibrium $(r, r, 0)$. The matrix of linearization is

$$L = \begin{pmatrix} \alpha + 3\sigma r^2 & \operatorname{Re} c & r \operatorname{Im} c \\ \operatorname{Re} c & \alpha + 3\sigma r^2 & -r \operatorname{Im} c \\ 2\gamma r - 2r^{-1} \operatorname{Im} c & -2\gamma r + 2r^{-1} \operatorname{Im} c & 2 \operatorname{Re} c \end{pmatrix}.$$

It is a daunting task to determine the eigenvalues of L. Instead, we follow Aronson et al. (1990, Proposition 5.1) and transform L to another matrix that has the same set of eigenvalues. Let

$$P = \begin{pmatrix} 1 & 1 & 0 \\ 1 & -1 & 0 \\ 0 & 0 & 1 \end{pmatrix}.$$

Then $L_1 = P^{-1}LP$, which is similar to L, has the form

$$L_1 = \begin{pmatrix} \alpha + 3\sigma r^2 + \operatorname{Re} c & 0 & 0 \\ 0 & \alpha + 3\sigma r^2 - \operatorname{Re} c & r \operatorname{Im} c \\ 0 & 4\gamma r - 4r^{-1} \operatorname{Im} c & -2 \operatorname{Re} c \end{pmatrix}.$$

This matrix has one eigenvalue

$$\alpha + 3\sigma r^2 + \operatorname{Re} c = -2(\alpha + \operatorname{Re} c).$$

It is negative when (10.7) is satisfied. The two other eigenvalues are negative if the trace of the lower 2×2 matrix, $-2(\alpha + 3 \operatorname{Re} c)$, is negative (this gives the second condition (10.8)) and its determinant is positive (this gives the last condition (10.9)).

To determine stability of the anti-phase solution (r, r, π), one needs to note that the matrix of linearization in this case has the same form as L with the only exception that c must be replaced by $-c$. Therefore, conditions (10.10), (10.11), and (10.12) can be obtained from (10.7), (10.8), and (10.9) by the replacement $c \to -c$. $\square$

Let us use $\operatorname{Re} c = |c| \cos \psi$ and $\operatorname{Im} c = |c| \sin \psi$ and rewrite the conditions (10.8) and (10.11) in the form

$$\alpha/|c| + \cos \psi > 0 ,$$
$$\alpha/|c| + 3 \cos \psi > 0 ,$$
$$(\alpha/|c| + \cos \psi)(\gamma/\sigma \sin \psi + \cos \psi) + 1 > 0 ,$$

and

$$\alpha/|c| - \cos \psi > 0 ,$$
$$\alpha/|c| - 3 \cos \psi > 0 ,$$
$$(\alpha/|c| - \cos \psi)(-\gamma/\sigma \sin \psi - \cos \psi) + 1 > 0 .$$

The sets satisfying these conditions are depicted in Figure 10.3 for the case $\gamma = 0$ (no shear). Notice that the parameter regions corresponding to the in-phase and anti-phase synchronization overlap near $\psi = \pm\pi/2$. A nonzero parameter γ distorts the picture, but it does not remove the coexistence of in-phase and anti-phase synchronization. We continue to analyze the conditions (10.8) and (10.11) in Chapter 13.

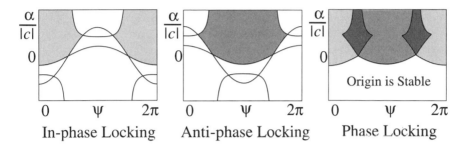

Figure 10.3. Parameter regions corresponding to various kinds of phase locking.

Each curve in Figure 10.3 denotes a set of parameters for which one of the six inequalities above becomes equality. The curves are bifurcational, since the qualitative behavior of the system changes when parameters pass through them. For example, when $\alpha/|c| + \cos\psi = 0$, an eigenvalue of the Jacobian matrix L (see the proof) crosses 0. It would be incorrect to assume that there is a saddle-node bifurcation in (10.13), even less so in (10.6), even though such a bifurcation is the most probable one in general systems having a zero eigenvalue. Careful analysis shows that (10.13) undergoes a pitchfork bifurcation, which corresponds to the Andronov-Hopf bifurcation in (10.6). We have already encountered this bifurcation (see remark on p. 304) when we studied the loss of stability at the origin $z_1 = z_2 = 0$. Other curves correspond to other bifurcations that give birth to various solutions, such as out of phase locking (i.e., the phase difference is a constant different from 0 or π) or phase trapping (i.e., frequency locking without phase locking). Some of the solutions are found analytically and numerically by Aronson et al. (1990). It is still an open problem to classify all dynamical regimes of (10.6) from the phase locking point of view.

We return to the pair of identical oscillators (10.6) in Chapter 13, where we study how the natural phase difference ψ depends on the synaptic organization of the network. That and Theorem 10.4 determine which synaptic organizations could produce in-phase and anti-phase synchronization and which could not.

10.4   Oscillatory Associative Memory

In this section we reveal the conditions under which the canonical model (10.1) can operate as a multiple attractor neural network (MA-type NN; see Section 3.2).

First, we assume that all $d_i = \sigma_i + i\gamma_i$ are real and negative. Without loss of generality we may take $d_i = -1$. Thus, we study a dynamical system of the form

$$z_i' = (\alpha_i + \omega_i)z_i - z_i|z_i|^2 + \sum_{j \neq i}^{n} c_{ij}z_j. \tag{10.14}$$

We take advantage of the fact that the system is invariant under rotations to present the following theorem, which was proved independently in slightly different settings by Hoppensteadt and Izhikevich (1996b) and Chakravarthy and Ghosh (1996).

Theorem 10.5 (Cohen-Grossberg Convergence Theorem for Oscillatory Neural Networks at Multiple Andronov-Hopf Bifurcations)   *If in the canonical model (10.14) all neurons have equal center frequencies* $\omega_1 = \cdots = \omega_n = \omega$ *and the matrix of synaptic connections* $C = (c_{ij})$ *is self-adjoint, i.e.,* $c_{ij} = \bar{c}_{ji}$, *then the neural network activity converges to a limit cycle. On the limit cycle all neurons have constant phase differences, which corresponds to synchronization of the network activity.*

We prove a similar result for a weakly connected network of limit cycle oscillators in Chapter 9 (Theorem 9.15). We call the result the Cohen-Grossberg theorem, since its proof is a modification of the one given by Cohen and Grossberg (1983) to prove absolute stability of a general network of nonlinear nonoscillatory neurons.

Proof. In the rotating coordinate system $u_i = e^{-i\omega\tau}z_i(\tau)$, (10.14) becomes

$$u_i' = \alpha_i u_i - u_i|u_i|^2 + \sum_{j=1}^{n} c_{ij}u_j, \quad i = 1, \ldots, n. \tag{10.15}$$

Note that the mapping $U : \mathbb{C}^{2n} \to \mathbb{R}$ given by

$$U(u_1, \ldots, u_n, \bar{u}_1, \ldots, \bar{u}_n) = -\sum_{i=1}^{n}\left(\alpha_i|u_i|^2 - \frac{1}{2}|u_i|^4 + \sum_{j=1}^{n} c_{ij}\bar{u}_i u_j\right)$$

is a global Liapunov function for (10.15). Indeed, it is continuous, bounded below (because it behaves like $\frac{1}{2}|u|^4$ for large u), and satisfies

$$u_i' = -\frac{\partial U}{\partial \bar{u}_i}, \quad \bar{u}_i' = -\frac{\partial U}{\partial u_i};$$

hence,

$$\frac{dU}{d\tau} = \sum_{i=1}^{n} \left(\frac{\partial U}{\partial u_i} u_i' + \frac{\partial U}{\partial \bar{u}_i} \bar{u}_i' \right) = -2 \sum_{i=1}^{n} |u_i'|^2 \leq 0 .$$

Notice that $dU/d\tau = 0$ precisely when $u_1' = \cdots = u_n' = 0$, i.e., at the equilibrium point of (10.15). Let $u^\star \in \mathbb{C}^n$ be such a point. Then, while the solution u of (10.15) converges to $u^\star$, the solution of (10.14) converges to the limit cycle $z = e^{\mathrm{i}\omega\tau} u^\star$. Obviously, any two oscillators have constant phase difference on this limit cycle. $\square$

It should be noted that the dynamics of (10.14) can converge to different limit cycles depending upon the initial conditions and the choice of the parameters $\alpha_1, \ldots, \alpha_n$. For fixed parameters there could be many such limit cycles corresponding to different memorized images.

From the theorem it follows that the canonical model (10.14) is an MA-type NN model (Hopfield 1982, Grossberg 1988), but instead of equilibrium points, the network activity converges to limit cycles (see Figure 10.4), as

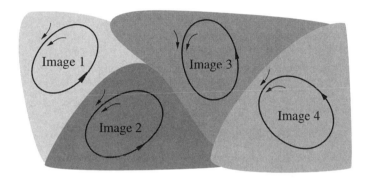

Figure 10.4. Schematic presentation of the phase space of the canonical model. Each memorized image is represented as a limit cycle attractor.

was postulated by Baird (1986). Whether or not this new feature renders the canonical model any advantages over the classical Hopfield model is still an open question.

We continue our studying of the canonical model in Chapter 13.

11
Multiple Cusp Bifurcation

In Section 5.3.3 we showed that any weakly connected neural network of the form

$$\dot{X}_i = F_i(X_i, \lambda) + \varepsilon G_i(X, \lambda, \rho, \varepsilon)$$

at a multiple cusp bifurcation is governed by the canonical model

$$x_i' = r_i + b_i x_i + \sigma_i x_i^3 + \sum_{j=1}^{n} c_{ij} x_j, \quad i = 1, \ldots, n, \qquad (11.1)$$

where $' = d/d\tau$, τ is slow time, x_i, r_i, b_i, c_{ij} are real variables, and $\sigma_i = \pm 1$. In this chapter we study some neurocomputational properties of this canonical model. In particular, we use Hirsch's theorem to prove that the canonical model can work as a globally asymptotically stable neural network (GAS-type NN) for certain choices of the parameters $b_1, \ldots, b_n$. We use Cohen and Grossberg's convergence theorem to show that the canonical model can also operate as a multiple attractor neural network (MA-type NN).

Since the pitchfork bifurcation is a particular case of the cusp bifurcation, we study the canonical model for the pitchfork bifurcation in this chapter, too. It is a special case of (11.1) for $r_i = 0$, $i = 1, \ldots, n$.

11.1 Preliminary Analysis

Before proceeding further, we explain the meaning of the data in (11.1). Each x_i depends on X_i and is a scalar that describes in some sense activity of the ith neuron, and the vector $x = (x_1, \ldots, x_n)^\top \in \mathbb{R}^n$ describes a physiological state of the network; The parameter $r_i \in \mathbb{R}$ is an external input from sensor organs to the ith neuron. It depends on λ and ρ. Each $b_i \in \mathbb{R}$ is an internal parameter, which also depends upon λ and ρ. The vector $(b_1, \ldots, b_n)^\top \in \mathbb{R}^n$ is a multidimensional bifurcation parameter; $C = (c_{ij}) \in \mathbb{R}^{n \times n}$ is a matrix of synaptic connections between neurons. There is strong neurobiological evidence that synapses are responsible for associative memorization and recall. We will see this realized in the canonical model (11.1), but from a rigorous mathematical point of view.

It should be stressed that (11.1) is an interesting dynamical system in itself without any connection to the WCNN theory. It exhibits useful behavior from a computational point of view and deserves to be studied for its own sake.

The simplicity of (11.1) in comparison with the original WCNN is misleading. It is very difficult to study it for an arbitrary choice of parameters. We will study it using bifurcation theory. Nevertheless, we can answer interesting questions about (11.1) only by making some assumptions about the parameters r_i, b_i, and C. In Section 11.1.2 we assume that the input from receptors is very strong, i.e., all r_i have large absolute values. In Section 11.1.3 we consider extreme choices for b_i. Section 11.3 is devoted to the study of (11.1) under the classical assumption that the matrix of synaptic connections C is symmetric. This restriction arises naturally when one considers Hebbian learning rules, which are discussed in Section 11.4.

In Section 11.5 we study bifurcations in the canonical model (11.1) when the external input $r_i = 0$ for all i. In this case (11.1) coincides with the canonical model for multiple pitchfork bifurcations in the WCNNs. Section 11.6 is devoted to studying the canonical model when only one or two images are memorized. In Sections 11.7 and 11.8 we illustrate the phenomena of bistability of perception and decision-making in analogy with problems in psychology.

We start studying (11.1) by asking the following question: What is the behavior of its solutions far from the origin $x = 0$, i.e., outside some ball $B_0(R) \subset \mathbb{R}^n$ with large radius R?

11.1.1 Global behavior

Let $B_0(R) \subset \mathbb{R}^n$ denote a ball with center at the origin and radius R. By the term "global behavior" we mean a flow structure of a flow outside the ball $B_0(R)$ for sufficiently large R.

A dynamical system is *bounded* if there is $R > 0$ such that $B_0(R)$ attracts all trajectories; i.e., for any initial condition $x(0)$ there exists t_0 such that $x(t) \in B_0(R)$ for all $t \geq t_0$. Obviously, all attractors of such a system lie inside $B_0(R)$.

Theorem 11.1 *A necessary and sufficient condition for the system (11.1) to be bounded is that $\sigma_1 = \cdots = \sigma_n = -1$; i.e., (11.1) must be*

$$x_i' = r_i + b_i x_i - x_i^3 + \sum_{j=1}^{n} c_{ij} x_j \, , \tag{11.2}$$

which corresponds to a multiple supercritical cusp bifurcation.

Proof. We are interested in the flow structure of (11.1) outside some ball $B_0(R)$ with sufficiently large radius R. Let $\delta = R^{-2}$ be a small parameter. After the rescaling $y = \sqrt{\delta} x$, $t = \delta^{-1} \tau$, (11.1) can be rewritten as

$$\dot{y}_i = \sigma_i y_i^3 + \delta(b_i y_i + \sum_{j=1}^{n} c_{ij} y_j + \sqrt{\delta} r_i),$$

which must be studied outside the unit ball $B_0(1)$.

Note that this is a δ-perturbation of the uncoupled system

$$\dot{y}_i = \sigma_i y_i^3, \qquad i = 1, \ldots, n \, . \tag{11.3}$$

Obviously, the unit ball $B_0(1)$ attracts all trajectories of (11.1) if and only if all $\sigma_i < 0$. Since the origin $y = 0$ is a hyperbolic equilibrium, any δ-perturbation of (11.3) has the same property, provided that δ is small enough. $\square$

So, the flow structure of (11.1) outside a ball with sufficiently large radius looks like that of (11.3), as depicted in Figure 11.1 for the cases $(\sigma_1, \sigma_2) = (-1, \mp 1)$.

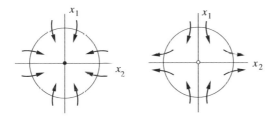

Figure 11.1. Global flow structures for the canonical model (11.1) for $n = 2$. When $\sigma_1 = \sigma_2 = -1$ (*left*), the activity is bounded. It is not bounded when $\sigma_2 = +1$ (*right*).

For the dynamics to be bounded is a desirable property in applications. Any initial condition $x(0)$ lies in a domain of attraction of some attractor that lies somewhere inside $B_0(R)$. Hence, for any $x(0)$ we have at least a hope of finding the asymptotic dynamics. From now on we will consider only (11.2) as the canonical model of the WCNN near a multiple cusp bifurcation point; i.e., we study the supercritical cusp bifurcation.

11.1.2   Strong Input from Receptors

What is the behavior of the canonical model (11.2) when the external input from receptors is very strong, i.e., when the parameter

$$R = \min_i |r_i|$$

is very large? Let $\delta = R^{-2/3}$ be a small parameter, and rescale variables by setting

$$y = \delta^{1/2} x, \qquad \tilde{r} = \delta^{3/2} r, \quad \text{and} \quad t = \delta^{-1} \tau.$$

Then (11.2) can be rewritten (after dropping $\sim$) as

$$\dot{y}_i = r_i - y_i^3 + \delta(b_i y_i + \sum_{j=1}^{n} c_{ij} y_j). \tag{11.4}$$

System (11.4) is a δ-perturbation of the uncoupled system

$$\dot{y}_i = r_i - y_i^3. \tag{11.5}$$

It is obvious that (11.5) has only one equilibrium, $y = (r_1^{1/3}, \ldots, r_n^{1/3})^\top \in \mathbb{R}^n$, for any external input $r_1, \ldots, r_n$. The Jacobian at that point is

$$L = \begin{pmatrix} -3r_1^{2/3} & 0 & \cdots & 0 \\ 0 & -3r_2^{2/3} & \cdots & 0 \\ \vdots & \vdots & \ddots & \vdots \\ 0 & 0 & \cdots & -3r_n^{2/3} \end{pmatrix}. \tag{11.6}$$

Note that according to the rescaling, all $|r_i| \geq 1$, and hence all diagonal elements in L are negative. Thus, the equilibrium point is a stable node.

All of the equilibrium points considered above are hyperbolic. Any δ-perturbations of (11.5) do not change the qualitative picture, provided that δ is small enough. So, the phase portrait of (11.2) for strong external inputs is qualitatively the same as that of (11.5).

Thus, when the input from receptors is strong (in comparison with the synaptic connections c_{ij} or the internal parameters b_i), then the network dynamics approach the unique equilibrium. At this equilibrium each neuron is either depolarized (excited) or hyperpolarized (inhibited) depending on the sign of the input r_i; see Figure 11.2.

11.1.3   Extreme Psychological Condition

It is convenient to assume that the parameters b_i, $i = 1, \ldots, n$ describes the psychological state of the network, because they affect the way the network reacts to external inputs. One might speculate that when

$$\beta = \min_i |b_i|$$

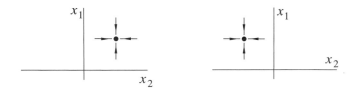

Figure 11.2. Canonical model (11.2) has a unique equilibrium when $|r_i| \gg 1$, $i = 1, 2$. When $r_1 > 0$, $r_2 > 0$ (*left*), all neurons are excited. When $r_2 < 0$ (*right*), neuron x_2 is inhibited.

is very large, the network is working in an "extreme psychological condition".

We use the same method of analysis of (11.2) as in previous sections. Let $\delta = \beta^{-1}$ be a small parameter. By rescaling

$$y = \sqrt{\delta}x , \qquad \tilde{b} = \delta b , \quad \text{and} \quad t = \delta^{-1}\tau$$

and dropping ˜, we can rewrite (11.2) as

$$\dot{y}_i = b_i y_i - y_i^3 + \delta(\sum_{j=1}^{n} c_{ij} y_j + \sqrt{\delta} r_i) .$$

This weakly connected system is a δ-perturbation of the uncoupled system

$$\dot{y}_i = b_i y_i - y_i^3 .$$

Here each equation has either one or three equilibrium points, depending upon the sign of $b_i \neq 0$. The points are $x_i = 0$ for any b_i and in addition, $x_i = \pm\sqrt{b_i}$ for $b_i > 0$. The Jacobian matrix at an equilibrium is

$$\begin{pmatrix} b_1 - 3y_1^2 & 0 & \cdots & 0 \\ 0 & b_2 - 3y_2^2 & \cdots & 0 \\ \vdots & \vdots & \ddots & \vdots \\ 0 & 0 & \cdots & b_n - 3y_n^2 \end{pmatrix} .$$

Obviously, the equilibrium point $y = (y_1, \ldots, y_n)^\top \in \mathbb{R}^n$ is a stable node if

$$y_i = \begin{cases} 0 & \text{if } b_i < 0 , \\ \pm\sqrt{b_i} & \text{if } b_i > 0 . \end{cases}$$

It is a saddle if some (but not all) of these conditions on y_i are violated, and it is an unstable node if all of the conditions are violated. Note that the presence of saddles and the unstable node is possible only if there exists at least one positive b_i. If all $b_i < 0$, then there is only one equilibrium point $y = 0 \in \mathbb{R}^n$, which is the stable node (see Figure 11.3). In any case, all equilibria are hyperbolic. Therefore, the canonical model (11.2) has hyperbolic equilibria of the same stability type.

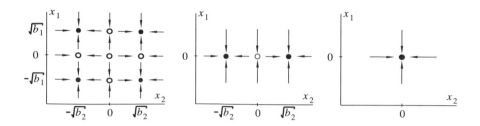

Figure 11.3. Phase portrait of (11.2) working in the extreme psychological regime $|b_i| \gg 1$: All $b_i > 0$ (*left*); $b_1 < 0$, $b_2 > 0$ (*middle*); and all $b_i < 0$ (*right*)

11.2   Canonical Model as a GAS-Type NN

Systems that have only one asymptotically stable equilibrium point and do not have any other attractors are called *globally asymptotically stable* systems (Hirsch 1989). Such systems are good candidates for GAS-type NNs (see Section 3.2). We have already seen that the canonical model (11.2) is globally asymptotically stable when the input r from receptors is strong. (The values of the other parameters, viz., b and C, are irrelevant). This is also true when the model is considered for negative b_i with large absolute values.

In the previous section we assumed that all $|b_i|$ are large. The remarkable fact is that for (11.2) to be globally asymptotically stable it suffices to require that $b = (b_1, \ldots, b_n)^\top$ take only intermediate values comparable to those of entries of C. A more accurate statement is the following:

Theorem 11.2 *The canonical model*

$$x'_i = r_i + b_i x_i - x_i^3 + \sum_{j=1}^{n} c_{ij} x_j$$

is globally asymptotically stable if

$$b_i < - \left(c_{ii} + \frac{1}{2} \sum_{j \neq i} |c_{ij} + c_{ji}| \right) \tag{11.7}$$

for all $i = 1, \ldots, n$.

Proof. We will use Hirsch's theorem (Hirsch 1989). Let L be the Jacobian of the canonical model at a point $x = (x_1, \ldots, x_n)^\top \in \mathbb{R}^n$. Hirsch's theorem claims that if there is a constant $-\delta < 0$ such that

$$\langle L\xi, \xi \rangle \leq -\delta \langle \xi, \xi \rangle$$

for all $\xi = (\xi_1, \ldots, \xi_n)^\top \in \mathbb{R}^n$, then the model is globally asymptotically stable. Here $\langle \xi, \eta \rangle$ denotes the inner (dot) product of vectors ξ and η. It is easy to check that

$$
\begin{aligned}
\langle L\xi, \xi \rangle &= \sum_{i=1}^{n}(b_i - 3x_i^2)\xi_i^2 + \sum_{i,j=1}^{n} c_{ij}\xi_i\xi_j \\
&= \sum_{i=1}^{n}(b_i + c_{ii} - 3x_i^2)\xi_i^2 + \frac{1}{2}\sum_{\substack{i,j=1 \\ i \neq j}}^{n}(c_{ij} + c_{ji})\xi_i\xi_j \\
&\leq \sum_{i=1}^{n}(b_i + c_{ii})\xi_i^2 + \frac{1}{4}\sum_{\substack{i,j=1 \\ i \neq j}}^{n}|c_{ij} + c_{ji}|(\xi_i^2 + \xi_j^2) \\
&= \sum_{i=1}^{n}\left(b_i + c_{ii} + \frac{1}{2}\sum_{j \neq i}|c_{ij} + c_{ji}|\right)\xi_i^2 \\
&< -\delta\sum_{i=1}^{n}\xi_i^2,
\end{aligned}
$$

where $-\delta = \max_i(b_i + c_{ii} + \frac{1}{2}\sum_{j \neq i}|c_{ij} + c_{ji}|)$. We used here the inequality

$$
\xi_i\xi_j \leq \frac{1}{2}(\xi_i^2 + \xi_j^2).
$$

Inequality (11.7) guarantees that $-\delta < 0$, and hence all the conditions of Hirsch's theorem are satisfied. This completes the proof. $\square$

Note that (11.7) is much more appealing than the requirement that the absolute values of b_i be arbitrarily large. Thus, even for "reasonable" values of the internal parameters b_i, the dynamics of the canonical model are globally asymptotically stable.

Another remarkable fact is that the external input $r \in \mathbb{R}^n$ does not come into the condition (11.7). What it does affect is the location of the unique attractor of the network. Therefore, the canonical model for b_i satisfying (11.7) can work as a GAS-type NN.

11.3   Canonical Model as an MA-Type NN

In this section we study the dynamics of the canonical model when the synaptic matrix $C = (c_{ij})$ is symmetric. The symmetry of C is a strong requirement, but it arises naturally if one considers the Hebbian learning rules, which will be discussed in the next section. Such systems are widely studied in the neural network literature (see the review by S. Grossberg 1988).

Theorem 11.3 (Cohen-Grossberg) *If the matrix of synaptic connections C is symmetric, then the canonical model*

$$x_i' = r_i + b_i x_i - x_i^3 + \sum_{j=1}^{n} c_{ij} x_j$$

is a gradient system.

Proof. To prove the theorem it suffices to present a function $U : \mathbb{R}^n \to \mathbb{R}$ satisfying

$$x_i' = -\frac{\partial U}{\partial x_i} .$$

It is easy to check that

$$U(x) = -\sum_{i=1}^{n} (r_i x_i + \frac{1}{2} b_i x_i^2 - \frac{1}{4} x_i^4) - \frac{1}{2} \sum_{i,j=1}^{n} c_{ij} x_i x_j$$

is such a function for the model. Note that far away from the origin $U(x)$ behaves like $\frac{1}{4} \sum_{i=1}^{n} x_i^4$, and hence $U(x)$ is bounded below. $\square$

Being a gradient system imposes many restrictions on the possible dynamics of the canonical model. For example, its dynamics cannot be oscillatory or chaotic. Indeed, it is easy to see that

$$\frac{dU}{dt} = \sum_{i=1}^{n} \frac{\partial U}{\partial x_i} x_i' = -\sum_{i=1}^{n} |x_i'|^2 \le 0$$

and that $dU/dt = 0$ if and only if all $x_i' = 0$. Therefore, the activity of the neural network converges only to equilibria. This property is considered to be very useful from a computational point of view, which we discuss later.

To what type of NNs does the canonical model belong? It is clear from previous sections and from the Cohen-Grossberg theorem that for one choice of the parameters the model has many attractors and hence is a candidate for an MA-type NN, whereas for other choices of the parameters it is globally asymptotically stable and hence is the GAS-type NN. We will show later that the canonical model can also stand somewhere between MA and GAS types and hence can be considered as a new NN type.

We are interested in the basic principles of human brain function. Hence, we will study only the qualitative behavior of the canonical model and neglect quantitative features. The main tools in the analysis below come from bifurcation theory. Unfortunately, comprehensive analysis of the model for arbitrary r_i, b_i, and C is formidable unless we impose some additional restrictions onto the parameter spaces. Among them there are two we discuss now. First of all, we will study the canonical models when

$$b_1 = \cdots = b_n = b .$$

Thus, instead of n bifurcation parameters $b_1, \ldots, b_n$ we have only one $b \in \mathbb{R}$.

The second assumption concerns the matrix of synaptic connections C, which is responsible for learning in the NNs.

11.4   Learning

11.4.1   Learning Rule for Coefficient c_{ij}

Not much is known about learning in the human brain, but our major hypotheses about the learning dynamics appear to be consistent with observations. We assume that

- Learning results from modifying synaptic connections between neurons (Hebb 1949).

- Learning is local; i.e., the modification depends upon activities of pre- and postsynaptic neurons and does not depend upon activities of the other neurons.

- The modification of synapses is slow compared with characteristic times of neuron dynamics.

- If either pre- or postsynaptic neurons or both are silent, then no synaptic changes take place except for exponential decay, which corresponds to forgetting.

These assumptions in terms of the WCNN

$$\dot{x}_i = f_i(x_i, \lambda) + \varepsilon g_i(x_1, \ldots, x_n, \lambda, \rho, \varepsilon)$$

have the following implications: The first hypothesis states that learning is described by modification of the coefficients $c_{ij} = \partial g_i / \partial x_j$. Recall that the actual synaptic connections have order ε. We denote them by $w_{ij} = \varepsilon c_{ij}$. The second hypothesis says that for fixed i and j the coefficient w_{ij} is modified according to equations of the form

$$w'_{ij} = h(w_{ij}, x_i, x_j) \ .$$

We introduce the slow time $\tau = \varepsilon t$ to account for the third hypothesis. We say that a neuron is silent if its activity is at an equilibrium point. Then the fourth hypothesis says that

$$h(w_{ij}, 0, x_j) = h(w_{ij}, x_i, 0) = h(w_{ij}, 0, 0) = \tilde{h}(w_{ij}) = -\gamma w_{ij} + \delta w_{ij}^2 + \ldots$$

for all x_i, x_j, so that h has the form

$$h(w_{ij}, x_i, x_j) = -\gamma w_{ij} + \beta x_i x_j + \delta_1 w_{ij} x_i + \delta_2 w_{ij} x_j + \delta w_{ij}^2 + \ldots \ .$$

Now we use the fact that the synaptic coefficient w_{ij} is of order ε. From Theorem 5.5 we know that the activities of neurons are of order $\sqrt{\varepsilon}$. After rescaling by $w_{ij} = \varepsilon c_{ij}$, $x_i = \sqrt{\varepsilon} y_i$, $i = 1, \ldots, n$, we obtain the learning rule

$$c'_{ij} = -\gamma c_{ij} + \beta y_i y_j + \mathcal{O}(\sqrt{\varepsilon}), \qquad (11.8)$$

which we refer to as the *Hebbian* synaptic modification rule. Note that although we consider general functions h, after the rescaling only two constants, γ and β, are significant to leading order. They are the rate of memory fading and the rate of synaptic plasticity, respectively. We assume that the fading rate γ and plasticity rate β are positive and the same for all synapses. It is also feasible to study the case $\gamma = \gamma_{ij}$ and $\beta = \beta_{ij}$.

Notice that if the neural network activity $y(t) = \xi \in \mathbb{R}^n$ is a constant, then

$$c_{ij} \to \frac{\beta}{\gamma} \xi_i \xi_j \ .$$

This motivates the following definition of the Hebbian learning rule for synaptic matrix C.

11.4.2   Learning Rule for Matrix C

Suppose $\xi^1, \ldots, \xi^m \subset \mathbb{R}^n$ are m key patterns to be memorized and reproduced by a neural network. Here each $\xi^s = (\xi_1^s, \ldots, \xi_n^s)^\top \in \mathbb{R}^n$, $s = 1, \ldots, m$, is a vector, and $m \leq n$. Let constants β_s measure "strength", or "quality", of the memory about the patterns ξ^s; Then the matrix C is said to be formed according to a Hebbian learning rule if

$$c_{ij} = \frac{1}{n} \sum_{s=1}^{m} \beta_s \xi_i^s \xi_j^s \ , \qquad 1 \leq i, j \leq n \ . \qquad (11.9)$$

The Hebbian learning rule (11.9) can be rewritten in the more convenient form

$$C = \frac{1}{n} \sum_{s=1}^{m} \beta_s \xi^s (\xi^s)^\top \ , \qquad (11.10)$$

where $^\top$ means transpose.

It is easy to see that the synaptic matrix C constructed according to (11.10) is symmetric. It is also true that any symmetric matrix C can be represented as (11.10) for some, possibly nonunique, choice of the orthogonal vectors $\xi^1, \ldots, \xi^m$. Thus, we have proved the following:

Proposition 11.4 *The matrix of synaptic connections C is symmetric if and only if there is a set of orthogonal patterns $\xi^1, \ldots, \xi^m$ such that C is constructed according to the Hebbian learning rule (11.10).*

Note that the orthogonal vectors $\xi^1, \ldots, \xi^m$ are eigenvectors of C. If we normalize them such that

$$|\xi^s|^2 = \langle \xi^s, \xi^s \rangle = \sum_{i=1}^{n} (\xi_i^s)^2 = n \, ,$$

then the constants β_s are eigenvalues of C, and (11.10) is the spectral decomposition of C. We assume that $\beta_1 \geq \cdots \geq \beta_m > 0$. If $m < n$, then there is an $(n - m)$-dimensional eigenspace $\ker C \subset \mathbb{R}^n$ corresponding to the zero eigenvalue. We denote this eigenvalue by $\beta_{m+1} = 0$.

To summarize, we can say that the Hebbian learning rule for orthogonal patterns gives a way of constructing the matrix of synaptic connections such that each pattern is an eigenvector of the matrix corresponding to a positive eigenvalue.

In our analysis below we will assume that the learning rule is Hebbian. This assumption imposes significant restrictions on possible dynamic behavior of the canonical model. For example, it follows from Section 11.3 that the model is a gradient dynamical system.

In order to make all computations without resorting to computer simulations, we also assume that $|\xi_i^s| = 1$ for all i; i.e.,

$$\xi^s = (\pm 1, \ldots, \pm 1)^\top, \qquad s = 1, \ldots, m \, .$$

For these purposes we introduce the set

$$\Xi^n = \{\xi \in \mathbb{R}^n, \ \xi = (\pm 1, \ldots, \pm 1)^\top\} \subset \mathbb{R}^n. \tag{11.11}$$

We will also need for our analysis an orthogonal basis for $\mathbb{R}^n$ that contains vectors only from Ξ^n. This basis always exists if $n = 2^k$ for some integer $k > 0$. The assumption $\xi^s \in \Xi^n$ might look artificial, but it is very important in neurocomputer applications and in digital circuit design.

Recall that we are interested in qualitative behavior of the canonical models. All attractors that we will study below are hyperbolic. Hence, if we perturb the parameters $r_1, \ldots, r_n, \ b_1, \ldots, b_n$, and C, i.e., if we violate the assumptions made above, the qualitative behavior will be the same provided that the perturbations are not very large.

11.5 Multiple Pitchfork Bifurcation

We start the bifurcation analysis of the canonical model for a multiple cusp singularity in WCNNs for the special case when there are no receptor inputs, i.e., when $r_1 = \cdots = r_n = 0$. Thus, we are interested in qualitative behavior of the dynamical system

$$x_i' = bx_i - x_i^3 + \sum_{j=1}^{n} c_{ij}x_j \quad i = 1, \ldots, n \, . \tag{11.12}$$

It is easy to check that the canonical model (11.12) describes a WCNN with $\mathbb{Z}_2$ symmetry $x \to -x$ near a *multiple pitchfork* bifurcation point (see Section 5.3.4). Symmetry means that for $b > 0$ each neuron is essentially a bistable element with two states: excitation ($x_i = \sqrt{b}$) and inhibition ($x_i = -\sqrt{b}$).

Since the canonical model for the pitchfork bifurcation is a special case of that for the cusp bifurcation, all facts that we derived in the previous sections are applicable to (11.12) too. In particular, (11.12) is a candidate for a GAS- or MA-type NN. We cannot say at this point that (11.12) can operate as these NN types because we do not know the locations of the attractors and how they relate to the memorized images. We study these issues below.

11.5.1   Stability of the Origin

Note that (11.12) always has an equilibrium point $x_1 = \cdots = x_n = 0$. It follows from Section 11.1.3, that the origin is the only equilibrium point, which is a stable node, for $b \ll -1$ or for b satisfying (11.7). We also know that for $b \gg 1$ the canonical model has many attractors (see Figure 11.3). What we do not know is the behavior of the model for intermediate values of b.

Thus, we have the following questions: What happens while b is increasing? How does the origin lose its stability? How many and of what type are the new equilibrium points? What is the relationship between them and the synaptic matrix $C = (c_{ij})$? These and other questions are studied in this section.

Below, we consider the canonical model (11.12) for a general matrix C. In our analysis we do not require symmetry of C. Let L be the Jacobian of the right-hand side of (11.12) at the origin. It is easy to see that

$$L = bI + C ,$$

where I is the unit matrix. Let $\beta_1, \ldots, \beta_m$ be the (distinct) eigenvalues of C ordered such that

$$\operatorname{Re}\beta_1 \geq \cdots \geq \operatorname{Re}\beta_m$$

counted with multiplicity. Obviously, L has m eigenvalues,

$$\lambda_s = b + \beta_s , \qquad s = 1, \ldots, m ,$$

with the same eigenvectors as those of C. The matrix L has all eigenvalues with negative real parts, and hence the origin is a stable equilibrium point for (11.12) if and only if $b < -\operatorname{Re}\beta_1$ (see Figure 11.4a).

Bifurcations of the Origin

If β_1 is a real eigenvalue with multiplicity one, then the canonical model (11.12) at the origin undergoes a pitchfork bifurcation when b crosses $-\beta_1$.

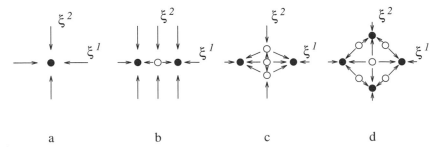

Figure 11.4. Phase portrait of the canonical model (11.12) of a WCNN near a multiple pitchfork bifurcation point for different values of the bifurcation parameter b. (a). $b < -\beta_1$. (b). $-\beta_1 < b < -\beta_2$. (c). $-\beta_2 < b < -\beta_2 + (\beta_1 - \beta_2)/2$. (d). $-\beta_2 + (\beta_1 - \beta_2)/2 < b$

For b slightly larger than $-\beta_1$ the origin is a saddle surrounded by two sinks (see Figure 11.4b) and those are the only equilibrium points for (11.12).

If (β_1, β_2) is a pair of complex conjugate eigenvalues with multiplicity one, then we can observe the Andronov-Hopf bifurcation for $b = -\operatorname{Re}\beta_1$.

For $b > -\beta_1$ it is possible to observe the birth of a pair of saddles or an unstable limit cycle every time b crosses $-\operatorname{Re}\beta_s$, where β_s is an eigenvalue with multiplicity one.

For the eigenvalues with multiplicity more than one bifurcations could be more complicated. We will consider some of them later.

Self-Ignition

Recall that each neuron is bistable only for $b > 0$. For negative b there is only one stable state $x_i = 0$ and hence it is "passive". But when the neurons are connected, they acquire a new property: bistability for $-\beta_1 < b < 0$. This is the property that each neuron alone cannot have. Thus a network of "passive" elements can exhibit "active" properties. This is called self-ignition and has already been studied in Section 10.2 for oscillatory neural networks. We encounter this phenomenon frequently in our analysis of brain function.

11.5.2 Stability of the Other Equilibria

It is noteworthy that we have not restricted the synaptic matrix C yet. All the bifurcations discussed above take place for any C. In return for this generality, we cannot trace the new equilibrium points and study their stability. Fortunately, we can do it if we assume that the synaptic matrix C is constructed according to the Hebbian learning rule

$$C = \frac{1}{n} \sum_{s=1}^{m} \beta_s \xi^s (\xi^s)^\top \,, \qquad \beta_1 \geq \cdots \geq \beta_m > 0 \,,$$

and that the memorized images $\xi^1, \ldots, \xi^m \in \Xi^n$ are orthogonal. For simplicity we assume that all β_s are different. At the end of this section we will discuss the case $\beta_1 = \cdots = \beta_m$. Let $x_i = y_s \xi_i^s$ for $i = 1, \ldots, n$. Then

$$x_i' = y_s' \xi_i^s = b y_s \xi_i^s - y_s^3 (\xi_i^s)^3 + \beta_s y_s \xi_i^s.$$

After (dot) multiplication by ξ_i^s, we have

$$y_s' = (b + \beta_s) y_s - y_s^3. \tag{11.13}$$

This equation has only one equilibrium point, $y_s = 0$, for $b < -\beta_s$, and for $b > -\beta_s$ there are three points, $y_s = 0$, $y_s = \pm \sqrt{b + \beta_s}$. Hence the original system (11.12) has two new equilibrium points, $x = \pm \sqrt{b + \beta_s}\, \xi^s$, after b crosses $-\beta_s$.

Note that the pair of new equilibrium points lies on the line spanned by the memorized pattern ξ^s. Every attractor lying on or near span (ξ^s) is called an attractor corresponding to the pattern ξ^s. When the network activity $x(t)$ approaches such an attractor, we say that the NN has recognized the memorized image ξ^s.

To our surprise, only the pair $x = \pm \sqrt{b + \beta}\, \xi^1$ is a pair of stable nodes, whereas the others $x = \pm \sqrt{b + \beta_s}\, \xi^s$, $s \geq 2$, are pairs of saddles, at least when b is near $-\beta_s$ (see Figure 11.4c).

Let us study the stability of $x = \pm \sqrt{b + \beta_k}\, \xi^k$ for some $k = 1, \ldots, m$. The matrix of linearization L at this point is

$$L = (b - 3(b + \beta_k)) I + C.$$

It has eigenvalues

$$\lambda_s = -2(b + \beta_k) + \beta_s - \beta_k, \qquad s = 1, \ldots, m+1,$$

where $\beta_{m+1} = 0$ corresponds to ker C. Note that $\lambda_1 \geq \cdots \geq \lambda_m > \lambda_{m+1}$. The maximum eigenvalue $\lambda_1 = -2(b + \beta_k) + \beta_1 - \beta_k$ is always negative only for $k = 1$. For $k \geq 2$ the inequality $\lambda_1 < 0$ gives us the condition

$$b > -\beta_k + \frac{\beta_1 - \beta_k}{2}. \tag{11.14}$$

One could say that in the life of the equilibrium point $\sqrt{b + \beta_k}\, \xi^k$ there are two major events: birth (when $b = -\beta_k$) and maturation ($b = -\beta_m + (\beta_1 - \beta_m)/2$) when the point becomes a stable node. For $k = 2$, see Figures 11.4d and 11.5.

It is easy to see that when $b = -\beta_k$, the eigenvalues of L are

$$\lambda_1 \geq \cdots \geq \lambda_{k-1} \geq 0 = \lambda_k \geq \lambda_{k+1} \geq \cdots \geq \lambda_m.$$

So, $\sqrt{b + \beta_k}\, \xi^k$ is the saddle such that $k - 1$ directions corresponding to $\xi^1, \ldots, \xi^{k-1}$ are unstable (see Figure 11.6a). Every time b crosses $-\beta_k +$

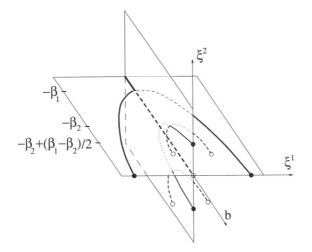

Figure 11.5. Bifurcation diagram

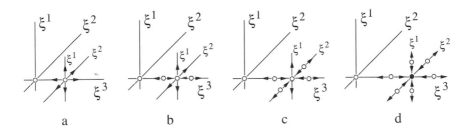

a   b   c   d

Figure 11.6. Every equilibrium point $\pm\sqrt{b + \beta_k}\xi^k$ becomes an attractor after the sequence of the pitchfork bifurcations. Every time b crosses $-\beta_k + (\beta_s - \beta_k)/2$, $s < k$, the ξ^s-direction becomes stable.

$(\beta_s - \beta_k)/2$, $s < k$, there is a pitchfork bifurcation. As a result, the direction corresponding to ξ^s becomes stable, and there appears a new pair of saddles lying in span (ξ^s, ξ^k) (see Figure 11.6b, c, and d). To summarize, we can say that for $b < \beta_1$ the only equilibrium point is the origin, but for $b > -\beta_m + (\beta_1 - \beta_m)/2$ there are m pairs of stable nodes corresponding to the memorized images $\xi^1, \ldots, \xi^m$ and many saddles lying in between these nodes.

Spurious Memory

Recall that we referred to the attractors that do not correspond to any of the memorized images as being *spurious memory*. Is there any spurious memory in (11.12)? The answer is YES. It happens for large b. Indeed, when $b > 0$, all eigenvalues of the matrix of linearization at the origin are positive, not only $\lambda_1, \ldots, \lambda_m$. The new unstable directions correspond to

$\ker C$ (of course, if $m < n$). It is easy to check that for $0 < b < \frac{\beta_1}{2}$ all the equilibrium points lying in $\ker C$ are saddles (except the origin, which is an unstable node), whereas for $b > \frac{\beta_1}{2}$ there are $2(n - m)$ stable nodes among them.

In order to avoid spurious memory, one should keep the bifurcation parameter below the critical value $\frac{\beta_1}{2}$. Actually, it is more reasonable to demand that b be negative. By this means we guarantee that nothing interesting is going on in the directions orthogonal to all the memorized images. But we must be cautious, because not all equilibrium points corresponding to memorized patterns are stable for $b < 0$. Indeed, the stability condition (11.14) for $b < 0$ can be satisfied only if

$$\beta_k > \frac{\beta_1}{3} \ .$$

Thus, all memorized images are stable nodes, and successful recognition is possible if the weight β_m of the weakest image is greater than one-third of that of the strongest one.

Obviously, we do not have this kind of problem when $\beta_1 = \cdots = \beta_m = \beta > 0$. For $b < -\beta$ the NN is globally asymptotically stable. For $b = -\beta$ there is a multiple pitchfork bifurcation with the birth of $2m$ stable nodes corresponding to the memorized images. For $-\beta < b < 0$ these nodes are the only attractors,[1] and behavior of the NN (11.12) is very simple in the directions orthogonal to the span $(\xi^1, \ldots, \xi^m)$. Thus, the system (11.12) can work as a typical MA-type NN. If the initial condition $x(0)$ is an input from receptors, then the activity $x(t)$ of (11.12) approaches the closest attractor, which corresponds to one of the previously memorized images. It is believed that this simple procedure is a basis for construction of new generations of computers – neurocomputers.

Nevertheless, our opinion is that this is too far from the basic principles of how a real brain functions (despite the fact that we know almost nothing about these principles). In the next section we explore another, more realistic, approach.

11.6 Bifurcations for $r \neq 0$ (Two Memorized Images)

As we have already mentioned, comprehensive analysis of the canonical model

$$x_i' = r_i + bx_i - x_i^3 + \sum_{j=1}^{n} c_{ij} x_j \tag{11.15}$$

---

[1]Actually, the correct statement is that these are the only attractors that bifurcated from the origin. Whether there are other attractors or not is still an open question.

is formidable. Even for a symmetric synaptic matrix $C = (c_{ij})$ it is difficult, although there is a Liapunov function in that case. Hence, every piece of information about (11.15) obtained by analytical tools is precious.

The next step in studying (11.15) is to assume that the number of memorized images $m \leq 2$. The key result in this direction is the reduction lemma, which enables us to reduce the number of independent variables to 2.

11.6.1   The Reduction Lemma

Suppose that $\xi^1, \ldots, \xi^n \in \Xi^n$ form an orthogonal basis for $\mathbb{R}^n$ and that ξ^1 and ξ^2 coincide with memorized images, where Ξ is defined in (11.11). Let

$$y_s = \frac{1}{n}\langle x, \xi^s \rangle = \frac{1}{n}\sum_{i=1}^{n} x_i \xi_i^s$$

be the projection of $x \in \mathbb{R}^n$ onto $\frac{1}{n}\xi^s$. Obviously, x can be represented as the sum

$$x = \sum_{s=1}^{n} y_s \xi^s . \tag{11.16}$$

A similar decomposition is possible for any input $r \in \mathbb{R}^n$: Let

$$a_s = \frac{1}{n}\langle r, \xi^s \rangle .$$

Then

$$r = \sum_{s=1}^{n} a_s \xi^s . \tag{11.17}$$

Let us prove the following.

Lemma 11.5 (Reduction Lemma) *If*

$$C = \frac{1}{n}(\beta_1 \xi^1 (\xi^1)^\top + \beta_2 \xi^2 (\xi^2)^\top)$$

for orthogonal $\xi^1, \xi^2 \in \Xi^n$, $b < 0$ and $a_3 = \cdots = a_n = 0$, then the plane $y_3 = \cdots = y_n = 0$ is a stable invariant manifold for (11.15) (see Figure 11.7) and activity on the manifold is governed by the system

$$\begin{cases} y_1' = a_1 + (b + \beta_1)y_1 - 3y_1 y_2^2 - y_1^3 \\ y_2' = a_2 + (b + \beta_2)y_2 - 3y_2 y_1^2 - y_2^3 \end{cases} . \tag{11.18}$$

The condition $a_3 = \cdots = a_n = 0$ means that the input from receptors r is in the span (ξ^1, ξ^2). For example, if we study the olfactory system, then with this restriction only two odors are inhaled and recognizable. One can say that the canonical model lives in a world of two images, since there are only two images to be memorized and recognized.

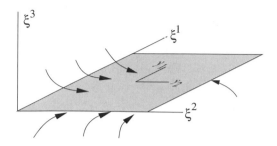

Figure 11.7. The plane spanned by ξ^1 and ξ^2 is stable and invariant when the canonical model lives in a world having two images.

Proof. Substituting (11.16) and (11.17) into (11.15) gives

$$\sum_{s=1}^{n} y'_s \xi_i^s = \sum_{s=1}^{n} a_s \xi_i^s + b \sum_{s=1}^{n} y_s \xi_i^s$$
$$- \sum_{s=1}^{n} \sum_{p=1}^{n} \sum_{q=1}^{n} y_s y_p y_q \xi_i^s \xi_i^p \xi_i^q + \sum_{s=1}^{n} y_s \beta_s \xi_i^s ,$$

where $\beta_3 = \cdots = \beta_n = 0$. Projecting both sides onto $\frac{1}{n}\xi^k$ gives

$$y'_k = a_k + (b + \beta_k) y_k - \sum_{s=1}^{n} \sum_{p=1}^{n} \sum_{q=1}^{n} y_s y_p y_q \left(\frac{1}{n} \sum_{i=1}^{n} \xi_i^k \xi_i^s \xi_i^p \xi_i^q \right) \qquad (11.19)$$

for $k = 1, \ldots, n$. Note that

$$\frac{1}{n} \sum_{i=1}^{n} \xi_i^k \xi_i^s \xi_i^p \xi_i^q = \begin{cases} 1, & \text{if } k = s, p = q \text{ or } k = p, s = q, \\ & \text{or } k = q, s = p, \\ d_{kspq}, & \text{if all indices are different,} \\ 0, & \text{otherwise,} \end{cases}$$

where $d_{kspq} \in \mathbb{R}$ are some constants. We used the assumption that $\xi_i^s = \pm 1$ for any s and i.

If the number of the memorized images m were greater than 2, then all equations in (11.19) would contain the constants d_{kspq}. It is possible to eliminate them if we consider (11.19) on the plane $y_3 = \cdots = y_n = 0$ for $m \leq 2$. Indeed, the product $y_s y_p y_q$ is always zero unless $1 \leq s, p, q \leq 2$. The inequality guarantees that at least two indices coincide. Hence, the sum $\frac{1}{n} \sum_{i=1}^{n} \xi_i^k \xi_i^s \xi_i^p \xi_i^q$ is either 1 or 0. It is 1 when all the indices are equal (this gives y_k^3) or when k is equal to only one of the three indices s, p, q (there are three such possibilities, whence the term $3 y_k y_{3-k}^2$). Thus, the system (11.19) on the plane can be rewritten as (11.18).

We still must show that the plane is a stable invariant manifold. From the lemma's conditions we know that $a_3 = \cdots = a_n = 0$ and $\beta_3 = \cdots = \beta_n = 0$.

Let us fix y_1 and y_2 and consider them as parameters. Keeping only linear terms, we can rewrite (11.19) for $k \geq 3$ as

$$y_k' = by_k - 3y_k(y_1^2 + y_2^2) + \text{higher order terms}, \qquad k = 3, \ldots, n \ .$$

The plane is invariant because $y_3' = \cdots = y_n' = 0$ on it. It is stable because $b - 3(y_1^2 + y_2^2) < b < 0$, which follows from the lemma's condition that $b < 0$. $\square$

It is still an open question whether the invariant plane is globally asymptotically stable or not. Our conjecture is that for $b < 0$ it is true, but we do not need this for our analysis below.

If $\beta_1 = \beta_2 = \beta$, then it is easy to check that (11.18) can be rewritten as

$$\begin{cases} u' = s + (b + \beta)u - u^3 \\ v' = c + (b + \beta)v - v^3 \end{cases}, \qquad (11.20)$$

where $u = y_1 + y_2$, $v = y_1 - y_2$, $s = a_1 + a_2$, and $c = a_1 - a_2$. The advantage of (11.20) is that it is uncoupled, and each equation can be studied independently.

11.6.2   Recognition: Only One Image Is Presented

Suppose only one image is presented for memorization and recognition. Without loss of generality we may assume that it is ξ^1, i.e.,

$$r_i = a\xi_i^1 .$$

Assuming that all the conditions of the reduction lemma are satisfied and that $\beta_1 = \beta_2 = \beta$, we can rewrite (11.15) as (11.20). Note that $s = c = a$. Thus, the activity on the (u, v) plane is the direct product of two identical equations

$$z' = a + (b + \beta)z - z^3 , \qquad z \in \mathbb{R} . \qquad (11.21)$$

If $b + \beta < 0$, then there is only one equilibrium point for any a. The activity on the (u, v) plane is qualitatively the same as that of the canonical model for a WCNN near a multiple pitchfork bifurcation point (11.12) for $b + \beta_1 < 0$, which is depicted in Figure 11.4a.

Weak Input

Suppose $b + \beta > 0$. There are three equilibrium points in (11.21) when $|a| < a^*$, where

$$a^* = 2 \left(\frac{b + \beta}{3} \right)^{\frac{3}{2}} ;$$

see Figure 11.8. Hence, (11.20) has nine equilibrium points. Again, there is not any qualitative distinction between the phase portrait of (11.12)

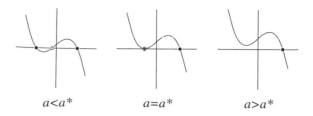

Figure 11.8. The graph of $a + (b+\beta)z - z^3$ for $b + \beta > 0$ and various values of a

depicted in Figure 11.4d and that of (11.20), which we depict in Figure 11.9a for $a > 0$. We see that $|a| < a^\star$ is too weak to produce any qualitative changes in the dynamics of the canonical model (11.15) in comparison with (11.12). Nevertheless, it is easy to see that the domain of attraction of the equilibrium point corresponding to the presented image ξ^1 is much bigger than the attraction domains of the other equilibrium points. By the term attraction domain *size* we mean here the distance from the attractor to the closest saddle. We use this definition in order to be able to compare domains that have infinite volumes.

Strong Input

When the parameter a crosses $\pm a^\star$, there is a saddle-node bifurcation in (11.21); see Figure 11.8. In system (11.20) one can observe two saddle-node bifurcations and one codimension-2 bifurcation (see Figure 11.9b). All of them take place simultaneously due to the fact that (11.20) is a direct product of two identical equations of the form (11.21). We consider these bifurcations elsewhere when we study the canonical model for WCNNs near a multiple saddle-node bifurcation point.

   If the input $r = a\xi^1$ is sufficiently strong (i.e., if $|a| > a^\star$), then there is only one equilibrium point, which is a stable node (see Figure 11.8 and Figure 11.9c). The equilibrium point is globally asymptotically stable in this case.

   We see that the canonical model (11.15) can work as GAS-type NN when the input strength a is strong enough, viz.,

$$|a| > a^\star = 2\left(\frac{b+\beta}{3}\right)^{\frac{3}{2}}.$$

We have performed all the analysis above for the case of one presented and two memorized images.

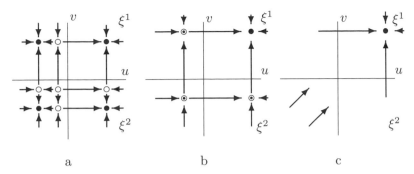

a   b   c

Figure 11.9. Phase portrait of the canonical model on the stable invariant plane spanned by ξ^1 and ξ^2. The first image is presented as an input into the network. (a). Input is weak, i.e., $|a| < a^*$. (b). For $|a| = a^*$ there are saddle-node bifurcations. (c). For $|a| > a^*$ the canonical model is globally asymptotically stable.

11.6.3   Recognition: Two Images Are Presented

Without loss of generality we may assume in the case of two presented images that

$$r = a_1 \xi^1 + a_2 \xi^2$$

for $a_1, a_2 > 0$. If $\beta_1 = \beta_2 = \beta$ and all the conditions of the reduction lemma are satisfied, then the canonical model (11.15) can be reduced to the two-dimensional system (11.20):

$$\begin{cases} u' = s + (b + \beta)u - u^3 \\ v' = c + (b + \beta)v - v^3 \end{cases}.$$

We cannot reduce (11.20) to a one-dimensional system because in general $s \neq c$. The constant $s = a_1 + a_2$ has the obvious meaning of overall *strength* of the input from receptors, whereas $c = a_1 - a_2$ is the *contrast* of the input. When $c > 0$ ($c < 0$) we say that ξ^1 (ξ^2) is dominant.

In order to determine the qualitative behavior of (11.20) we have to compare s and c with the bifurcation value a^*. When both s and c are less than a^*, the qualitative phase portrait of (11.20) depicted in Figure 11.10a coincides with that of (11.12) depicted in Figure 11.4d provided that $b + \beta > 0$.

An interesting behavior arises when the overall input from receptors s is strong enough, i.e., when $s > a^*$. Then, (11.20) generically has either one or three equilibrium points (see Figure 11.10b). Its behavior is determined by the equation

$$v' = c + (b + \beta)v - v^3, \qquad v \in \mathbb{R}. \tag{11.22}$$

Obviously, the activity of (11.22) depends crucially not only upon which image is dominant but also upon how dominant it is. If $|c| < a^*$, then there is a coexistence between these two images (see Figure 11.10b). Both

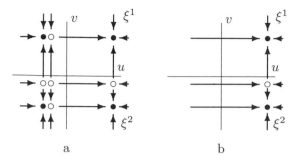

a   b

Figure 11.10. Phase portrait for the canonical model on the stable invariant plane spanned by ξ^1 and ξ^2. The input is a mixture of two images ξ^1 and ξ^2. (a). Overall input is weak. (b). Strong input and weak contrast. There is a coexistence of two attractors.

equilibrium points are stable. If $|c| > a^\star$, then only one image survives, the dominant image.

One possible explanation of the coexistence of two attractors corresponding to two different images is that the NN cannot distinguish between them when the contrast $|c|$ is small. One could say that the two-attractor state corresponds to the *I do not know* answer. We prefer another explanation, suggested by the psychological experiment described in the next section.

11.7   Bistability of Perception

In the previous sections we showed that if the conditions of the reduction lemma are satisfied and the overall input from receptors is strong $(s > a^\star)$, then the canonical model behaves qualitatively like the equation

$$v' = c + bv - v^3 , \qquad v \in \mathbb{R} ,$$

where $c = a_1 - a_2$ is the contrast between two images $a_1\xi^1$ and $a_2\xi^2$ and b is a real parameter (we incorporated β into b, so b can be positive or negative).

We have already mentioned that if the contrast is weak ($|c| < a^\star$), then (11.22) has two attractors corresponding to the previously memorized images ξ^1 and ξ^2.

First of all, note that the coexistence of two attractors contradicts the GAS-type NN paradigm, which requires that the NN have only one attractor. We must accept the fact that the brain is a very complicated system having many attractors. Its dynamic behavior depends not only upon the input r, the synaptic memory C, and the psychological state b, but also upon short-term past activity (which sometimes is called short-term memory (Grossberg 1988)). In our case this is the initial condition $x(0)$. Obviously, which attractor will be selected by the NN depends upon the initial state. Simultaneous existence of several attractors for the input that is a mixture of images suggests the following hypothesis: The NN perceives the ambiguous input according to the network's past short-term activity $x(0)$.

The behavior of the *artificial* NN (11.15) is similar to the behavior of the real human brain in the following psychological experiments as found by Attneave (1971) and others. The fourth figure from the left in Figure 11.11 was shown to be perceived with equal probability as the human face or body. If the figure is included in a sequence, then its perception depends upon the direction in which the sequence is viewed.

This phenomenon was studied from a catastrophe theory point of view (Poston and Stewart 1978, Stewart and Peregoy 1983), and it was shown (theoretically and experimentally) that there is a one-dimensional section of a cusp catastrophe in the human perception of the figures.

The remarkable fact is that the WCNN approximated by (11.22) also exhibits the cusp catastrophe. Suppose ξ^1 and ξ^2 represent the body and the face images, respectively. If we fix $b > 0$ and vary the image contrast $c = a_1 - a_2$, then the artificial NN also has the same bistable perception of the presented images $a_1\xi^1$ and $a_2\xi^2$ (see the bottom row in Figure 11.11).

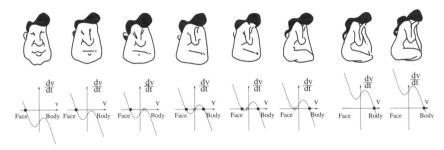

Figure 11.11. Bistability of perception. (adapted from Poston and Stewart 1978)

What we have not explained yet is the switching of our attention (say, from body to face and back) while we observe an ambiguous picture. These oscillations in our perception cannot be explained by catastrophe theory. We can tackle this problem by embedding the WCNN (5.1) into the λ-space, i.e., by allowing the internal parameter λ to vary. This idea was used by Haken (1988) and Ditzinger and Haken (1989) in their analysis of a similar system.

The actual switching of attention is not periodic, but chaotic. Richards et al. (1994) estimated the dimension of such a chaotic attractor in the human brain. Depending on the picture, the dimension varies from 3.4 to 4.8, which suggests that the perceptual oscillations can be modeled by a low-dimensional dynamics system.

As we can see, the canonical model (11.15) can work as MA- and GAS-type NNs simultaneously. Indeed, its activity crucially depends upon the input $r \in \mathbb{R}^n$ from receptors. If the input is strong enough and there is no ambiguity, then (11.15) has only one attractor and hence works as a GAS-type NN. If the input is weak or ambiguous, then (11.15) can have many attractors and hence can work as an MA-type neural network.

We think that the real brain might use similar principles. Consider, for example, the olfactory system. It is believed that each inhaled odor has its own attractor – a stable limit cycle. The analysis of the canonical model (11.15) suggests that when an animal inhales a mixture of odors, the appropriate limit cycles become stable so that there is a one-to-one correspondence between the inhaled odors and the attractors. Similar results were obtained by studying another NN (Izhikevich and Malinetskii 1993), but the attractors there were chaotic.

11.8   Quasi-Static Variation of a Bifurcation Parameter

In the two preceding sections we studied the behavior of the canonical model (11.15) for fixed b. We varied the contrast c and saw that there were two attractors when the contrast was weak. The NN recognized one of the two presented images according to the initial conditions, not to dominance of one of them over the other. Of course, the attraction domain of the stronger image was bigger than that of the weaker one, but the network could not determine which image was dominant. One possibility of making such a determination would be to collect statistics over many trials for random initial conditions. There is another possibility of determining which image is dominant. We have to fix the contrast c and vary the bifurcation parameter b very slowly so that we can neglect the transient processes and assume that $x(t)$ is arbitrarily close to the attractor. Such variation of the parameter is called *quasi-static*.

Recall that for b sufficiently small the canonical model (11.15) is globally asymptotically stable; i.e., it has only one attractor, which is a stable node. For large b, system (11.15) has many attractors. Suppose we start from small b. Then for any initial condition $x(0)$ the activity $x(t)$ approaches the unique attractor, and after some transient process $x(t)$ is in a small neighborhood of the attractor. Let us increase b quasi-statically. The activity $x(t)$ remains in the small neighborhood, provided that the attractor is hyperbolic.

Suppose the input is a combination of two previously memorized images ξ^1 and ξ^2. Suppose also that all conditions of the reduction lemma are satisfied, and $\beta_1 = \beta_2 = \beta$ and $s = a_1 + a_2$ are large enough. Then the qualitative behavior of the canonical model (11.15) is governed by the dynamical system

$$v' = c + (b + \beta)v - v^3 , \quad v \in \mathbb{R} .$$

Suppose $c = 0$; i.e., the input images have equal strength. As we expected, (11.22) has only one attractor $v = 0$ for small b. Here *small* means $b < -\beta$. If we increase b quasi-statically, the activity $v(t)$ is always in a neighborhood of the origin, provided that $b < -\beta$.

When $b = -\beta$, the NN must choose one of the stable branches of the pitchfork bifurcation diagram depicted in Figure 11.12a. The nonzero contrast c is a perturbation (or an imperfection, see Golubitsky and Shaeffer 1979) of the nonhyperbolic equilibrium at the pitchfork bifurcation (see Figure 11.12b and c). No matter how small the contrast c is, the NN correctly chooses the corresponding branch, provided that the quasi-static increasing of b is slow enough. The case when the first image ξ^1 is dominant is depicted in Figure 11.12b. The stroboscopic presentation of the phenomenon is depicted in Figure 11.12d.

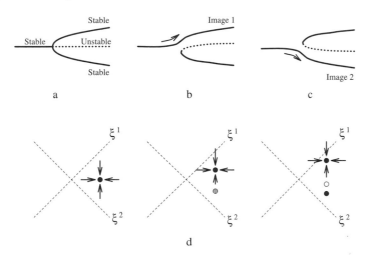

Figure 11.12. Bifurcation diagrams for quasi-static variation of parameter b. (a). The contrast $c = 0$. (b). The first image is dominant. (c). The second image is dominant. (d). Stroboscopic presentation of the phenomenon for $b < -\beta, b = -\beta$ and $b > -\beta$, respectively.

One can speculate that when the internal parameter b crosses a bifurcation value, the NN "thinks". Choosing one of the stable branches could be called "decision-making". Prolonged staying near the unstable branch could be called the "I don't know" state. Thus, we speculate that some types of nonhyperbolic behavior exhibited by the canonical model near bifurcation points are intimately connected to such vague psychological processes as recognition and thinking.

The considerations above are in the spirit of the nonhyperbolic NN approach. The future of the network's activity is determined by the local processes at the nonhyperbolic equilibrium when b crosses the bifurcation value. We discuss and generalize such NNs in Section 3.1.

12

Quasi-Static Bifurcations

In this chapter we analyze the canonical models (6.17) and (6.24) for singularly perturbed WCNNs at quasi-static saddle-node and quasi-static cusp bifurcations. In particular, we consider the canonical models in the special case

$$a_1 = \cdots = a_n = a > 0 .$$

An appropriate change of coordinates and taking the limit as $\varepsilon \to 0$ transform the canonical models to

$$\begin{cases} x_i' = -y_i + r_i x_i + x_i^2 + \sum_{j=1}^{n} c_{ij} x_j \\ y_i' = x_i \end{cases} , \qquad i = 1, \ldots, n , \qquad (12.1)$$

and

$$\begin{cases} x_i' = -y_i + r_i x_i \pm x_i^3 + \sum_{j=1}^{n} c_{ij} x_j \\ y_i' = x_i \end{cases} , \qquad i = 1, \ldots, n , \qquad (12.2)$$

respectively. Here $r_i \in \mathbb{R}$ describes input to the ith relaxation neuron. We use the notation $x = (x_1, \ldots, x_n)^\top \in \mathbb{R}^n$, $y = (y_1, \ldots, y_n)^\top \in \mathbb{R}^n$, and $C = (c_{ij}) \in \mathbb{R}^{n \times n}$.

12.1 Stability of the Equilibrium

Note that the canonical models (12.1) and (12.2) always have a unique equilibrium point, namely, the origin $(x, y) = (0, 0) \in \mathbb{R}^n \times \mathbb{R}^n$. In this section we study the stability of the origin, as determined by the Jacobian matrix

$$L = \begin{pmatrix} R \mid C & -I \\ I & 0 \end{pmatrix}, \qquad (12.3)$$

where I is the unit $n \times n$ matrix, $C = (c_{ij})$, and

$$R = \begin{pmatrix} r_1 & 0 & \cdots & 0 \\ 0 & r_2 & \cdots & 0 \\ \vdots & \vdots & \ddots & \vdots \\ 0 & 0 & \cdots & r_n \end{pmatrix}.$$

Recall that a matrix is called *hyperbolic* if all its eigenvalues have nonzero real parts. It is *stable* if all its eigenvalues have negative real parts. An eigenvalue with the largest (most positive) real part is the *leading* (dominant) eigenvalue.

Theorem 12.1 *The Jacobian matrix $L \in \mathbb{R}^{2n \times 2n}$ defined in (12.3) has the following properties:*

(a) L is nonsingular.
(b) L is stable if and only if $R + C$ is stable.
(c) L is hyperbolic if and only if $R + C$ is hyperbolic.
(d) If $R+C$ has a zero eigenvalue, then L has a pair of purely imaginary eigenvalues $\pm i$.
(e) If $R + C$ has a pair of purely imaginary eigenvalues, then L has two pairs of purely imaginary eigenvalues.

Proof. Suppose $Lv = \mu v$ for some $\mu \in \mathbb{C}$ and a nonzero vector $v \in \mathbb{C}^{2n}$. We use the notation

$$v = \begin{pmatrix} v_1 \\ v_2 \end{pmatrix},$$

where $v_1, v_2 \in \mathbb{C}^n$. Then

$$\begin{aligned} (L - \mu I)v &= \begin{pmatrix} R + C - \mu I & -I \\ I & -\mu I \end{pmatrix} \begin{pmatrix} v_1 \\ v_2 \end{pmatrix} \\ &= \begin{pmatrix} (R + C - \mu I)v_1 - v_2 \\ v_1 - \mu v_2 \end{pmatrix} = \begin{pmatrix} 0 \\ 0 \end{pmatrix}. \end{aligned}$$

Thus, we have

$$\begin{aligned} (R + C - \mu I)v_1 - v_2 &= 0, \\ v_1 - \mu v_2 &= 0. \end{aligned} \qquad (12.4)$$

(a) If $\mu = 0$, then the second equation implies $v_1 = 0$. The first equation implies $v_2 = 0$; therefore, $v = 0$. This means that L cannot have a zero eigenvalue, and hence it is always nonsingular.

(*b*) From part (*a*) we know that $\mu \neq 0$. Then $v_2 = \mu^{-1}v_1$, and the eigenvector v of L has the form

$$v = \begin{pmatrix} v_1 \\ \mu^{-1}v_1 \end{pmatrix}.$$

The first equation in (12.4) gives

$$(R + C - (\mu + \mu^{-1})I)v_1 = 0$$

for some nonzero $v_1 \in \mathbb{R}^n$. Hence the matrix

$$R + C - (\mu + \mu^{-1})I \tag{12.5}$$

is singular. Its eigenvalues are

$$\lambda_i - (\mu + \mu^{-1}), \quad i = 1, \ldots, k,$$

where $\lambda_1, \ldots, \lambda_k$ are those of $R + C$. Since it is singular, at least one of the eigenvalues of (12.5) should be zero. Hence μ is a solution of one of the equations

$$\lambda_i = \mu + \mu^{-1}, \quad i = 1, \ldots, k.$$

It is easy to check that if $\operatorname{Re}\lambda_i < 0$, then $\operatorname{Re}\mu < 0$ and vice versa.

(*c*) The equation above reveals the relationship between eigenvalues of L and $R + C$. If L is nonhyperbolic, then from part (*a*) it follows that μ is purely imaginary. Hence λ is purely imaginary or zero. Conversely, if λ is zero or purely imaginary, then μ is purely imaginary.

(*d*) and (*e*) follow from the equation above. □

Corollary 12.2 *The equilibrium point loses its stability through a (possibly multiple) Andronov-Hopf bifurcation.*

We study the Andronov-Hopf bifurcation in Section 12.3. In particular, we are interested in when the bifurcation is subcritical or supercritical and how this depends on the input.

Next, we analyze some neurophysiological consequences of Theorem 12.1.

12.2   Dale's Principle and Synchronization

One application of (6.1) is in modeling weakly connected networks of relaxation neural oscillators. In this case x_i and y_i denote rescaled activities of local populations of excitatory and inhibitory neurons, respectively. Each coefficient c_{ij} describes the strength of synaptic connections from x_j to x_i.

We say that x_j is depolarized when $x_j > 0$. Notice that if $c_{ij} > 0$ ($c_{ij} < 0$), then depolarization of x_j facilitates (impedes) that of x_i. We call such synapses excitatory (inhibitory). Copious neurophysiological data suggest that excitatory neurons have only excitatory synapses. This observation is usually referred to as *Dale's principle*; see the discussion in Section 1.2.2. In our case it implies that $c_{ij} \geq 0$ for all i and j (where $c_{ij} = 0$ corresponds to absence of a synapse from x_j to x_i).

In this section we show that Dale's principle imposes some restriction on the local dynamics of the canonical models when the origin loses stability. In particular, we prove that the neural oscillators can synchronize.

Theorem 12.3 *Suppose the synaptic matrix $C = (c_{ij})$ satisfies Dale's principle ($c_{ij} \geq 0$ for all i and j). Then generically*

- *The equilibrium of (12.1) and (12.2) loses stability via an Andronov-Hopf bifurcation.*

- *The network's local activity is in-phase synchronized; i.e., any two neural oscillators have nearly zero phase difference.*

The idea of the proof was suggested by S. Treil (Michigan State University).

Proof. Let

$$\rho = \min_{1 \leq i \leq n} (r_i + c_{ii})$$

and consider the matrix A defined by

$$A = R + C - \rho I .$$

Dale's principle ensures that A has nonnegative entries. Now we can apply the Perron-Frobenius theorem (Gantmacher 1959) to A to show that the leading eigenvalue λ of A is real and nonnegative and that the corresponding eigenvector u has only nonnegative entries. Typically, λ has multiplicity one and u has positive entries. The leading eigenvalue of $R + C$ is $\lambda + \rho$, which is real and also has multiplicity one. Theorem 12.1 guarantees that when the equilibrium loses stability, the Jacobian matrix L has only one pair of purely imaginary eigenvalues $\pm i$. Thus, the multiple Andronov-Hopf bifurcation is not typical in this sense.

From the proof of Theorem 12.1 it follows that the corresponding eigenvectors of L have the form

$$\begin{pmatrix} u \\ \mp iu \end{pmatrix} ,$$

where u was defined above. The local activity near the equilibrium is described by

$$\begin{pmatrix} x(t) \\ y(t) \end{pmatrix} = \begin{pmatrix} u \\ -iu \end{pmatrix} z(t) + \begin{pmatrix} u \\ +iu \end{pmatrix} \bar{z}(t) + \text{ higher-order terms,}$$

where $z(t) \in \mathbb{C}$ is small. The activity of each neural oscillator has the form

$$\begin{pmatrix} x_i(t) \\ y_i(t) \end{pmatrix} = 2u_i \begin{pmatrix} \operatorname{Re} z(t) \\ \operatorname{Im} z(t) \end{pmatrix} + \text{higher-order terms.} \tag{12.6}$$

We can express the activity of the ith oscillator through that of the jth oscillator by

$$\begin{pmatrix} x_i(t) \\ y_i(t) \end{pmatrix} = \frac{u_i}{u_j} \begin{pmatrix} x_j(t) \\ y_j(t) \end{pmatrix} + \text{higher-order terms,} \tag{12.7}$$

where $u_i/u_j > 0$ because u has positive entries. Hence the ith and jth oscillators have zero phase difference (up to some order). □

If the Andronov-Hopf bifurcation is supercritical; i.e., there is a birth of a stable limit cycle, then in-phase synchronization is asymptotic. More precisely, all local solutions have the form (12.6), where $z(t)$ is small and periodic and the higher-order terms remain sufficiently small as $t \to \infty$.

If the Andronov-Hopf bifurcation is subcritical (i.e., there is a death of an unstable limit cycle), then $z(t)$ in (12.6) grows as $t \to \infty$, and in-phase synchronization is only *local*. The higher-order terms in (12.6) can grow with time, and after a while they can be significant.

Remark. If Dale's principle is not satisfied, then:

- A multiple Andronov-Hopf bifurcation with precisely two pairs of purely imaginary eigenvalues is also generic. This follows from Theorem 12.1, part (e).

- Either in-phase or anti-phase synchronization is possible; i.e., the phase difference between any two oscillators could be nearly 0 or π. This follows from (12.7) because u_i/u_j is a scalar and could be positive or negative.

There have been many studies of existence and stability of in phase and anti-phase solutions in linearly coupled relaxation oscillators. See, for example, Belair and Holmes (1984), Storti and Rand (1986), Somers and Kopell (1993), Kopell and Somers (1995), and Mirollo and Strogatz (1990). Our results complement and extend those of these authors, since they perform global analysis of *two strongly* coupled oscillators, while we perform local analysis of n weakly coupled oscillators.

Remark. An important difference between weakly connected networks of relaxation and nonrelaxation oscillators is that in the former the phase differences are usually either 0 or π, but in the latter they may assume arbitrary values.

12.3   Further Analysis of the Andronov-Hopf Bifurcation

In this section we study the Andronov-Hopf bifurcation in canonical models when $R + C$ has a simple zero eigenvalue. Our major goal is to determine when it is subcritical or supercritical and how this depends on the matrix C and the inputs $r_1, \ldots, r_n$.

We begin with an analysis of the canonical model (12.2)

$$\begin{cases} x_i' = -y_i + r_i x_i + \sigma_i x_i^3 + \sum_{j=1}^{n} c_{ij} x_j \\ y_i' = x_i \end{cases} , \quad \sigma_i = \pm 1 , \quad i = 1, \ldots, n ,$$

because it is simpler than that of (12.1).

Let $v_1 = (v_{11}, \ldots, v_{1n})^\top \in \mathbb{R}^n$ be the normalized eigenvector of $R + C$ corresponding to the zero eigenvalue. Let $w_1 = (w_{11}, \ldots, w_{1n}) \in \mathbb{R}^n$ be dual to v_1; i.e.,

$$w_1 v_1 = \sum_{i=1}^{n} w_{1i} v_{1i} = 1$$

and w_1 is orthogonal to the other (generalized) eigenvectors of $R + C$.

Theorem 12.4 *If the parameter a defined by*

$$a = \frac{3}{4} \sum_{i=1}^{n} \sigma_i w_{1i} v_{1i}^3 \tag{12.8}$$

is positive (negative), then the Andronov-Hopf bifurcation in (12.2) is subcritical (supercritical).

Proof of the theorem is given in Appendix, Section 12.7.

In neural network studies it is frequently assumed that synapses are modified according to the Hebbian learning rule. This implies that the synaptic matrix C is symmetric. It is also reasonable to consider a network of approximately similar oscillators. They can have different quantitative features, but their qualitative behavior should be comparable. These two observations motivate the following result.

Corollary 12.5 *Suppose that*

1. *The synaptic matrix $C = (c_{ij})$ is symmetric.*

2. *All oscillators have the same type, i.e.,*

$$\sigma_1 = \cdots = \sigma_n = \sigma .$$

If $\sigma = +1$ ($\sigma = -1$), then the Andronov-Hopf bifurcation in (12.2) is always subcritical (supercritical).

Proof. If C is symmetric, then so is $R + C$. Since every symmetric matrix has orthogonal eigenvectors, we have $v_1 = w_1^\top$. Therefore, (12.8) can be rewritten as

$$a = \frac{3}{4}\sigma \sum_{i=1}^{n} v_{1i}^4 \ ,$$

and its sign is determined by σ. $\square$

We can relax assumption 1 in the corollary simply by requiring that v_1 be orthogonal to the other (generalized) eigenvectors.

Analysis of the canonical model (12.1)

$$\begin{cases} x_i' = -y_i + r_i x_i + x_i^2 + \sum_{j=1}^{n} c_{ij} x_j \\ y_i' = x_i \end{cases} ,$$

is more complicated than that of (12.2). Since we are interested mostly in the case when the synaptic matrix C is symmetric, we assume at the very start that it is. The case when C is not symmetric has yet to be studied.

Theorem 12.6 *The Andronov-Hopf bifurcation in (12.1) for a symmetric matrix $C = (c_{ij})$ is always subcritical.*

Proof of the theorem is given in Appendix, Section 12.7.

12.4   Nonhyperbolic Neural Networks

Let us return to the question of how the canonical models might perform pattern recognition tasks. First we outline the main idea, and then we present rigorous mathematical considerations.

Together with the canonical models for a multiple quasi-static saddle-node bifurcation

$$\begin{cases} x_i' = -y_i + r_i x_i + x_i^2 + \sum_{j=1}^{n} c_{ij} x_j \\ y_i' = x_i \end{cases} \tag{12.9}$$

and pitchfork bifurcation

$$\begin{cases} x_i' = -y_i + r_i x_i + x_i^3 + \sum_{j=1}^{n} c_{ij} x_j \\ y_i' = x_i \end{cases} , \tag{12.10}$$

we may as well consider the canonical models for multiple subcritical pitchfork bifurcation

$$x_i' = r_i x_i + x_i^3 + \sum_{j=1}^{n} c_{ij} x_j \ , \tag{12.11}$$

or the canonical model for a multiple subcritical Andronov-Hopf bifurcation

$$z_i' = r_i z_i + d_i z_i |z_i|^2 + \sum_{j=1}^{n} c_{ij} z_j , \quad \operatorname{Re} d_i > 0 . \tag{12.12}$$

Notice that these models have an equilibrium – the origin – for any choice of parameters. Stability of the origin is determined by the Jacobian matrix $R + C$, where

$$R = \begin{pmatrix} r_1 & 0 & \cdots & 0 \\ 0 & r_2 & \cdots & 0 \\ \vdots & \vdots & \ddots & \vdots \\ 0 & 0 & \cdots & r_n \end{pmatrix}$$

and $C = (c_{ij})$ is the synaptic matrix. Both R and C are complex-valued for (12.12) and real-valued for the other canonical models. The origin is stable if all eigenvalues of $R + C$ have negative real parts and is unstable otherwise.

The origin loses its stability via subcritical pitchfork bifurcation (for (12.11)) or Andronov-Hopf bifurcation (for the other models). In some neighborhood of the bifurcation point the direction along the eigenvector corresponding to the leading eigenvalue of $R + C$ becomes unstable, and the activity vector moves along this direction. After a while it leaves a small neighborhood of the origin, and an observer notices some macroscopic changes in dynamics of the canonical models (see Figure 3.5). Thus the local event – loss of stability by the origin – produces a global effect. This is the key idea of the nonhyperbolic NN approach, which we described in Section 3.2. Below, we explain in detail the idea outlined above.

12.5   Problem 1

Given an input vector $r^k = (r_1^k, \ldots, r_n^k) \in \mathbb{R}^n$ we can construct diagonal matrix R^k by

$$R^k = \begin{pmatrix} r_1^k & 0 & \cdots & 0 \\ 0 & r_2^k & \cdots & 0 \\ \vdots & \vdots & \ddots & \vdots \\ 0 & 0 & \cdots & r_n^k \end{pmatrix} .$$

We use the following notation: $\lambda_k \in \mathbb{R}$ denotes the leading eigenvalue of the matrix $R^k + C$, i.e., the eigenvalue with the largest real part. The vector $u^k \in \mathbb{R}^n$ denotes an eigenvector of $R^k + C$ corresponding to λ_k if the leading eigenvalue is unique and simple, i.e., it has multiplicity one.

Suppose we are given a set of input vectors $\{r^1, \ldots, r^m\} \subset \mathbb{R}^n$ and a set of key patterns $\{v^1, \ldots, v^m\} \subset \mathbb{R}^n$ to be memorized. Consider the following problem:

PROBLEM 1. *Find a matrix $C \in \mathbb{R}^{n \times n}$ such that for all matrices $R^k + C$, $k = 1, \ldots, m$, the leading eigenvalues $\lambda_k = 0$ are simple and $u^k = v^k$.*

Suppose that given $r^1, \ldots, r^m$ and $v^1, \ldots, v^m$ there is such a matrix C. Then the canonical models can perform the pattern recognition tasks in the sense described next.

We say that the kth input r^k from external receptors is given if the parameters $r_1, \ldots, r_n$ in the canonical models are given by

$$r_i = r_i^k + \rho, \qquad i = 1, \ldots, n, \tag{12.13}$$

where $\rho \in \mathbb{R}$ is a scalar bifurcation parameter. Then, for $\rho < 0$ the equilibrium point of the canonical models (the origin) is stable; for $\rho = 0$ there is a bifurcation; and for $\rho > 0$ the equilibrium is unstable.

For the canonical model (12.11) the equilibrium loses stability through subcritical pitchfork bifurcation (see Section 11.5). For small positive ρ the canonical model dynamics approach the center manifold, which is tangent to the center subspace $E \in \mathbb{R}^n$ defined by

$$E = \text{span}\left\{u^k\right\},$$

where u^k is the eigenvector of $R^k + C$ corresponding to the leading eigenvalue λ_k. According to Problem 1, the vector u^k coincides with the memorized pattern v^k. Thus, when the kth input pattern is given, the activity of the canonical models is close to the linear subspace E that is determined by the memorized pattern v^k. A rough sketch of the local dynamics is depicted in Figure 12.1.

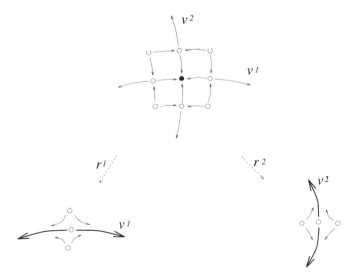

Figure 12.1. Depending upon the input (r^1 or r^2), the center manifold is tangent to the corresponding memorized vectors (v^1 or v^2).

In the canonical models (12.12), (12.9), and (12.10) the origin loses stability through Andronov-Hopf bifurcations (see the remark on p. 304 and Section 12.3). The center manifold is tangent to the center subspace defined by

$$E = \operatorname{span} \left\{ \begin{pmatrix} u^k \\ 0 \end{pmatrix}, \begin{pmatrix} 0 \\ \bar{u}^k \end{pmatrix} \right\}. \tag{12.14}$$

If the Andronov-Hopf bifurcation is supercritical for all inputs r^k, then the newborn stable limit cycles lie close to the corresponding center subspaces defined by (12.14). Thus, for each input r^k there is an attractor – stable limit cycle, which has a location prescribed by the memorized pattern v^k. This was observed in the olfactory bulb (see Section 3.1.1). The only difference is that the limit cycles in the canonical models have small radii. Finally, we note that the supercritical Andronov-Hopf bifurcations can be observed in the canonical model (12.2) for $\sigma_i = -1$, $i = 1, \ldots, n$. Therefore, (12.2) can function like a GAS-type neural network, but there are two steady states: an unstable equilibrium (the origin) and a stable limit cycle.

If the Andronov-Hopf bifurcation is subcritical for all inputs r^k, then the canonical models are like nonhyperbolic neural networks. Indeed, after the bifurcation, which is local, the dynamics leave some neighborhood of the equilibrium point along the direction determined by one of the memorized patterns $v^1, \ldots, v^m$.

For example, suppose the network recognizes v^1 and that $|v_1^1| \gg |v_2^1|$. Then the first neural oscillator oscillates with amplitude much larger than that of the second one. If the Andronov-Hopf bifurcation is supercritical, then the attractor has the same property. If the bifurcation is subcritical, then these oscillations are observed locally and might persist globally. In both cases, an experimenter on the olfactory bulb discovers that there is a spatial pattern of oscillations: various sites of the olfactory bulb oscillate with the same frequency but with different amplitudes (Skarda and Freeman 1987). We call such an activity the amplitude modulation (AM).

12.6   Problems 2 and 3

To the best of our knowledge, Problem 1 is still unresolved. We do not know any general method that allows one to construct such a synaptic matrix C, although for some special cases, such as $m = 1$, the construction is trivial.

Below we present alternative problems that can be easier to resolve.

PROBLEM 2. *Find matrix C such that each λ_k is simple and real and $u^k = v^k$ for all k.*

Since we do not require that $\lambda_k = 0$, the bifurcations occur for $\rho = -\lambda_k$, where ρ is defined in (12.13).

PROBLEM 3. *Find matrix C such that the leading eigenvalues λ_k of $R^k +$ C, $k = 1, \ldots, m$, are real and simple and the corresponding eigenvectors u^k are pairwise orthogonal.*

Note that we do not require here that $u^k = v^k$. The requirement that all u^k be orthogonal means that the response of the neural network on various inputs is as different as possible, even when the inputs to be memorized are similar (like a cat and a dog).

12.7 Proofs of Theorems 12.4 and 12.6

The proofs of the theorems use the center manifold reduction, which we discussed in Section 2.1.1. Below we adapt the center manifold results to our $2n$-dimensional systems.

The canonical models (12.1) and (12.2) can be written concisely in the form

$$Z' = F(Z), \tag{12.15}$$

where $Z = (x, y) \in \mathbb{R}^{2n}$, $F : \mathbb{R}^{2n} \to \mathbb{R}^{2n}$, and $F(0) = 0$. The Jacobian matrix $L = DF$ at the equilibrium is given by (12.3). From Theorem 12.1 it follows that the equilibrium $Z = 0$ loses stability via an Andronov-Hopf bifurcation when $R + C$ has one simple zero eigenvalue and the other eigenvalues lie in the left half-plane. Let v_1 be the eigenvector of $R + C$ corresponding to the zero eigenvalue. Then L has a pair of purely imaginary eigenvalues $\pm i$ with the corresponding eigenvectors

$$\begin{pmatrix} v_1 \\ \mp i v_1 \end{pmatrix}. \tag{12.16}$$

To determine the type of bifurcation that occurs, we restrict (12.15) to the center manifold, which is tangent to the center subspace:

$$\begin{aligned} E^c &= \operatorname{span}\left\{ \begin{pmatrix} v_1 \\ 0 \end{pmatrix}, \begin{pmatrix} 0 \\ v_1 \end{pmatrix} \right\} \\ &= \left\{ \begin{pmatrix} v_1 \\ 0 \end{pmatrix} x + \begin{pmatrix} 0 \\ v_1 \end{pmatrix} y \mid x, y \in \mathbb{R} \right\}, \end{aligned} \tag{12.17}$$

where x and y can be treated as coordinates on E^c. On the manifold (12.15) has the normal form

$$\begin{cases} x' = -y + f(x, y) \\ y' = x + g(x, y) \end{cases}, \tag{12.18}$$

where f and g denote the nonlinear terms in x and y. Then the Andronov-Hopf bifurcation is subcritical (supercritical) if the parameter

$$\begin{aligned} a = \ &\tfrac{1}{16}\left(f_{xxx} + f_{xyy} + g_{xxy} + g_{yyy} + f_{xy}(f_{xx} + f_{yy}) \right. \\ &\left. -g_{xy}(g_{xx} + g_{yy}) - f_{xx}g_{xx} + f_{yy}g_{yy} \right) \end{aligned} \tag{12.19}$$

is positive (negative). Derivation of (12.19) can be found, for example, in Guckenheimer and Holmes (1983). Note that during the center manifold reduction it suffices to compute f and g only up to third-order terms in x and y.

To perform the reduction we must introduce some objects : Let $E^s \subset \mathbb{R}^{2n}$ be the stable subspace spanned by the (generalized) eigenvectors of L corresponding to the eigenvalues of L having negative real parts. Thus, we have the splitting

$$\mathbb{R}^{2n} = E^c \oplus E^s .$$

Let $\Pi_c : \mathbb{R}^{2n} \to E^c$ and $\Pi_s : \mathbb{R}^{2n} \to E^s$ be projectors such that

$$\ker \Pi_c = E^s \quad \text{and} \quad \ker \Pi_s = E^c .$$

If w_1 is a dual vector to v_1, then Π_c is given by

$$\Pi_c = \begin{pmatrix} v_1 w_1 \\ v_1 w_1 \end{pmatrix} , \tag{12.20}$$

where $v_1 w_1$ denotes the $n \times n$ matrix defined by the tensor product $v_1 \otimes w_1$. Note also that Π_c and Π_s commute with L. The center manifold theorem ensures that there is a mapping $\Psi : E^c \to E^s$ with

$$\Psi(0) = 0 \quad \text{and} \quad D\Psi(0) = 0$$

such that the manifold $\mathcal{M}$ defined by

$$\mathcal{M} = \{v + \Psi(v) \mid v \in E^c\}$$

is invariant and locally attractive. The reduced system has the form

$$v' = \Pi_c F(v + \Psi(v)) , \tag{12.21}$$

and (12.18) is just (12.21) written in local coordinates on E^c.

The initial portion of the Taylor's expansion of the function $\Psi(v)$, which defines the center manifold, can be determined from the equation

$$D\Psi(v)v' = \Pi_s F(v + \Psi(v)) , \tag{12.22}$$

where v' is defined in (12.21).

We do not need (12.22) for proving Theorem 12.4 because the canonical model (12.2) does not have quadratic terms in x and y. But we use (12.22) in proof of Theorem 12.6.

12.7.1 Proof of Theorem 12.4

From (12.17) we see that

$$v' = \begin{pmatrix} v_1 \\ 0 \end{pmatrix} x' + \begin{pmatrix} 0 \\ v_1 \end{pmatrix} y' . \tag{12.23}$$

Since Π_c and L commute, $\Pi_c \Psi = 0$, and $\Pi_c v = v$ for $v \in E^c$, we have

$$\Pi_c L(v + \Psi(v)) = L\Pi_c(v + \Psi(v)) = Lv . \tag{12.24}$$

Therefore, the right-hand side of (12.21) is

$$L \begin{pmatrix} v_1 \\ 0 \end{pmatrix} x + L \begin{pmatrix} 0 \\ v_1 \end{pmatrix} y + \Pi_c \begin{pmatrix} (\sigma_1(v_{11}x)^3, \ldots, \sigma_n(v_{1n}x)^3) + \text{h.o.t.} \\ 0 \end{pmatrix}$$

$$= \begin{pmatrix} 0 \\ v_1 \end{pmatrix} x + \begin{pmatrix} -v_1 \\ 0 \end{pmatrix} y + \Pi_c \begin{pmatrix} v_1 x^3 \sum_{i=1}^n w_{1i}\sigma_i v_{1i}^3 + \text{h.o.t.} \\ 0 \end{pmatrix} .$$

Multiplying by $(w_1, 0)$ and $(0, w_1)$ gives

$$\begin{cases} x' & = & -y & + x^3 \sum_{i=1}^n \sigma_i w_{1i} v_{1i}^3 + \text{h.o.t.} \\ y' & = & x \end{cases} .$$

The parameter a defined in (12.19) is

$$a = \frac{3}{4} \sum_{i=1}^n \sigma_i w_{1i} v_{1i}^3 .$$

□

12.7.2   Proof of Theorem 12.6

If the connection matrix C is symmetric, then so is $R+C$. Therefore, $R+C$ has n orthogonal eigenvectors $v_1, \ldots, v_n$ forming a basis for $\mathbb{R}^n$. As before, the vector v_1 corresponds to the zero eigenvalue of $R + C$ and $v_2, \ldots, v_n$ correspond to the other eigenvalues $\lambda_2, \ldots, \lambda_n$, which are negative. Using the proof of Theorem 12.1 we can define the stable subspace of L by

$$E^s - \text{span} \left\{ \begin{pmatrix} v_2 \\ 0 \end{pmatrix}, \begin{pmatrix} 0 \\ v_2 \end{pmatrix}, \ldots, \begin{pmatrix} v_n \\ 0 \end{pmatrix}, \begin{pmatrix} 0 \\ v_n \end{pmatrix} \right\}$$

and

$$\Pi_s = \begin{pmatrix} \sum_{k-2}^n v_k v_k^T \\ \sum_{k=2}^n v_k v_k^T \end{pmatrix} .$$

To determine the parameter u we must find the quadratic terms of $\Psi(v)$. Let

$$\Psi \left(\begin{pmatrix} v_1 \\ 0 \end{pmatrix} x + \begin{pmatrix} 0 \\ v_1 \end{pmatrix} y \right) = x^2 \begin{pmatrix} p_1 \\ q_1 \end{pmatrix} + xy \begin{pmatrix} p_2 \\ q_2 \end{pmatrix} + y^2 \begin{pmatrix} p_3 \\ q_3 \end{pmatrix} + \text{h.o.t.} ,$$

where

$$p_i = \sum_{k=2}^n p_{ik} v_k \quad \text{and} \quad q_i = \sum_{k=2}^n q_{ik} v_k$$

for $i = 1, 2, 3$. Since

$$\begin{cases} x' = -y + \text{h.o.t.} \\ y' = x + \text{h.o.t.} \end{cases},$$

the left-hand side of (12.22) is

$$D\Psi(v)v' = xy \begin{pmatrix} 2p_3 - 2p_1 \\ 2q_3 - 2q_1 \end{pmatrix} + (x^2 - y^2) \begin{pmatrix} p_2 \\ q_2 \end{pmatrix} + \text{h.o.t.}$$

Since Π_s commutes with L, and $\Pi_s v = 0$, we have

$$\begin{aligned}
\Pi_s L(v + \Psi(v)) &= L\Pi_s(v + \Psi(v)) = L\Psi(v) \\
&= x^2 \begin{pmatrix} \sum_{k=2}^{n} \lambda_k v_k p_{1k} - q_1 \\ p_1 \end{pmatrix} \\
&\quad + xy \begin{pmatrix} \sum_{k=2}^{n} \lambda_k v_k p_{2k} - q_2 \\ p_2 \end{pmatrix} \\
&\quad + y^2 \begin{pmatrix} \sum_{k=2}^{n} \lambda_k v_k p_{3k} - q_3 \\ p_3 \end{pmatrix} + \text{h.o.t.}
\end{aligned}$$

Thus, the right-hand side of (12.22) is

$$L\Psi(v) + \Pi_s \begin{pmatrix} (v_{1i}x)_i^2 \\ 0 \end{pmatrix} = L\Psi(v) + x^2 \begin{pmatrix} \sum_{k=2}^{n} v_k \sum_{i=1}^{n} v_{ki} v_{1i}^2 \\ 0 \end{pmatrix}.$$

Combining like terms in the three equations above and considering projections on each v_k, we obtain the system

$$\begin{pmatrix} p_{2k} \\ q_{2k} \end{pmatrix} = \begin{pmatrix} \lambda_k p_{1k} - q_{1k} \\ p_{1k} \end{pmatrix} + \begin{pmatrix} \sum_{i=1}^{n} v_{ki} v_{1i}^2 \\ 0 \end{pmatrix}$$

$$2 \begin{pmatrix} p_{3k} \\ q_{3k} \end{pmatrix} - 2 \begin{pmatrix} p_{1k} \\ q_{1k} \end{pmatrix} = \begin{pmatrix} \lambda_k p_{2k} - q_{2k} \\ p_{2k} \end{pmatrix}$$

$$- \begin{pmatrix} p_{2k} \\ q_{2k} \end{pmatrix} = \begin{pmatrix} \lambda_k p_{3k} - q_{3k} \\ p_{3k} \end{pmatrix},$$

which is be solved for p_{ik}, q_{ik}, $i = 1, 2, 3$; $k = 2, \ldots, n$. Below we will use only the values of p_{1k} and p_{3k}, which are given by

$$p_{1k} = -p_{3k} = -\frac{2\lambda_k}{4\lambda_k^2 + 9} \sum_{i=1}^{n} v_{ki} v_{1i}^2. \tag{12.25}$$

Now let us determine the reduced system (12.21). Its left-hand side is given by (12.23). Using (12.24), we can find its right-hand side: it is

$$\begin{aligned}
&L \begin{pmatrix} v_1 \\ 0 \end{pmatrix} x + L \begin{pmatrix} 0 \\ v_1 \end{pmatrix} y \\
&+ \Pi_c \begin{pmatrix} (v_{1i}x + \sum_{k=2}^{n} v_{ki}(x^2 p_{1k} + xy p_{2k} + y^2 p_{3k}) + \text{h.o.t.})_i^2 \\ 0 \end{pmatrix} \\
&= \begin{pmatrix} 0 \\ v_1 \end{pmatrix} x + \begin{pmatrix} -v_1 \\ 0 \end{pmatrix} y \\
&+ \begin{pmatrix} v_1 \sum_{i=1}^{n} v_{1i} \{ v_{1i}^2 x^2 + 2 v_{1i} x \sum_{k=2}^{n} v_{ki}(x^2 p_{1k} + xy p_{2k} + y^2 p_{3k}) \} + \text{h.o.t.} \\ 0 \end{pmatrix}.
\end{aligned}$$

Multiplying by $(v_1^\top, 0)$ and $(0, v_1^\top)$ gives

$$
\begin{cases}
x' = -y + x^2 \sum_{i=1}^n v_{1i}^3 + x^3 \sum_{i=1}^n \sum_{k=2}^n 2v_{1i}^2 v_{ki} p_{1k} \\
\quad + xy^2 \sum_{i=1}^n \sum_{k=2}^n 2v_{1i}^2 v_{ki} p_{3k} \ + \ \text{h.o.t.} \\
y' = x
\end{cases}
.
$$

It follows from (12.19) that

$$
a = \frac{1}{4} \sum_{i=1}^n v_{1i}^2 \sum_{k=2}^n v_{ki}(3p_{1k} + p_{3k}) \, .
$$

Using (12.25) we see that

$$
a = -\sum_{k=2}^n \left(\sum_{i=1}^n v_{ki} v_{1i}^2 \right)^2 \frac{\lambda_k}{4\lambda_k^2 + 9} > 0 \, ,
$$

since each $\lambda_k < 0$. $\square$

13
Synaptic Organizations of the Brain

In this chapter we study the relationship between synaptic organizations and dynamical properties of networks of neural oscillators. In particular, we are interested in which synaptic organizations can memorize and reproduce phase information. Most of our results are obtained for neural oscillators near multiple Andronov-Hopf bifurcations.

13.1   Introduction

Neurophysiological studies of various brain structures show that there is a pattern of local synaptic circuitry in many parts of the brain: Local populations of excitatory and inhibitory neurons have extensive and strong synaptic connections between each other, so that action potentials generated by the former excite the latter, which in turn, reciprocally inhibit the former (see Figure 13.1). They can consist of motoneurons and Renshaw interneurons in the spinal cord, mitral and granule cells in the olfactory bulb, pyramidal cells and thalamic interneurons in the cortico-thalamic system, pyramidal and basket cells in the hippocampus, etc. Such pairs of interacting excitatory and inhibitory populations of neurons, which we called *neural oscillators* in Section 1.4.4, can also be found in the cerebellum, olfactory cortex, and neocortex.

The neural oscillators within one brain structure can be connected into a network because the excitatory (and sometimes inhibitory) neurons can have synaptic contacts with other, distant, neurons. For example, in the

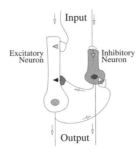

Figure 13.1. Schematic representation of the neural oscillator. It consists of excitatory (white) and inhibitory (shaded) populations of neurons. For simplicity only one neuron from each population is pictured. White arrow denotes excitatory synaptic connections; black, inhibitory; and gray, undetermined.

olfactory bulb the mitral cells have contacts with other mitral cells (see Figure 13.2), whereas the granule cells apparently do not make any distant contacts, they do not even have axons. Their only purpose is to provide a reciprocal dendro-dendritic inhibition for the mitral cells. Sometimes inhibitory neurons can also have long axons; for example, the periglomerular cells in the olfactory bulb.

Though on the local level all neural oscillators appear to be similar, the type of connections between them may differ. In this case we say that the networks of such oscillators have different *synaptic organizations*. For instance, in Figure 13.2 the contacts between mitral cells and distant granule cells (dashed line) might or might not exist. These cases correspond to various synaptic organizations and hence to various dynamical properties of the network. The notion of synaptic organization is closely related to the notion of anatomy of the brain. Thus, in this chapter we study relationships between anatomy and functions of the brain. For example, we show that some synaptic organizations allow the network to memorize time delays, or phase deviation information, whereas the others do not allow such a possibility.

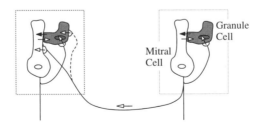

Olfactory Bulb

Figure 13.2. The neural oscillators (dotted boxes) are connected into a network. The mitral cell makes contacts with other mitral cells and may have contacts with other granule cells.

For the sake of clarity we always depict only two neural oscillators and the synaptic connections only in one direction, as in Figure 13.2. It is implicitly assumed that the network consists of many neural oscillators and the synaptic connections of the same type exist between any two oscillators and in all directions, unless stated otherwise.

13.2 Neural Oscillators

In this chapter we assume that activities of both excitatory and inhibitory neurons are described by one-dimensional variables, say $x \in \mathbb{R}$ and $y \in \mathbb{R}$, respectively. This is a technical assumption made to simplify our computations and to allow us to use Dale's principle, which we discussed in Section 1.2.2.

We assume that excitatory neuron activity is governed by the equation

$$\dot{x} = f(x, \rho_x) \, ,$$

where $\rho_x \in \mathbb{R}$ describes the input to the neuron from the inhibitory neuron y and from other neurons of the network. The function f satisfies $f'_\rho \geq 0$, which means that increasing the input ρ_x increases the activity x. Similarly, we assume that inhibitory neuron dynamics is governed by

$$\dot{y} = g(\rho_y, y) \, ,$$

where $\rho_y \in \mathbb{R}$ describes the input from x and other neurons, and $g'_\rho \geq 0$.

If we neglect inputs from other neurons, we may take (without loss of generality) $\rho_x = -y$ and $\rho_y = x$ to comply with Dale's principle. In this case a neural oscillator has the form

$$\begin{cases} \dot{x} = f(x, -y) \\ \dot{y} = g(x, \, y) \end{cases} , \qquad (13.1)$$

A typical example of such a neural oscillator is the Wilson-Cowan's model (Section 1.4.4). We cannot use Dale's principle to determine the sign of $\partial f / \partial x$ or $\partial g / \partial y$ because these are not synaptic coefficients, but nonlinear feedback quantities.

In this chapter we distinguish two cases:

1. The neural oscillator (13.1) is at a supercritical Andronov-Hopf bifurcation.

2. The neural oscillator (13.1) has a stable hyperbolic limit cycle attractor.

The first case is easier than the second one in many respects.

Suppose the pair x_i, y_i denotes activity of the ith neural oscillator for each $i = 1, \ldots, n$. A weakly connected network of such oscillators is a dynamical system of the form

$$
\begin{cases}
\dot{x}_i = f(x_i, \rho_{xi}) , & \rho_{xi} = -y_i + \varepsilon \sum_{j=1}^{n} \left(s_{ij1} x_j - s_{ij2} y_j \right) \\
\dot{y}_i = g(\rho_{yi}, y_i) , & \rho_{yi} = x_i + \varepsilon \sum_{j=1}^{n} \left(s_{ij3} x_j - s_{ij4} y_j \right)
\end{cases} , \qquad (13.2)
$$

where synaptic constants $s_{ijk} > 0$ for all i, j and k, and $\varepsilon \ll 1$. One can think of (13.2) as being a generalization of the Wilson-Cowan model (1.3).

It should be noted that we do not require that the connections within an oscillator be weak. For our analysis we use weakness of connections between neural oscillators.

We can rewrite the system (13.2) in the form

$$
\begin{cases}
\dot{x}_i = f(x_i, -y_i) + \varepsilon p_i(x, y, \varepsilon) \\
\dot{y}_i = g(x_i, y_i) + \varepsilon q_i(x, y, \varepsilon)
\end{cases} , \qquad (13.3)
$$

where the functions p and q have the following special form

$$
p_i(x, y, 0) = f_\rho'(x_i, -y_i) \sum_{j=1}^{n} \left(s_{ij1} x_j - s_{ij2} y_j \right) , \qquad (13.4)
$$

$$
q_i(x, y, 0) = g_\rho'(x_i, y_i) \sum_{j=1}^{n} \left(s_{ij3} x_j - s_{ij4} y_j \right) . \qquad (13.5)
$$

Obviously, system (13.2) is not the most general, though the majority of models of oscillatory neural networks can be so written. Besides, almost all results that we discuss in this chapter were obtained for more general WCNN of the form (13.3) for arbitrary functions p and q (Hoppensteadt and Izhikevich 1996a,b).

13.3   Multiple Andronov-Hopf Bifurcation

The supercritical Andronov-Hopf bifurcation corresponds to the birth or disappearance of a small amplitude limit cycle attractor. It is a local bifurcation at an equilibrium. Without loss of generality we may assume that each neural oscillator (13.1) has $(0,0)$ as an equilibrium. So we have

$$
f(0,0) = 0 ,
$$
$$
g(0,0) = 0 .
$$

It is convenient to denote the Jacobian matrix at the equilibrium by

$$
L = \begin{pmatrix} a_1 & -a_2 \\ a_3 & -a_4 \end{pmatrix} ; \qquad (13.6)
$$

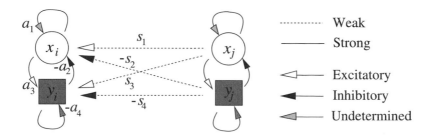

Figure 13.3. A network of two neural oscillators.

see Figure 13.3. The neural oscillator is at an Andronov-Hopf bifurcation (see Section 2.2.4) when the Jacobian matrix L satisfies

$$\operatorname{tr} L = a_1 - a_4 = 0 \qquad \text{and} \qquad \det L = -a_1 a_4 + a_2 a_3 > 0 .$$

We associate with each neural oscillator its natural frequency

$$\Omega = \sqrt{\det L} = \sqrt{a_2 a_3 - a_1 a_4}$$

in this case.

Dale's principle implies that a_2 and a_3 in (13.6) are nonnegative constants. Since $a_1 = a_4$ the condition $\det L > 0$ implies that both a_2 and a_3 are strictly positive. Neither $a_1 \geq 0$ nor $a_1 < 0$ contradicts Dale's principle. To distinguish these cases, we say that a neural oscillator is of *Type A* in the first case and of *Type B* in the second case. The differences between Type A and B neural oscillators (see Figure 13.4) were not essential for our

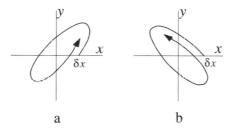

a   b

Figure 13.4. Differences in dynamic behavior of type A and B neural oscillators.

mathematical analysis of multiple Andronov-Hopf bifurcations so far. Nevertheless, they are crucial when we classify synaptic organizations. Both types were studied by Hoppensteadt and Izhikevich (1996a,b), who showed that Type A neural oscillators have more interesting properties: In some sense they are "smarter". For the sake of simplicity we consider only Type A below; that is, we implicitly assume that $a_1 = a_4 \geq 0$.

13.3.1   Converted Synaptic Coefficients

From Theorem 5.8 it follows that the WCNN (13.2) near multiple Andronov-Hopf bifurcation is governed by the canonical model

$$z_i' = (\alpha_i + \mathrm{i}\omega_i)z_i + (\sigma + \mathrm{i}\gamma)z_i|z_i|^2 + \sum_{j \neq i}^{n} c_{ij}z_j . \tag{13.7}$$

In this chapter we derive the exact relationship between complex-valued synaptic coefficients c_{ij} and actual synaptic functions p_i and q_i from (13.3). This is important for neurobiological interpretations of the theory developed below. It will allow us to interpret all results obtained by studying (13.3) in terms of the original WCNN (13.2), i.e., in terms of excitatory and inhibitory populations of neurons and interactions between them.

Lemma 13.1 *The relationship between c_{ij} in (13.7) and*

$$S_{ij} = \left(\begin{array}{cc} s_1 & -s_2 \\ s_3 & -s_4 \end{array} \right)_{ij}$$

in (13.2) is given by

$$c_{ij} = u_1 s_1 + u_2 s_2 + u_3 s_3 + u_4 s_4 , \tag{13.8}$$

where

$$u_1 = \frac{a_2}{2} - \mathrm{i}\frac{a_2 a_4}{2\Omega} , \qquad u_2 = u_3 = \mathrm{i}\frac{a_2 a_3}{2\Omega} , \qquad u_4 = -\frac{a_3}{2} - \mathrm{i}\frac{a_3 a_4}{2\Omega}$$

are some complex numbers, which we show as vectors in $\mathbb{R}^2$ in Figure 13.5.

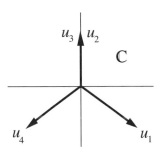

Figure 13.5. Complex numbers u_k from Lemma 13.1.

Proof. The eigenvectors of L are

$$v = \left(\begin{array}{c} 1 \\ \frac{a_4 - \mathrm{i}\Omega}{a_2} \end{array} \right) \qquad \text{and} \qquad \bar{v} = \left(\begin{array}{c} 1 \\ \frac{a_4 + \mathrm{i}\Omega}{a_2} \end{array} \right). \tag{13.9}$$

The dual vectors are

$$w = \frac{1}{2}\left(1 - i\frac{a_4}{\Omega}, \ i\frac{a_2}{\Omega}\right) \quad \text{and} \quad \bar{w} = \frac{1}{2}\left(1 + i\frac{a_4}{\Omega}, \ -i\frac{a_2}{\Omega}\right).$$

From (5.39), (13.4) and (13.5) it follows that

$$c_{ij} = w \begin{pmatrix} a_2 & 0 \\ 0 & a_3 \end{pmatrix} S_{ij} v \,,$$

which can easily be written in the linear form (13.8). □

Expression (13.8) allows us to transfer results between WCNN (13.2) and canonical model (13.7). Notice that the synaptic connections from the ith to the jth neural oscillator are entries of the 2×2 matrix S_{ij}, and hence are 4-dimensional. The complex-valued synaptic coefficient c_{ij} has dimension 2. Thus, there is an apparent reduction of dimension. To stress this fact we sometimes call c_{ij} the "converted" synaptic coefficient.

13.3.2 Classification of Synaptic Organizations

According to Lemma 10.2 the value $\psi_{ij} = \text{Arg}\, c_{ij}$, which we called the natural phase difference, encodes phase information. Let us determine possible values of ψ_{ij} for various synaptic organizations S_{ij} satisfying Dale's principle. It is easy to do this using Lemma 13.1 and the vectors from Figure 13.5, since c_{ij} is a linear combination of the vectors with appropriate positive coefficients determined by the synaptic organization. Complete classification of c_{ij} for all S_{ij} is summarized in Figure 13.6.

Using the classification in Figure 13.6 we can solve a number of problems. Knowing the phase difference between two neural oscillators we could find possible synaptic organization that produce the difference. Knowing changes in synapses we could find changes in phase differences and vice versa, etc. We will use this classification below when we analyze possible synaptic organizations from the point of view of memorization of phase information.

The value $\psi_{ij} = \text{Arg}\, c_{ij}$ describes a possible phase difference between a pair of neural oscillators having equal center frequencies and connected in one direction: from the jth to the ith oscillator. If the oscillators are interconnected in both directions, the phase difference depends on c_{ij} and c_{ji}, and it usually differs from either ψ_{ij} or ψ_{ji}. We study some aspects of this problem in Section 13.3.3. If there are more than two interconnected neural oscillators, the situation becomes even more complicated. In both cases one may use the classification in Figure 13.6 only to determine c_{ij}, but not to determine the actual phase difference between the ith and the jth neural oscillators.

Notice the similarity between the natural phase difference for excitatory $\rightarrow$ inhibitory and for inhibitory $\rightarrow$ excitatory synaptic organizations.

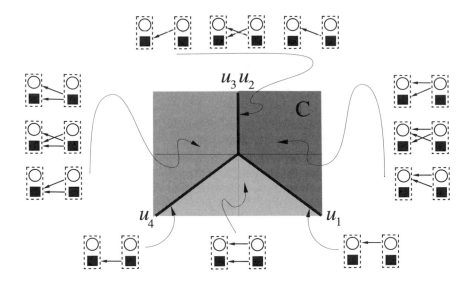

Figure 13.6. Possible values of synaptic connections c_{ij} for different synaptic configurations S_{ij} satisfying Dale's principle. For synaptic configurations that are not explicitly depicted the possible values of c_{ij} may occupy all shaded areas. (Hoppensteadt and Izhikevich 1996a).

We encountered such a feature when we studied weakly coupled Wilson-Cowan's oscillators having limit cycle attractors (see Figure 9.6). Later in this chapter we show that this is a universal phenomenon that is relatively independent of the equations that describe neural oscillators.

We see that some synaptic organizations can reproduce a range of phase differences, while others reproduce a unique phase difference. For example, if there is only one connection, say excitatory → inhibitory, then the phase difference is $\pi/2$ (the synaptic organization is at left hand side of the top row in Figure 13.6). This means that the left oscillator completes a quarter of its cycle by the time the right oscillator starts a new one. For an observer it may seem that the left neural oscillator is ahead of the right ones, that is, that there is a negative time delay. At first glance this might contradict one's intuition. In fact, there is a simple explanation to this event: The excitatory neuron of the right neural oscillator excites the inhibitory neuron of the left one and thereby facilitates reciprocal inhibition to its excitatory neuron. This expedites completion of the cycle. Next time, the left neural oscillator starts its cycle before the right one, thus producing the impression that the time delay is negative. Similar explanations are possible for other synaptic organizations.

Phase Differences in Chains

The fact that a postsynaptic neural oscillator can be ahead of a presynaptic one can be presented in a more dramatic form by considering chains of such oscillators, which we discussed in Section 9.4.5.

Indeed, consider an array of neural oscillators with the synaptic organization depicted in Figure 13.7. In measuring activity of each oscillator in

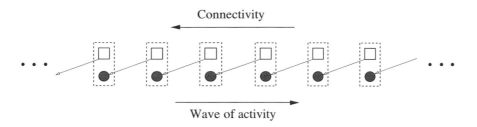

Figure 13.7. An array of neural oscillators with a synaptic organization capable of producing reverse waves of activity.

such arrays, one can observe a running wave. Since the phase difference between each pair of neural oscillators in the array is $\pi/2$ (the left neural oscillator is a quarter cycle ahead of the right one), the wave moves to the right. This may motivate an experimenter to look for synaptic connections in the opposite direction, that is, from left to right, though there may not be any.

This example emphasizes the importance of distinguishing synaptic organizations and how they affect dynamical properties of networks of neural oscillators. We have discussed only one organization, although similar considerations are applicable to any of them. In all cases the phase difference crucially depends on the synaptic organizations, not on the transmission delays.

Vacuous Connections

Let us treat (13.8) as a linear mapping from $\mathbb{R}^4$ to $\mathbb{R}^2$. Obviously, it has a kernel, which is two-dimensional in this case. Therefore, the synaptic connections S_{ij} between two neural oscillators can be in the kernel so that the converted synaptic coefficient $c_{ij} = 0$, which means that the oscillators do not communicate. Although physiologically present, such connections are functionally ineffective. Dale's principle prohibits the existence of such vacuous connections in most of synaptic organizations. Nevertheless, they are possible in some synaptic organizations (Hoppensteadt and Izhikevich 1996b).

Theorem 13.2 (Vacuous Connections) *There are nonzero synaptic configurations S_{ij} between the ith and the jth neural oscillators such that $c_{ij} = 0$. Such configurations can be found that satisfy Dale's principle.*

Proof. We can take a linear combination of u_1 and u_4 so that the real part (projection on the horizontal axis) of $u_1 s_1 + u_4 s_4$ is zero and add u_2 or u_3 or both to kill the imaginary part (projection on the vertical axis). $\square$

The corresponding synaptic organizations are depicted in Figure 13.8. One is warned that not all neural oscillators connected as shown in the

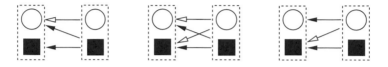

Figure 13.8. Synaptic organizations that can have vacuous connections, provided that the efficacy of synapses is chosen appropriately.

figure produce this effect: The strengths of synaptic connections between the neural oscillators must be chosen appropriately.

13.3.3 Synchronization of Two Identical Neural Oscillators

Let us combine the classification in Figure 13.6 with Theorem 10.4, which determines the conditions under which a pair of identical oscillators

$$\begin{cases} z_1' = (\alpha + i\omega)z_i + (\sigma + i\gamma)z_1|z_1|^2 + cz_2 \\ z_2' = (\alpha + i\omega)z_2 + (\sigma + i\gamma)z_2|z_2|^2 + cz_1 \end{cases}$$

is in-phase or anti-phase synchronized. In-phase synchronization ($z_1 = z_2$) occurs when

$$\alpha + \operatorname{Re} c > 0 , \quad \alpha + 3\operatorname{Re} c > 0$$

and

$$(\operatorname{Re} c + \alpha)(\gamma/\sigma \operatorname{Im} c + \operatorname{Re} c) + |c|^2 > 0 .$$

Anti-phase synchronization ($z_1 = -z_2$) occurs when

$$\alpha - \operatorname{Re} c > 0 , \quad \alpha - 3\operatorname{Re} c > 0$$

and

$$(\operatorname{Re} c - \alpha)(\gamma/\sigma \operatorname{Im} c + \operatorname{Re} c) + |c|^2 > 0 .$$

For the sake of simplicity we consider Type A neural oscillators. First, suppose that $\gamma = 0$ (no shear) and $\alpha > 0$ (the oscillators are pacemakers).

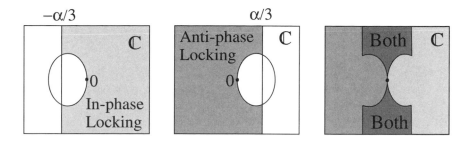

Figure 13.9. The complex plane $\mathbb{C}$ partitioned into regions of various kinds of synchronization.

If we treat α as a parameter, then the complex plane $\mathbb{C}$ is partitioned into regions where in-phase, anti-phase, or both synchronizations are possible (see Figure 13.9).

Classification of synaptic organizations with respect to phase locking they can reproduce is depicted in Figure 13.10, which is obtained by superposition of Figure 13.6 and the right-hand side of Figure 13.9. Synaptic

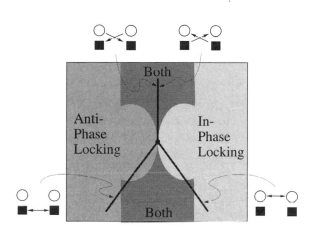

Figure 13.10. Classification of synaptic organizations with respect to the phase locking pattern they can reproduce. Synaptic organizations that are not explicitly shown can be obtained as linear combinations of the corresponding vectors.

organizations corresponding to connections between excitatory (or between inhibitory) neurons can exhibit both in- and anti-phase synchronization if the strength of connection $|c|$ is chosen appropriately (not too weak, not too strong). Increasing α expands the ellipse and thereby decreases the size of the segment in which both types of synchronization are possible.

If the parameter α is negative, which corresponds to passive oscillators, the regions of in- and anti-phase locking do not overlap in this case. Moreover, not all synaptic organizations can exhibit phase locking; see Figure 13.11.

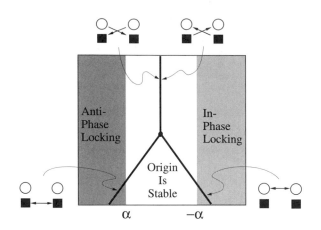

Figure 13.11. Classification of synaptic organizations for $\alpha < 0$.

If there is a nonzero shear parameter γ, the ellipses are enlarged and rotated; see Figure 13.12. When $|\gamma/\sigma| > 2\sqrt{2}$, the ellipses are transformed

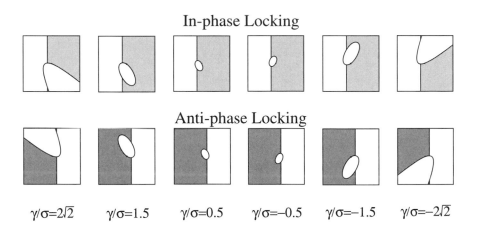

Figure 13.12. A nonzero shear parameter γ can augment and rotate the ellipses.

into hyperbolas. We see in Figure 13.13 that negative γ hampers in-phase locking for neural oscillators connected via excitatory neurons. It also hinders anti-phase locking for synaptic organizations corresponding to excita-

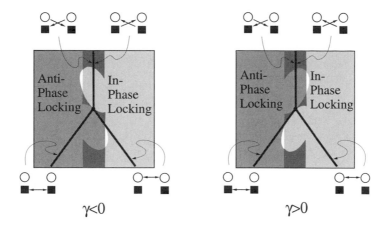

Figure 13.13. Classification of synaptic organizations for a nonzero shear parameter γ.

tory $\rightarrow$ inhibitory or inhibitory $\rightarrow$ excitatory type of connections. In both cases the desired locking is not achieved if the connection strength is too weak. Moreover, the in-phase locking can never be achieved for the excitatory $\rightarrow$ excitatory synaptic organization if $\gamma/\sigma > 2\sqrt{2}$. In this case the behavior of two weakly coupled neural oscillators at the Andronov-Hopf bifurcation point resembles that of weakly coupled Wilson-Cowan oscillators having a limit cycle attractor, which we studied in Section 9.2.2. Indeed, in both cases the inhibitory $\rightarrow$ excitatory and the excitatory $\rightarrow$ inhibitory synaptic organizations exhibit in-phase locking, whereas the excitatory $\rightarrow$ excitatory and inhibitory $\rightarrow$ inhibitory synaptic organizations exhibit anti-phase locking.

Similar considerations, with obvious modifications, are applicable to the case $\gamma > 0$. We see that a nonzero shear parameter γ affects dramatically the occurrence of synchronization for some synaptic organizations.

13.4 Learning Dynamics

Little is known about learning dynamics in the human brain. In this section we continue with the same assumptions as those in Section 11.4, which lead to the following modification rules for components of the matrix S_{ij}.

$$
\begin{aligned}
s_1' &= -bs_1 + \beta_1 x_i x_j + \mathcal{O}(\sqrt{\varepsilon})\,, \\
s_2' &= -bs_2 + \beta_2 x_i y_j + \mathcal{O}(\sqrt{\varepsilon})\,, \\
s_3' &= -bs_3 + \beta_3 y_i x_j + \mathcal{O}(\sqrt{\varepsilon})\,, \\
s_4' &= -bs_4 + \beta_4 y_i y_j + \mathcal{O}(\sqrt{\varepsilon})\,,
\end{aligned}
$$

where x and y are rescaled (divided by $\sqrt{\varepsilon}$) activities of excitatory and inhibitory neurons, respectively. We refer to the system above as being the Hebbian synaptic modification rule for a weakly connected network of neural oscillators. Recall that the constants b and β are the rate of memory fading and the rate of synaptic plasticity, respectively. We assume that the fading rate b is positive and the same for all synapses. The plasticity rates can differ for different synapses. To distinguish them we write β_{ijk} for $i, j = 1, \ldots, n; \ k = 1, 2, 3, 4$.

In order to understand how learning influences the dynamics of the canonical model we must calculate the changes in the converted synaptic coefficient c_{ij}.

Lemma 13.3 *The Hebbian learning rule for a weakly connected network of neural oscillators can be written in the form*

$$c'_{ij} = -bc_{ij} + k_{ij2}z_i\bar{z}_j + k_{ij3}\bar{z}_iz_j , \qquad (13.10)$$

where

$$k_{ij2} \;=\; \kappa(a_2\beta_1 - a_3\beta_2) + \frac{a_3}{a_2}\bar{\kappa}(a_2\beta_3 - a_3\beta_4) , \qquad (13.11)$$

$$k_{ij3} \;=\; \kappa(a_2\beta_1 - a_3\beta_3) + \frac{a_3}{a_2}\bar{\kappa}(a_2\beta_2 - a_3\beta_4) , \qquad (13.12)$$

$$\kappa \;=\; \frac{1}{2}(1 - i\frac{a_4}{\Omega}) ,$$

where we write β_k for β_{ijk}, $i, j = 1, \ldots, n, \ k = 1, 2, 3, 4$.

Proof. Using (13.8) we see that

$$c'_{ij} = u_1s'_1 + u_2s'_2 + u_3s'_3 + u_4s'_4 .$$

Using the Hebbian learning rule we obtain

$$c'_{ij} = -bc_{ij} + u_1\beta_1x_ix_j + u_2\beta_2x_iy_j + u_3\beta_3y_ix_j + u_4\beta_4y_iy_j . \qquad (13.13)$$

Using (5.37) and (13.9) we obtain the following expression for the rescaled (divided by $\sqrt{\varepsilon}$) variables x and y.

$$\begin{pmatrix} x_i \\ y_i \end{pmatrix} = \begin{pmatrix} 1 \\ \frac{a_4 - i\Omega}{a_2} \end{pmatrix} e^{i\Omega t}z_i + \begin{pmatrix} 1 \\ \frac{a_4 + i\Omega}{a_2} \end{pmatrix} e^{-i\Omega t}\bar{z}_i + \mathcal{O}(\sqrt{\varepsilon}) .$$

Substituting this into (13.13) yields

$$c'_{ij} = -bc_{ij} + k_{ij1}e^{2i\Omega t}z_iz_j + k_{ij2}z_i\bar{z}_j + k_{ij3}\bar{z}_iz_j + k_{ij4}e^{-2i\Omega t}\bar{z}_i\bar{z}_j + \mathcal{O}(\sqrt{\varepsilon}) , \qquad (13.14)$$

where $k_{ij1}, k_{ij2}, k_{ij3}$, and k_{ij4} are some coefficients. We are not interested in k_{ij1} and k_{ij4} because after averaging, all terms containing $e^{\mathrm{i}m\Omega t}$ with $m \neq 0$ disappear, and we will have

$$c'_{ij} = -bc_{ij} + k_{ij2}z_i\bar{z}_j + k_{ij3}\bar{z}_iz_j + \mathcal{O}(\sqrt{\varepsilon}).$$

Taking the limit $\varepsilon \to 0$ gives (13.10). It is easy to check that k_{ij2} and k_{ij3} are given as shown in (13.11). $\square$

Remark. The fourth hypothesis about the learning dynamics (forgetting when pre- or postsynaptic neurons are silent; see Section 11.4) is redundant when we study weakly connected oscillators. Indeed, if we drop the assumption, then there are extra linear and nonlinear terms in equations for s'_k. This results in additional terms in (13.14), which eventually vanish after the averaging. The resulting equation for c'_{ij} has the same form as (13.10).

Note that we assumed little about the actual learning dynamics. Nevertheless, the family of possible learning rules (13.10) that satisfy our assumptions is apparently narrow. In the next section we show that to be "useful", the learning rule (13.10) must satisfy the additional conditions $\mathrm{Im}\, k_{ij2} = 0$ and $k_{ij3} = 0$. Using this and (13.11) we can determine what restrictions must be imposed on the plasticity rates $\beta_1, \beta_2, \beta_3, \beta_4$, and hence onto the possible organization of the network so that it can memorize phase information, which we discuss next.

13.4.1 Memorization of Phase Information

We develop here the concept of memorization of *phase differences*. By this we understand the following: If during a learning period neuron A excites neuron B such that B generates an action potential with time delay δ, then changes occur such that whenever A generates an action potential, then so does B with the same time delay δ; see Figure 13.14.

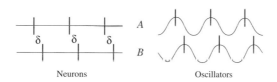

Figure 13.14. Time delays (*left*) are phase differences (*right*) when neurons generate spikes repeatedly.

Since in the real brain neurons tend to generate the action potentials repeatedly, instead of the time delay we will be interested in the phase difference between the neurons A and B. So, if during a learning period, two neural oscillators generate action potentials with some phase difference, then after the learning is completed, they can reproduce the same phase difference.

Whether memorization of phase differences is important or not is a neurophysiological question. We suppose here that it is important. Then, we would like to understand what conditions must be imposed on a network's architecture to ensure that it can memorize phase differences.

The memorization of phase information in terms of the canonical model means the following: Suppose during a learning period the oscillator activities $z_i(\tau)$ are given such that the phase differences $\operatorname{Arg} z_i \bar{z}_j$ are kept fixed. We call the pattern of the phase differences the *image* to be memorized. Suppose also that the synaptic coefficients c_{ij} are allowed to evolve according to the learning rule (13.10). Then we say that the canonical model *memorized* the image if there is an attractor in the z-space $\mathbb{C}^n$ such that when the activity $z(\tau)$ is on the attractor, the phase differences between the oscillators coincide with those to be learned.

Theorem 13.4 *Consider the weakly connected network of oscillators governed by*

$$z_i' = (\alpha_i + i\omega_i)z_i + (\sigma_i + i\gamma_i)z_i|z_i|^2 + \sum_{j=1}^n c_{ij}z_j, \quad i = 1, \dots, n,$$

together with the learning rule (13.10). Suppose the neural oscillators have equal frequencies ($\omega_1 = \cdots = \omega_n = \omega$), the Andronov-Hopf bifurcation is supercritical ($\sigma_i < 0$) and there is no shear ($\gamma_i = 0$). Such network can memorize phase differences of at least one image if and only if

$$k_{ij2} > 0 \qquad and \qquad k_{ij3} = 0 ; \tag{13.15}$$

i.e., the learning rule (13.10) has the form

$$c_{ij}' = -bc_{ij} + k_{ij}z_i\bar{z}_j , \qquad i \neq j , \tag{13.16}$$

where $k_{ij} = \operatorname{Re} k_{ij2}$, $i, j = 1, \dots, n$, are positive real numbers.

Proof. Let us introduce the new rotating coordinate system $e^{i\omega\tau}z_i(\tau)$. In the new coordinates the canonical model becomes

$$z_i' = \alpha_i z_i + \sigma_i z_i|z_i|^2 + \sum_{j=1}^n c_{ij}z_j, \quad i = 1, \dots, n . \tag{13.17}$$

Sufficient condition. First, we prove that (13.15) is a sufficient condition. Our goal is to show that after learning is completed, the dynamical system (13.17) has an attractor such that the phase differences on the attractor coincide with those of the memorized pattern.

Let (13.17) be in the learning mode such that the phase differences $\varphi_i(\tau) - \varphi_j(\tau) = \text{Arg}\, z_i \bar{z}_j$ are kept fixed. Then, according to the learning rule (13.16) the coefficients c_{ij} approach $\frac{k_{ij}}{b} z_i \bar{z}_j$ and hence ψ_{ij} approaches $\text{Arg}\, z_i \bar{z}_j$, where $c_{ij} = |c_{ij}| e^{i\psi_{ij}}$. Note that ψ_{ij} satisfies

$$\psi_{ij} = -\psi_{ji} \qquad \text{and} \qquad \psi_{ij} = \psi_{ik} + \psi_{kj} \qquad (13.18)$$

for any i, j, and k. We must show that after learning is completed, the neural network can reproduce a pattern of activity having the memorized phase differences ψ_{ij}.

We have assumed that during learning all activities $z_i \neq 0$, so that the phases φ_i of the oscillators are well-defined. It is easy to see that after learning is completed we have $c_{ij} \neq 0$ for $i \neq j$.

Consider (13.17) in polar coordinates:

$$\begin{cases} r_i' = \alpha_i r_i + \sigma_i r_i^3 + \sum_{j=1}^n |c_{ij}| r_j \cos(\varphi_j + \psi_{ij} - \varphi_i) \\ \varphi_i' = \frac{1}{r_i} \sum_{j=1}^n |c_{ij}| r_j \sin(\varphi_j + \psi_{ij} - \varphi_i) \end{cases} . \qquad (13.19)$$

Let us show that the radial components, determined by

$$r_i' = \alpha_i r_i + \sigma_i r_i^3 + \sum_{j=1}^n |c_{ij}| r_j \cos(\varphi_j + \psi_{ij} - \varphi_i) , \qquad i = 1, \ldots, n , \quad (13.20)$$

are bounded. Indeed, let $B(0, R) \subset \mathbb{R}^n$ be a ball at the origin with arbitrarily large radius $R > 0$. Consider the flow of (13.20) outside the ball. After the rescaling $r_i \to R r_i$, $\tau \to R^{-2}\tau$, the system (13.20) becomes

$$r_i' = \sigma_i r_i^3 + \mathcal{O}(R^{-2}) , \qquad i = 1, \ldots, n ,$$

which is an R^{-2}-perturbation of

$$r_i' = \sigma_i r_i^3 , \qquad i = 1, \ldots, n . \qquad (13.21)$$

Since all $\sigma_i < 0$, the activity vector of (13.21) is inside a unit ball $B(0, 1)$ for any initial conditions after some finite transient. Any perturbation of (13.21) has the same property. Therefore, after the finite transition interval the activity vector of (13.20) is inside $B(0, R)$ for any initial conditions and any values of $\varphi_1, \ldots, \varphi_n$. Hence, all attractors of (13.19) lie inside $B(0, R) \times \mathbb{T}^n$.

Now consider the dynamical system

$$\varphi_i' = \frac{1}{r_i} \sum_{j=1}^n |c_{ij}| r_j \sin(\varphi_j + \psi_{ij} - \varphi_i) \qquad (13.22)$$

for fixed $(r_1, \ldots, r_n)$. Let us check that the line

$$\varphi_i = \varphi_1 + \psi_{ik} , \qquad i > 1 , \qquad (13.23)$$

parametrized by φ_1 is a global invariant manifold. Indeed, using (13.18) we have

$$\varphi_j + \psi_{ij} - \varphi_i = \varphi_1 + \psi_{j1} + \psi_{ij} - \varphi_1 - \psi_{i1} = \psi_{ij} - \psi_{ij} = 0 \qquad (13.24)$$

for all i and j, and hence

$$\varphi_i' = 0$$

on the line (13.23). Therefore, it is invariant. If we used any other φ_k instead of φ_1 in (13.23), we would obtain the same invariant manifold.

In order to study the stability of the manifold, consider the auxiliary system (13.22) where φ_1 is fixed. Since all attractors of (13.19) are inside $B(0, R) \times \mathbb{T}^n$, we may assume that $r_i < R$ for all i. The Jacobian matrix of (13.22), say $J = (J_{ij})_{i,j>1}$, is diagonal-dominant because

$$J_{ij} = \begin{cases} \frac{r_j}{r_i}|c_{ij}| & i \neq j, \\ -\frac{1}{r_i}\sum_{m=1}^n |c_{im}|r_m & i = j \end{cases}$$

and

$$J_{ii} + \sum_{j \neq i}^n J_{ij} = -\frac{r_1}{r_i}|c_{i1}| < -r_1\mu_1 < 0 ,$$

where

$$\mu_k = \frac{1}{R} \min_{i>1} |c_{i1}| .$$

This means that all eigenvalues of J have negative real parts; Hence, every

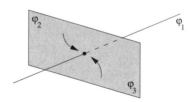

Figure 13.15. Every point on the line (13.23) parametrized by φ_1 is stable.

point on the line (13.23) is an asymptotically stable equilibrium point for (13.22) when φ_1 is fixed; see Figure 13.15. Since

$$(\varphi_2(\tau), \ldots, \varphi_n(\tau)) \to (\varphi_1 + \psi_{21}, \ldots, \varphi_1 + \psi_{n1}) ,$$

independently from values of $r_1, \ldots, r_n$, the flow of the $2n$-dimensional system (13.19) is directed everywhere toward the invariant manifold defined by (13.23), at least for $(r_1, \ldots, r_n) \neq (0, \ldots, 0)$. Hence, the manifold contains an attractor of (13.19). Moreover, it is possible to prove that the complement of its domain of attraction has measure zero; i.e., this is the only attractor for (13.17). Note that on the manifold the phase differences satisfy

$$\varphi_i - \varphi_j = \psi_{ij} \ .$$

Thus, (13.15) is a sufficient condition for memorization and recall of phase differences.

It should be stressed that the oscillators have constant phase differences on the manifold even when the attractor is not an equilibrium point. It is still an open problem whether the system $r_i' = \alpha_i r_i + \sigma_i r_i^3 + \sum_{j=1}^n |c_{ij}| r_j$ can sustain nontrivial dynamics or not. For example, if its attractor is chaotic, then one can observe an interesting phenomenon: The oscillator's amplitudes $r_1(\tau), \ldots, r_n(\tau)$ have chaotic activity, whereas their phases $\varphi_1(\tau), \ldots, \varphi_n(\tau)$ have constant differences ψ_{ij}. Thus, the synchronization of *strongly* connected oscillators does not necessarily mean that the entire network's activity is on a limit cycle.

Necessary condition. Next, we show that conditions (13.15) are necessary. Since the pattern of activity to be memorized and the values of $\alpha_1, \ldots, \alpha_n$ are not specified, it is assumed that the network can learn and reproduce phase differences of any activity pattern $z^* = (z_1^*, \ldots, z_n^*)^\top$ for any choice of $\alpha_1, \ldots, \alpha_n$.

The phase difference between the ith and jth oscillators during the learning period is $\mathrm{Arg}\, z_i^* - \mathrm{Arg}\, z_j^* = \mathrm{Arg}\, z_i^* \bar{z}_j^*$. Hence, the same value must be reproduced after the learning is completed. From Lemma 10.2 it follows that the network can always reproduce the phase differences $\varphi_i - \varphi_j = \psi_{ij} = \mathrm{Arg}\, c_{ij}$. Therefore, the equality $\psi_{ij} = \mathrm{Arg}\, z_i^* \bar{z}_j^*$ must be satisfied for any z^*. This is possible only if (13.15) holds. Hence (13.15) is necessary. □

Note the similarity of (13.16) and the Hebbian rule (11.8). The only difference is that in (13.16) the variables c and z are complex-valued.

Let us rewrite (13.16) in polar coordinates: If $c_{ij} = |c_{ij}| e^{\mathrm{i}\psi_{ij}}$ and $z_i = r_i e^{\mathrm{i}\varphi_i}$, then

$$\begin{cases} |c_{ij}|' = -b|c_{ij}| + k_{ij} r_i r_j \cos(\varphi_i - \varphi_j - \psi_{ij}) \\ \psi_{ij}' = \frac{1}{|c_{ij}|} k_{ij} r_i r_j \sin(\varphi_i - \varphi_j - \psi_{ij}) \end{cases} .$$

From the second equation it is clear that

$$\psi_{ij} \rightarrow \varphi_i - \varphi_j$$

as we expected on the basis of Lemma 10.2. Notice that if $\psi_{ij} = \varphi_i - \varphi_j$, then $\cos(\varphi_i - \varphi_j - \psi_{ij}) = 1$, and first equation coincides with the Hebbian learning rule.

Since we know how k_{ij2} and k_{ij3} depend upon the original WCNNs we can restate the results of Theorem 13.4 in terms of (13.2). Almost all results discussed in the next section are straightforward consequences of the following result.

Corollary 13.5 *A weakly connected network of neural oscillators can memorize phase differences if and only if the plasticity rates satisfy*

$$a_2\beta_1 = a_3\beta_3 , \qquad a_2\beta_2 = a_3\beta_4 \qquad (13.25)$$

and

$$k_{ij} = a_2\beta_1 - a_3\beta_2 > 0 \qquad (13.26)$$

for each i and j.

The proof follows from application of condition (13.15) in Theorem 13.4 to the representation (13.11) in Lemma 13.3.

13.4.2   Synaptic Organizations and Phase Information

In this section we apply Corollary 13.5 to various synaptic organizations. Suppose that $\beta_{ij1} = 0$ for some $i \neq j$. That is, there is no modification of synapses between the jth and ith excitatory neurons except for fading (atrophy). So, even if a synapse s_{ij1} between x_j and x_i existed at the beginning, it would atrophy with time. Thus, without loss of generality we may assume that $\beta_{ij1} = 0$ indicates that formation and growth of synapses from x_j onto x_i is impossible. The same consideration can be applied to β_{ij2}, β_{ij3}, and β_{ij4}. In Figure 13.16 we draw arrows from one neuron to another only when the corresponding plasticity rate is nonzero, i.e., only

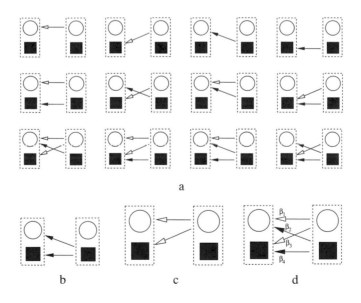

Figure 13.16. Synaptic organizations that can (b, c, and d) and cannot (a) memorize phase information.

if synaptic contact between the two neurons is possible. Different choices of the arrows correspond to different synaptic organizations of the neural network.

Corollary 13.6 *The synaptic organizations depicted in Figure 13.16a cannot memorize phase information.*

Proof. According to condition (13.25), if one of the plasticity rates is zero, then so should be the other one corresponding to it. Thus, the arrows must be in pairs; i.e., if a neuron has synaptic contacts with some neural oscillator, then it must have access to both excitatory and inhibitory neurons of the neural oscillator. Obviously, none of the architectures in Figure 13.16a satisfies this condition. □

Corollary 13.7 *The synaptic organization depicted in Figure 13.16b can memorize phase information only if there are two oscillators and the phase difference to be memorized is close to π.*

Proof. From the proof of Theorem 13.4 it follows that $\operatorname{Arg} c_{ij} = -\operatorname{Arg} c_{ji}$; that is, c_{ij} and c_{ji} must lie on two symmetric rays from the origin (see Figure 13.17). In order to satisfy Dale's principle both beams must be

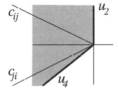

Figure 13.17. c_{ij} and c_{ji} must be inside the shaded area between u_4 and u_2 in the proof of Corollary 13.7.

inside the shaded area between u_4 and u_2 (see the classification in Figure 13.6). Obviously, the phase difference to be memorized must be sufficiently close to π (within ($\operatorname{Arg} u_4$)-neighborhood).

Suppose the network consists of more than two oscillators. Then the phase difference between the first and second and between the second and third neural oscillators should be close to π. Hence the phase difference between the first and third oscillators is close to 0. Thus c_{13} and c_{31} violate Dale's principle. □

Since networks seldom have only two elements, we can conclude that this synaptic organization is not much better than those depicted in Figure 13.16a. Nevertheless, this result is applicable to the case when there are two neural oscillators representing two distinct brain structures.

Corollary 13.8 *The synaptic organization in Figure 13.16c can memorize arbitrary phase information.*

Proof. Since $\beta_2 = 0$, we have $k_{ij} = a_2\beta_1$. When two excitatory neurons generate action potentials simultaneously, the strength of excitatory synaptic connections between them increases (Hebb 1949), which corresponds to $\beta_1 > 0$. Therefore $k_{ij} > 0$, and the result follows from Theorem 13.4. $\square$

Corollary 13.9 *Consider the synaptic organization in Figure 13.16d.*

- *If $a_2\beta_1 > a_3\beta_2$, then the synaptic organization can learn arbitrary phase information.*

- *If $a_2\beta_1 < a_3\beta_2$, then it unlearns phase information.*

- *If $a_2\beta_1 = a_3\beta_2$, then it passively forgets phase information.*

Proof. The result follows directly from the expression

$$k_{ij} = a_2\beta_1 - a_3\beta_2$$

(see Corollary 13.5) and the learning rule

$$c'_{ij} = -bc_{ij} + k_{ij}z_i\bar{z}_j$$

(see Theorem 13.4). $\square$

What sign does β_2 have? Since the presynaptic neuron y_j is inhibitory, it is not clear what changes take place in the synapses from y_j to x_i (or y_i) when they fire simultaneously. We consider the two possible cases:

- Case $\beta_2 < 0$ corresponds to decreasing of strength of the inhibitory synapse.

- Case $\beta_2 > 0$ corresponds to increasing of strength of the inhibitory synapse.

In the first case, k_{ij} is always positive and the synaptic organization in Figure 13.16d can learn phase information. In the second case k_{ij} may become negative. Then the network memorizes not the presented image but its inverse, like a photographic negative. Sometimes it is convenient to think of this as unlearning of the image. So, the synaptic organization on Figure 13.16d is able not only to learn, but also to unlearn information simply by adjusting the plasticity rates β_1 and β_2.

The special choice of plasticity constants for which $k_{ij} = 0$ is interesting: Since the plasticity constants are not zero, there are undoubtedly some changes in synaptic coefficients s_{ijk} between the neurons. Nevertheless, the network as a whole does not learn anything because for $k_{ij} = 0$

$$c'_{ij} = -bc_{ij} .$$

Moreover, the network forgets (loses) information because $c_{ij} \to 0$ as τ increases. Hence, the fully connected synaptic organization can exhibit a broad dynamic repertoire.

The synaptic organizations depicted in Figure 13.16c and d indicate that to learn phase information the excitatory neurons must have long axons. The inhibitory neurons may serve as local-circuit inter-neurons (Shepherd 1983, Rakic 1976).

13.5   Limit Cycle Neural Oscillators

Next we consider the weakly connected network of neural oscillators (13.2), which we write in the form

$$
\begin{cases} \dot{x}_i = f(x_i, -y_i) + \varepsilon p_i(x, y, \varepsilon) \\ \dot{y}_i = g(x_i, \ y_i) \ + \varepsilon q_i(x, y, \varepsilon) \end{cases} , \tag{13.27}
$$

where the functions p and q have the following special form

$$
p_i(x, y, 0) = f'_\rho(x_i, -y_i) \sum_{j=1}^{n} (s_{ij1}x_j - s_{ij2}y_j) , \tag{13.28}
$$

$$
q_i(x, y, 0) = g'_\rho(x_i, \ y_i) \sum_{j=1}^{n} (s_{ij3}x_j - s_{ij4}y_j) . \tag{13.29}
$$

We suppose here that each of the neural oscillators

$$
\begin{cases} \dot{x}_i = f(x_i, -y_i) \\ \dot{y}_i = g(x_i, \ y_i) \end{cases} \tag{13.30}
$$

has a hyperbolic limit cycle attractor $\gamma \in \mathbb{R}^2$, which without loss of generality we assume has period 2π.

13.5.1   Phase Equations

Let $\theta_i \in \mathbb{S}^1$ denote the phase of the ith oscillator on γ. Then, the oscillatory weakly connected network (13.27) has a local phase model (see Chapter 9)

$$
\dot{\theta}_i = 1 + \varepsilon h_i(\theta, \varepsilon) , \qquad \theta = (\theta_1, \dots, \theta_n) \in \mathbb{T}^n ,
$$

where h_i are some 2π-periodic functions. As in Chapter 9, we write

$$
\theta_i(t) = 2\pi t + \varphi_i(\tau) ,
$$

where $\varphi_i(\tau) \in \mathbb{S}^1$ is the phase deviation due to weak connections and $\tau = \varepsilon t$ is the slow time. From Malkin's Theorem (Theorem 9.2) it follows that

$$
\varphi'_i = H_i(\varphi - \varphi_i, \varepsilon) , \qquad \varphi = (\varphi_1, \dots, \varphi_n) \in \mathbb{T}^n ,
$$

where H_i are some periodic functions that we determine below.

Connection Functions H_i

The linearization of (13.30) at the limit cycle γ has the form

$$DF(\gamma(t)) = \begin{pmatrix} f_x'(\gamma(t)) & -f_\rho'(\gamma(t)) \\ g_\rho'(\gamma(t)) & g_y'(\gamma(t)) \end{pmatrix} .$$

Let $Q = (Q_1, Q_2)^\top \in \mathbb{R}^2$ be a periodic solution to the adjoint linear problem $\dot{Q} = -\{DF(\gamma(t))\}^\top Q$; that is,

$$\begin{cases} \dot{Q}_1 & = & -f_x'(\gamma(t))\,Q_1 - g_\rho'(\gamma(t))\,Q_2 \\ \dot{Q}_2 & = & f_\rho'(\gamma(t))\,Q_1 - g_y'(\gamma(t))\,Q_2 \end{cases} . \qquad (13.31)$$

We normalize Q so that it satisfies

$$Q_1(t)\,f(\gamma(t)) + Q_2(t)\,g(\gamma(t)) = 1$$

for some (and hence for all) $t \in [0, 2\pi]$. According to Malkin's Theorem the connection functions H_i are given by

$$H_i(\varphi - \varphi_i, 0) = \frac{1}{2\pi} \int_0^{2\pi} (Q_1(t), Q_2(t)) \begin{pmatrix} p_i(\gamma(t + \varphi - \varphi_i)) \\ q_i(\gamma(t + \varphi - \varphi_i)) \end{pmatrix} dt .$$

Taking into account the additive form of the functions p_i and q_i in (13.28) and (13.29), we represent H_i as

$$H_i(\varphi - \varphi_i, 0) = \sum_{j=1}^n H_{ij}(\chi_{ij}) ,$$

where $\chi_{ij} = \varphi_j - \varphi_i \in \mathbb{S}^1$ is the phase difference between the ith and the jth oscillators. Each function H_{ij} has the form

$$H_{ij}(\chi) = s_1 H_{ee}(\chi) + s_2 H_{ei}(\chi) + s_3 H_{ie}(\chi) + s_4 H_{ii}(\chi) , \qquad (13.32)$$

(where we omit $_{ij}$ from s_{ijk} and χ_{ij} for clarity), and

$$H_{ee}(\chi) = \frac{1}{2\pi} \int_0^{2\pi} Q_1(t)\,f_\rho'(\gamma(t))\,x(t + \chi)\,dt \qquad (13.33)$$

$$H_{ei}(\chi) = -\frac{1}{2\pi} \int_0^{2\pi} Q_1(t)\,f_\rho'(\gamma(t))\,y(t + \chi)\,dt \qquad (13.34)$$

$$H_{ie}(\chi) = \frac{1}{2\pi} \int_0^{2\pi} Q_2(t)\,g_\rho'(\gamma(t))\,x(t + \chi)\,dt \qquad (13.35)$$

$$H_{ii}(\chi) = -\frac{1}{2\pi} \int_0^{2\pi} Q_2(t)\,g_\rho'(\gamma(t))\,y(t + \chi)\,dt , \qquad (13.36)$$

where $x(t + \chi)$ and $y(t + \chi)$ are the coordinates on the limit cycle $\gamma(t + \chi)$. The functions H_{ee}, H_{ei}, H_{ie} and H_{ii} have the same meaning as vectors u_1, u_2, u_3 and u_4 in Lemma 13.1; see Figure 13.18. They describe the effect of synaptic connection strengths (or weights) s_1, s_2, s_3 and s_4, respectively. Knowing f and g in (13.30) one can directly evaluate these functions numerically, as we did in Figure 9.6.

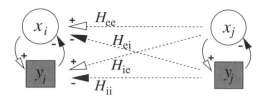

Figure 13.18. Functions H_{ee}, H_{ei}, H_{ie} and H_{ii}.

13.5.2   Synchronization of Two Identical Oscillators

Consider a network comprising two identical neural oscillators that is described by (13.27) for $n = 2$. In this case the phase model has the form

$$\begin{cases} \varphi_1' = H(\varphi_2 - \varphi_1) \\ \varphi_2' = H(\varphi_1 - \varphi_2) \end{cases},$$

where H is defined by (13.32), and terms of order $\mathcal{O}(\varepsilon)$ are ignored.

Proposition 13.10 *Two coupled identical oscillators have both in-phase and anti-phase solutions, either of which may be unstable.*

Proof. Let us denote $\chi = \varphi_2 - \varphi_1$. Then the phase model can be written concisely as

$$\chi' = H(-\chi) - H(\chi) . \tag{13.37}$$

In-phase and anti-phase solutions correspond to the equilibria $\chi = 0$ and $\chi = \pi$, respectively. Obviously, $\chi = 0$ is an equilibrium. Since H is 2π-periodic, $\chi = \pi$ is an equilibrium too. $\square$

We note that the proposition does not tell anything about the stability of the in-phase and anti-phase solutions, and out-of-phase solutions might also exist for (13.37). The problem of stability, which we discuss below, is a difficult one and it is not yet completely resolved.

The in-phase solution $\chi = 0$ is stable if $H'(0) > 0$, since in this case

$$\frac{\partial}{\partial \chi} \left(H(-\chi) - H(\chi) \right) \Big|_{\chi - 0} = -H'(0) - H'(0) = -2H'(0) < 0 .$$

Similarly, the anti-phase solution $\chi = \pi$ is stable if $H'(\pi) > 0$. If $H' = 0$, then the solution can be stable or unstable depending on higher derivatives of H. This case is considered elsewhere.

In general, H' may assume arbitrary values at 0 or π. However, Dale's principle imposes certain restrictions on the sign of H'. From (13.32) it follows that

$$H'(\chi) = s_1 H_{ee}'(\chi) + s_2 H_{ei}'(\chi) + s_3 H_{ie}'(\chi) + s_4 H_{ii}'(\chi) ,$$

where $s_1, s_2, s_3,$ and s_4 are positive parameters. If at least two numbers among $H'_{ee}, H'_{ei}, H'_{ie}$ and H'_{ii} have opposite signs, then a suitable choice of s_1, s_2, s_3 and s_4 can make the solution χ stable or unstable without violation of Dale's principle. The theorem below is based on the fact that H'_{ei} and H'_{ie} always have the same sign at $\chi = 0$.

Theorem 13.11 *Consider the synaptic organizations in Figure 13.19. Either all of these have stable in-phase synchronized solutions or none do.*

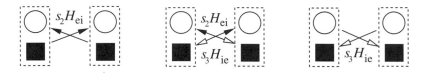

Figure 13.19. Synaptic organizations that have in-phase solution of the same stability type.

In other words, if any one of them has a stable in-phase synchronized solution, then so do the others.

Proof. Let us show that

$$H'_{ei}(0) = H'_{ie}(0) + \mathcal{O}(\varepsilon) \ .$$

The theorem will follow immediately. Indeed, the stability of the in-phase solution is determined by the sign of

$$H'(0) = s_2 H'_{ei}(0) + s_3 H'_{ie}(0) = (s_2 + s_3) H'_{ei}(0) + \mathcal{O}(\varepsilon)$$

which depends upon $H'_{ei}(0) \neq 0$, but not upon s_2 and s_3. (We do not consider the nonhyperbolic case $H'_{ei}(0) = 0$ here).

We use the following identities, which are easy to verify:

$$\frac{\partial x(t + \chi)}{\partial \chi}\bigg|_{\chi=0} = \dot{x}(t) = f(\gamma(t)) + \mathcal{O}(\varepsilon) \ ,$$

$$\frac{\partial y(t + \chi)}{\partial \chi}\bigg|_{\chi=0} = \dot{y}(t) = g(\gamma(t)) + \mathcal{O}(\varepsilon) \ ,$$

$$\frac{dg(\gamma(t))}{dt} = g'_\rho(\gamma(t)) f(\gamma(t)) + g'_y(\gamma(t)) g(\gamma(t)) + \mathcal{O}(\varepsilon) \ .$$

From the second line in (13.31) it follows that

$$Q_1(t) f'_\rho(\gamma(t)) = \dot{Q}_2(t) + Q_2(t) g'_y(\gamma(t)) \ .$$

If $a(t)$ and $b(t)$ are any smooth 2π-periodic functions, then from

$$\int_0^{2\pi} \dot{a}(t) b(t) \, dt = a(t) b(t) \bigg|_0^{2\pi} - \int_0^{2\pi} a(t) \dot{b}(t) \, dt$$

it follows that

$$\int_0^{2\pi} \dot{a}(t)b(t)\,dt = -\int_0^{2\pi} a(t)\dot{b}(t)\,dt \ .$$

The identities above show that within $\mathcal{O}(\varepsilon)$

$$
\begin{aligned}
H'_{\mathrm{ei}}(0) \ &= \ -\frac{1}{2\pi}\int_0^{2\pi} Q_1(t)\,f'_\rho(\gamma(t))\,g(\gamma(t))\,dt \\[2mm]
&= \ -\frac{1}{2\pi}\int_0^{2\pi} \left(\dot{Q}_2(t) + Q_2(t)\,g'_y(\gamma(t))\right) g(\gamma(t))\,dt \\[2mm]
&= \ \frac{1}{2\pi}\int_0^{2\pi} Q_2(t)\left(g'_\rho(\gamma(t))\,f(\gamma(t)) + g'_y(\gamma(t))\,g(\gamma(t))\right) dt \\[2mm]
&\quad -\frac{1}{2\pi}\int_0^{2\pi} Q_2(t)\,g'_y(\gamma(t))\,g(\gamma(t))\,dt \\[2mm]
&= \ \frac{1}{2\pi}\int_0^{2\pi} Q_2(t)g'_\rho(\gamma(t))\,f(\gamma(t))\,dt \\[2mm]
&= \ H'_{\mathrm{ie}}(0) \ ,
\end{aligned}
$$

which completes the proof. □

Proving an analogue of this theorem for anti-phase locking and for other synaptic organizations is an important but unsolved problem.

References

Abbott L.F. and Marder E. (1995) Activity-Dependent Regulation of Neuronal Conductances. In Arbib MA (Ed) Brain Theory and Neural Networks, The MIT press, Cambridge, MA.

Abbott L.F. and van Vreeswijk C. (1993) Asynchronous states in a network of pulse-coupled oscillators. Phys. Rev. E 48:1483–1490.

Abeles M. (1988) Neural Codes For Higher Brain Functions. In H.J. Markowitsch (ed.) Information Processing by the Brain, Hans Huber Publishers, Toronto.

Alexander J.C., Doedel E.J., and Othmer H.G. (1990) On the Resonance Structure in a Forced Excitable System. SIAM Journal on Applied Mathematics, 50:1373–1418.

Andronov A. and Leontovich E. (1939) Some cases of the dependence of the limit cycles upon parameters. Uch. Zap. Gork. Univ. 6:3-24, (in Russian).

Arnold V.I. (1961) Small divisors, I. On mappings of a circle onto itself. Izv. Akad. Nauk SSSR, Ser. Math. 25:21–86.

Arnold V.I. (1982) Geometrical Methods in the Theory of Ordinary Differential Equations. Springer-Verlag, New York, [Russian original: Additional Chapters of the Theory of Ordinary Differential Equations, Moscow 1977].

Arnold V.I., Afrajmovich V.S., Il'yashenko Yu.S., and Shil'nikov L.P. (1994) Bifurcation Theory In: Arnold V.I. (eds) Dynamical Systems V. Bifurcation theory and catastrophe theory. Springer-Verlag, New York.

Aronson D.G., Ermentrout G.B., and Kopell N. (1990) Amplitude response of coupled oscillators. Physica D 41:403–449.

Attneave F. (1971) Multistability in Perception. Scientific American 225: 63–71.

Babloyantz A. and Lourenco C. (1994) Computation with chaos: A paradigm for cortical activity, Proceedings of National Academy of Sciences USA 83:3513–3517.

Baer S.M. and Erneux T. (1986) Singular Hopf Bifurcation to Relaxation Oscillations. SIAM Journal on Applied Mathematics, 46:721–739.

Baer S.M. and Erneux T. (1992) Singular Hopf Bifurcation to Relaxation Oscillations II. SIAM Journal on Applied Mathematics, 52:1651–1664.

Baer S.M., Erneux T., and Rinzel J. (1989) The slow passage through a Hopf bifurcation: delay, memory effects, and resonances. SIAM Journal on Applied Mathematics 49:55–71.

Baer S.M., Rinzel J., and Carrillo H. (1995) Analysis of an autonomous phase model for neuronal parabolic bursting. Journal of Mathematical Biology 33:309–333.

Baesens C., Guckenheimer J., Kim S., and MacKay R.S. (1991) Three coupled oscillators: mode-locking, global bifurcations and toroidal chaos. Physica D 49:387–475.

Baird B. (1986) Nonlinear dynamics of pattern formation and pattern recognition in the rabbit olfactory bulb. Physica D 22:150–175.

Bar-Eli K. (1985) On the stability of coupled chemical oscillators. Physica D 14:242–252.

Basar E. (1990) Chaos in Brain Function. Springer-Verlag, New York.

Bedrov Y.A., Akoev G.N., and Dick O.E. (1992) Partition of the Hodgkin-Huxley type model parameter space into regions of qualitatively different solutions. Biological Cybernetics 66:413–418.

Belair J. and Holmes P. (1984) On linearly coupled relaxation oscillations. Quarterly of Appl. Math 42:193–219.

Bertram R., Butte M.J., Kiemel T., and Sherman A. (1995) Topological and phenomenological classification of bursting oscillations. Bulletin of Mathematical Biology 57:413–439.

Blechman I.I. (1971) Synchronization of Dynamical Systems. [in Russian: "Sinchronizatzia Dinamicheskich Sistem", Science, Moscow].

Bliss T.V.P. and Lynch M.A. (1988) Long-term potentiation of synaptic transmission in the hippocampus: Properties and Mechanisms. In Landfield P., Deadwyler S. (eds.) Long-Term Potentiation: From Biophysics to Behavior, Alan R. Liss, New York.

Bronstein I.U. and Kopanskii A.Ya. (1994) Smooth Invariant Manifolds and Normal Forms. World Scientific, Singapore.

Bogoliubov N.N. and Mitropolsky Y.A. (1961) Asymptotic Methods in the Theory of Non-linear Oscillations. Gordon and Breach Science Publishers, New York.

Borisyuk R.M. (1991) Interacting neural oscillators can initiate the selective attention. In Holden A.V., Kryukov V.I. (eds) Neurocomputers and Attention. I. Neurobiology, synchronization and chaos, Manchester University Press, Manchester.

Borisyuk R.M., Kirillov A.B. (1992) Bifurcation analysis of a neural network model. Biological Cybernetics 66:319–325.

Borisyuk G.N., Borisyuk R.M., Khibnik A.I., and Roose D. (1995) Dynamics and bifurcations of two coupled neural oscillators with different connection types, Bulletin of Mathematical Biology 57:809–840.

Caianiello E.R., Ceccarelli M., and Marinaro M. (1992) Can spurious states be useful? Complex Systems, 6:1-12.

Carpenter G.A. (1979) Bursting phenomena in excitable membranes. SIAM Journal on Applied Mathematics 36:334–372.

Chakraborty T. and Rand R. (1988) The transition from phase locking to drift in a system of two weakly coupled Van der Pol oscillators. Intl. Journal Non-Linear Mechanics. 23:369–376.

Chakravarthy S.V. and Ghosh J. (1996) A complex-valued associative memory for storing patterns as oscillatory states. Biological Cybernetics 75:229–238.

Chawanya T., Aoyagi T., Nishikawa I., Okuda K., and Kuramoto Y. (1993) A model for feature linking via collective oscillations in the primary visual cortex. Biological Cybernetics 463–489.

Chay T.R. (1996a) Electrical bursting and luminal calcium oscillation in excitable cell models. Biological Cybernetics 75:419–431.

Chay T.R. (1996b) Modeling Slowly Bursting Neurons via Calcium Store and Voltage-Independent Calcium Current. Neural Computation 8:951–978.

Chillingworth D.R.J. (1976) Differential Topology With a View to Applications, Pitman Publishing, London, San Francisco, Melbourne.

Cohen A.H., Holmes P.J., and Rand R.H. (1982) The Nature of the Coupling Between Segmental Oscillators of the Lamprey Spinal Generator for Locomotion: A Mathematical Model. Journal of Mathematical Biology 13:345–369.

Cohen M.A. and Grossberg S. (1983) Absolute stability of global pattern formation and parallel memory storage by competitive neural networks. IEEE Transactions SMC-13, 815–826.

Connor J.A., Walter D. and McKown R. (1977) Modifications of the Hodgkin-Huxley Axon Suggested by Experimental Results from Crustacean Axons, Biophysical Journal 18:81–102.

Cymbalyuk G.S., Nikolaev E.V., and Borisyuk R.M. (1994) In-phase and antiphase self-oscillations in a model of two electrically coupled pacemakers. Biological Cybernetics 71:153–160.

Daido H. (1996) Onset of cooperative entrainment of limit-cycle oscillators with uniform all-to-all interactions: bifurcation of the order function. Physica D 91:24-66.

Dale H.H. (1935) Pharmacology and the nerve endings. Proc. Roy. Soc. Med. 28:319–332.

Destexhe A. and Babloyantz A. (1991) Deterministic Chaos in a Model of the Thalamo-Cortical System. In Self-Organization, Emerging Properties and Learning, A. Babloyantz, (ed.), Plenum Press, ARW Series, New York.

Ditzinger T. and Haken G. (1989) Oscillations in the Perception of Ambiguous Patterns. Biological Cybernetics 61:279–287.

Eckhaus W. (1983) Relaxation Oscillations Including a Standard Chase of French Ducks. Lecture Notes in Math., 985:432–449.

Eckhorn R., Bauer R., Jordan W., Brosch M., Kruse W., Munk M., and Reitboeck H.J. (1988) Coherent Oscillations: A mechanism of feature link in the visual cortex? Biological Cybernetics 60:121–130.

Elbert T., Ray W.J., Kowalik Z.J., Skinner J.E., Graf K.E., and Birbaumer N. (1994) Chaos and Physiology: Deterministic Chaos in Excitable Cell Assemblies.

Erdi P., Grobler T., Barna G., and Kaski K. (1993) Dynamics of the olfactory bulb: bifurcations, learning, and memory. Biological Cybernetics 69:57–66.

Ermentrout G.B. (1981) $n : m$ Phase-Locking of Weakly Coupled Oscillators. Journal of Mathematical Biology, 12:327–342.

Ermentrout G.B. (1994) An Introduction to Neural Oscillators. In Neural Modeling and Neural Networks, F Ventriglia, (ed.), Pergamon Press, Oxford, pp.79–110.

Ermentrout G.B. (1996) Type I Membranes, Phase Resetting Curves, and Synchrony. Neural Computation 8:979–1001.

Ermentrout G.B. and Kopell N. (1984) Frequency Plateaus in a Chain of Weakly Coupled Oscillators, I. SIAM Journal on Applied Mathematics 15:215–237.

Ermentrout G.B. and Kopell N. (1986) Parabolic Bursting in an Excitable System Coupled With a Slow Oscillation. SIAM Journal on Applied Mathematics 46:233–253.

Ermentrout G.B. and Kopell N. (1986) Subcellular Oscillations and Bursting. Mathematical Biosciences 78:265–291.

Ermentrout G.B. and Kopell N. (1990) Oscillator death in systems of coupled neural oscillators. SIAM Journal on Applied Mathematics 50:125–146.

Ermentrout G.B. and Kopell N. (1991) Multiple pulse interactions and averaging in systems of coupled neural oscillators. Journal of Mathematical Biology 29:195–217.

Ermentrout G.B. and Kopell N. (1994) Learning of phase lags in coupled neural oscillators. Neural Computation 6:225–241.

Ermentrout G.B., Crook S., and Bower J.M. (1997) Connectivity, Axonal Delay, and Synchrony in Cortical Oscillators. preprint.

Ezhov A.A. and Vvedensky V.L. (1996) Object Generation with Neural Networks (When Spurious Memories are Useful). Neural networks 9:1491–1495.

Farkas M. (1994) Periodic Motions. Springer-Verlag, New York.

Fenichel N. (1971) Persistence and smoothness of invariant manifolds for flows. Ind. Univ. Math. J., 21:193–225.

FitzHugh R. (1969) Mathematical Models of Excitation and Propagation in Nerve. In Biological Engineering, H.P. Schwan, (ed.), McGraw-Hill, New York, pp.1-85.

Flaherty J. and Hoppensteadt F.C. (1978) Frequency entrainment of forced van der Pol's oscillator. Stud. Appl. Math. 58:5–15.

Frankel P. and Kiemel T. (1993) Relative phase behavior of two slowly coupled oscillators. SIAM Journal on Applied Mathematics 53:1436–1446.

Freeman W.J. (1975) Mass Action in the Nervous System. New York, Academic Press.

Fukai T. (1996) Bulbocortical interplay in olfactory information processing via synchronous oscillations. Biological Cybernetics 74:309–317.

Gajic Z., Petkovski D., and Shen X. (1990) Singularly Perturbed and Weakly Coupled Linear Control Systems. Lecture Notes in Control and Information Sciences, Springer-Verlag.

Gantmacher (1959) Applications of the Theory of Matrices. Interscience, New York.

Gerstner W., van-Hemmen J.L., and Cowan J.D. (1996) What matters in Neuronal Locking? Neural Computation 8:1653 1676.

Gerstner W. and van Hemmen J.L. (1994) Coding and Information Processing in Neural Networks. In Models of Neural Networks II, Domany E., van Hemmen J.L. and Schulten K., (eds.), Springer-Verlag, New York.

Glass L. and Belair J. (1985) Continuation of Arnold tongues in mathematical models of periodically forced biological oscillators. In Othmer HG (ed.), Nonlinear Oscillations in Biology and Chemistry. Lecture Notes in Biomathematics, Springer-Verlag, pp. 232–243.

Golubitsky M., Schaeffer D.G. (1985) Singularities and Groups in Bifurcation Theory, Volume I, Springer-Verlag, New York.

Grasman J. (1987) Asymptotic Methods for Relaxation Oscillations and Applications. Springer-Verlag, New York.

Gray C.M. (1994) Synchronous oscillations in neuronal systems: mechanism and functions. Journal of Computational Neuroscience 1:11-38.

Gray C.M., Konig P., Engel A.K., and Singer W. (1989) Oscillatory responses in cat visual cortex exhibit inter-columnar synchronization which reflects global stimulus properties. Nature 338:334–337.

Golubitsky M. and Shaeffer D. (1979) A theory for imperfect bifurcation via singularity theory. Communications on Pure and Applied Mathematics, 32:21–98.

Grossberg S. (1988) Nonlinear neural networks: Principles, mechanisms, and architectures. Neural Networks 1:17–61.

Guckenheimer J. (1975) Isochrons and phaseless sets. Journal of Mathematical Biology 1:259–272.

Guckenheimer J. and Holmes D. (1983) Nonlinear Oscillations, Dynamical Systems, and Bifurcations of Vector Fields. Springer-Verlag, New York.

Guckenheimer J., Harris-Warrick R., Peck J., and Willms A. (1997) Bifurcations, Bursting and Spike Frequency Adaptation. Journal of Computational Neuroscience. to appear.

Haken H. (1988) Information and Self-Organization. A Macroscopic Approach to Complex Systems, Springer-Verlag, New York.

Hale J.K. (1969) Ordinary Differential Equations. Wiley, New York.

Hansel D., Mato G., and Meunier C. (1995) Synchrony in excitatory neural networks. Neural Computations 7:307–335.

Hassard B.D. (1978) Bifurcation of periodic solutions of the Hodgkin-Huxley Model for the Squid Giant Axon, Journal of Theoretical Biology, vol.71, pp. 401–420.

Hassard B.D., Kazarinoff N.D., and Wan Y.H. (1981) Theory and Applications of Hopf Bifurcation. Cambridge University Press, Cambridge.

Hebb D.O. (1949) The Organization of behavior. Wiley, New York.

Hirsch J.C., Fourment A., and Marc M.E. (1983) Sleep-related variations of membrane potential in the lateral geniculate body relay neurons of the cat. Brain Research 259:308–312.

Hirsch M.W. (1989) Convergent activation dynamics in continuous time networks. Neural Networks 2:331–349.

Hirsch M.W., Pugh C.C., and Shub M. (1977) Invariant Manifolds, Springer-Verlag, New York.

Hodgkin A.L. (1948) The local electric changes associated with repetitive action in a non-medulated axon. Journal of Physiology 107:165–181.

Hodgkin A.L. and Huxley A.F. (1954) A quantitative description of membrane current and application to conduction and excitation in nerve. Journal Physiol., 117:500–544.

Holden A.V., Hyde J., and Muhamad M. (1991) Equilibria. Periodicity, Bursting and Chaos in Neural Activity. In Proceedings of he 9th Summer Workshop on Mathematical Physics, vol.1 pp. 96–128.

Hopfield J.J. (1982) Neural networks and physical systems with emergent collective computational abilities. Proceedings of National Academy of Sciences USA 79:2554–2558.

Hoppensteadt F.C. (1986) An Introduction to the Mathematics of Neurons. Cambridge Univ. Press, Cambridge, U. K.

Hoppensteadt F.C. (1989) Intermittent chaos, self-organization, and learning from synchronous synaptic activity in model neuron networks. Proceedings of National Academy of Sciences USA 86:2991–2995.

Hoppensteadt F.C. (1991) The searchlight hypothesis. J Math Biol 29: 689–691.

Hoppensteadt F.C. (1993) Analysis and simulations of chaotic systems. Springer-Verlag, New York.

Hoppensteadt F.C. and Izhikevich E.M. (1995) Canonical models for bifurcations from equilibrium in weakly connected neural networks. World Congress on Neural Networks, Washington DC, I:80–83.

Hoppensteadt F.C. and Izhikevich E.M. (1996a) Synaptic Organizations and Dynamical Properties of Weakly Connected Neural Oscillators: I. Analysis of Canonical Model. Biological Cybernetics, 75:117–127.

Hoppensteadt F.C. and Izhikevich E.M. (1996b) Synaptic organizations and dynamical properties of weakly connected neural oscillators. II. Learning of phase information. Biological Cybernetics, 75:129–135.

Iooss G. and Adelmeyer M. (1992) Topics in Bifurcation Theory. Advanced series in nonlinear dynamics, vol.3, World Scientific.

Ishii S., Fukumizu K., and Watanabe S. (1996) A Network of Chaotic Elements for Information Processing. Neural Networks 9:25–40.

Izhikevich E.M. and Malinetskii G.G. (1992) A possible role of chaos in neurosystems. Dokl. Akad. Nauk 326:626-632 [translated in Sov. Phys. Docl. (1993) 37(10) October 1992:492–495].

Izhikevich E.M. and Malinetskii G.G. (1993) A neural network with chaotic behavior. preprint #17, The Institute of Applied Mathematics, Russian Academy of Sciences, Moscow (in Russian).

Izhikevich E.M. (1996) Bifurcations in brain dynamics. Ph.D. Thesis, Department of Mathematics, Michigan State University.

Izhikevich E.M. (1997) Neuro-computational properties near multiple cusp singularities. Neural Networks, submitted.

Johnston D. and Wu S.M. (1995) Foundations of Cellular Neurophysiology. The MIT Press.

Kaneko K. (1994a) Relevance of dynamic clustering to biological networks. Physica D 75:55–73.

Kaneko K. (1994b) Information cascade with marginal stability in a network of chaotic elements. Physica D 77:456–472.

Kazanovich Ya.B. and Borisyuk R.M. (1994) Synchronization in a neural network of phase oscillators with the central element. Biological Cybernetics 71:177–185.

Keener J.P. (1989) Frequency dependent decoupling of parallel excitable fibers. SIAM Journal on Applied Mathematics 49:210–230.

Kerszberg M. and Masson C. (1995) Signal-induced selection among spontaneous oscillatory patterns in a model of honeybee olfactory glomeruli. Biological Cybernetics 72:487–495.

Kilmer W. (1996) Global Inhibition for Selecting Modes of Attention. Neural Networks 9:567–573.

Kirchgraber U. and Palmer K.J. (1990) Geometry in the Neighborhood of Invariant Manifolds of Maps and Flows and Linearization. Longman Scientific & Technical, UK.

Kolovskii M.Z. (1960) On existence conditions of periodic solutions of systems of differential equations with discontinuous right-hand side having a small parameter. Applied Mathematics and Mechanics, in Russian, 24, n4.

Kopell N. (1986) Coupled oscillators and locomotion by fish. In Othmer HG (Ed) Nonlinear Oscillations in Biology and Chemistry. Lecture Notes in Biomathematics, Springer-Verlag.

Kopell N. (1995) Chains of Coupled Oscillators. In Arbib M.A. (Ed) Brain Theory and Neural Networks, The MIT press, Cambridge, MA.

Kopell N. and Ermentrout G.B. (1990) Phase transitions and other phenomena in chains of coupled oscillators. SIAM Journal on Applied Mathematics 50:1014–1052.

Kopell N. and Somers D. (1995) Anti-phase solutions in relaxation oscillators coupled through excitatory interactions. Journal of Mathematical Biology 33:261–280.

Kowalski J.M., Albert G.L., Rhoades B.K., and Gross G.W. (1992) Neuronal networks with spontaneous, correlated bursting activity: Theory and Simulations. Neural Networks 5:805–822.

Kryukov V.I. (1991) An attention model based on principle of dominanta. In Holden A.V., Kryukov V.I. (eds) Neurocomputers and Attention. I. Neurobiology, synchronization and chaos, Manchester University Press, Manchester, 319-352.

Kryukov V.I., Borisyuk G.N., Borisyuk R.M., Kirillov A.B., and Kovalenko Ye. I. (1990) Metastable and unstable states in the brain. In Dobrushin R.L., Kryukov V.I., Toom A.L. (eds) Stochastic cellular systems: ergodicity, memory, morphogenesis. Manchester University Press, chapter III, pp. 225-358.

Kuramoto Y. (1991) Collective synchronization of pulse-coupled oscillators and excitable units. Physica D 50:15–30.

Kuramoto Y. (1984) Chemical Oscillations, Waves, and Turbulence. Springer-Verlag, New York.

Kuznetsov Yu. (1995) Elements of Applied Bifurcation Theory. Springer-Verlag, New York.

Li Z., Hopfield J.J. (1989) Modeling the olfactory bulb and its neural oscillatory processings. Biological Cybernetics 61:379–392.

MacKay R.S. and Sepulchre J.-A. (1995) Multistability in networks of weakly coupled bistable units. Physica D 82:243–254.

Malkin I.G. (1949) Methods of Poincare and Liapunov in theory of non-linear oscillations. [in Russian: "Metodi Puankare i Liapunova v teorii nelineinix kolebanii" Gostexizdat, Moscow].

Malkin I.G. (1956) Some Problems in Nonlinear Oscillation Theory. [in Russian: "Nekotorye zadachi teorii nelineinix kolebanii" Gostexizdat, Moscow].

von der Malsburg C. (1995) Dynamic Link Architecture. In Arbib M.A. (Ed) Brain Theory and Neural Networks, The MIT press, Cambridge, MA.

von der Malsburg C. and Buhmann J. (1992) Sensory segmentation with coupled neural oscillators. Biological Cybernetics 67:233–242.

Marsden J.E. and McCracken (1976) The Hopf Bifurcation and Its Applications. Springer-Verlag, New York.

Mason A., Nicoll A. and Stratford K. (1991) Synaptic Transmission between Individual Pyramidal Neurons of the Rat Visual Cortex in vitro. The Journal of Neuroscience, 11:72–84.

McNaughton B.L., Barnes C.A., and Andersen P. (1981) Synaptic Efficacy and EPSP Summation in Granule Cells of Rat Fascia Dentata Studied in Vitro, Journal of Neurophysiology 46:952–966.

Miles R. and Wong R.K.S. (1986) Excitatory synaptic interactions between CA3 neurones in the guinea-pig hippocampus, Journal Phyoiol. 373: 397–418.

Mirollo R.E. and Strogatz S.H. (1990) Synchronization of pulse-coupled biological oscillators. SIAM Journal on Applied Mathematics 50:1645–1662.

Mishchenko E.F., Kolesov Yu.S., Kolesov A.Yu., and Rozov N.K. (1994) Asymptotic Methods in Singularly Perturbed Systems. Plenum Press, New York.

Morris C. and Lecar H. (1981) Voltage Oscillations in the Barnacle Giant Muscle Fiber, Biophysical Journal 35:193–213.

Moser J. (1966) On the theory of quasi-periodic motions. SIAM Review 8:145–172.

Nejshtadt A. (1985) Asymptotic investigation of the loss of stability by an equilibrium as a pair of eigenvalues slowly cross the imaginary axis. Usp. Mat. Nauk 40:190-191.

Neu J.C. (1979) Coupled chemical oscillators. SIAM Journal on Applied Mathematics 37:307–315.

Ornstein D.S. (1974) Ergodic Theory, Randomness, and Dynamical Systems. Yale University Press, New Haven and London.

Osborne N.N. (1983) Dale's Principle and Communications Between Neurones. Pergamon Press, Oxford, New York.

Palis J. and de Melo W. (1982) Geometric Theory of Dynamical Systems: An introduction. Springer-Verlag, New York.

Peponides G.M. and Kokotovic P.V. (1983) Weak connections, time scales, and aggregation of nonlinear systems. IEEE Trans. on Systems, Man, and Cybernetics, SMC-13:527–532.

Pernarowski M., Miura R.M., and Kevorkian J. (1992) Perturbation techniques for models of bursting electrical activity in pancreatic β-cells. SIAM Journal on Applied Mathematics 52:1627–1650.

Pitkovski A.S. and Kurths J. (1994) Collective behavior of ensembles of globally coupled maps. Physica D 76:411–419.

Poston T. and Stewart I. (1978) Nonlinear modeling of multistable perception. Behavioral Science 23:318–334.

Rakic P. (1976) Local circuit neurons. MIT Press, Cambridge, Mass.

Rand R.H., Cohen A.H., and Holmes P.J. (1987) Systems of coupled oscillators as models of central pattern generators. In Cohen A.H., Grillner S., and Rossignol S. (eds) Neural Control of Rhythmic Movements in Vertebrates, New York, Wiley, pp.369–413.

Rapaport A. (1952) Ignition phenomenon in random nets. Bulletin of Mathematical Biophysics 14:35–44.

Richards W., Wilson H.R., and Sommer M.A. (1994) Chaos in percepts? Biological Cybernetics 70:345–349.

Rinzel J. and Ermentrout G.B. (1989) Analysis of Neural Excitability and Oscillations. In Koch C., Segev I. (eds) Methods in Neuronal Modeling, The MIT Press, Cambridge, Mass.

Rinzel J. and Lee Y.S. (1986) On different mechanisms for membrane potential bursting. In Othmer H.G. (Ed) Nonlinear Oscillations in Biology and Chemistry. Lecture Notes in Biomathematics, Springer-Verlag.

Rinzel J. and Lee Y.S. (1987) Dissection of a model for neuronal parabolic bursting. Journal of Mathematical Biology, 25:653–675.

Rinzel J. and Miller R.N. (1980) Numerical Calculation of Stable and Unstable Periodic Solution to the Hodgkin-Huxley Equations. Mathematical Biosciences 49:27–59.

Ruelle D. (1989) Elements of Differentiable Dynamics and Bifurcation Theory. Academic Press, inc.

Sanders J.A. and Verhulst F. (1985) Averaging Methods in Nonlinear Dynamical Systems, Springer-Verlag, New York.

Sayer R.J., Friedlander M.J., and Redman S.J. (1990) The time course and amplitude of EPSPs evoked at synapses between pairs of CA3/CA1 neurons in the hippocampal slice. Journal Neuroscience. 10:826–836.

Schecter S. (1987) The saddle-node separatrix-loop bifurcation. SIAM Journal Math. Anal. 18:1142–1156.

Schuster H.G. and Wagner P. (1990) A model for neuronal oscillations in the visual cortex. I Mean field theory and derivation of the phase equations. Biological Cybernetics 64:77–82.

Shepherd G.M. (1976) Models of LCN function in the olfactory bulb. In Rakic P. (ed.), Local circuit neurons, MIT Press, Cambridge, Mass.

Shepherd G.M. (1979) The Synaptic Organization of the Brain. Oxford University Press, New York.

Shepherd G.M. (1983) Neurobiology. Oxford University Press, New York.

Skarda C.A. and Freeman W.J. (1987) How brain makes chaos in order to make sense of the world. Behav Brain Sci 10:161–195.

Skinner F.K., Kopell N., and Marder E. (1994) Mechanism for Oscillations and Frequency Control in Reciprocally Inhibitory Model Neural Networks. Journal of Computational Neuroscience 1:69–87.

Smale S. (1974) A mathematical model of two cells via Turing's equation. Lectures in Applied Mathematics 6:15-26.

Smolen P., Terman D., and Rinzel J. (1993) Properties of a bursting model with two slow inhibitory variables. SIAM Journal on Applied Mathematics 53:861–892.

Somers D. and Kopell N. (1993) Rapid synchronization through fast threshold modulation. Biological Cybernetics 68:393–407.

Soto-Trevino C., Kopell N., and Watson D. (1996) Parabolic bursting revisited. Journal of Mathematical Biology 35:114–128

Steriade M. (1978) Cortical long axoned-cells and putative interneurons during the sleep-waking cycle. Behav. Brain Science 3:465–514.

Steriade M. and Llinas R.R. (1988) The Functional States of the Thalamus and the Associated Neuronal Interplay. Physiological Reviews 68:649–741.

Steriade M., Jones E.G., and Llinas R.R. (1990) Thalamic Oscillations and Signaling. A Neurosciences Institute Publication. John Wiley & Sons, New York.

Steriade M., McCormick D.A., and Sejnowski T.J. (1993) Thalamo-cortical Oscillations in the Sleeping and Aroused Brain. Science 262:679–685.

Stewart I.N. and Peregoy P.L. (1983) Catastrophe theory modeling in psychology. Psychological Bulletin, 94:336–362.

Stollenwerk N. and Pasemann F. (1996) Control strategies for chaotic neuromodules. Intl. Journal of Bifurcation and Chaos 6:693–703.

Storti D.W. and Rand R.H. (1986) Dynamics of two strongly coupled relaxation oscillators. SIAM Journal on Applied Mathematics 46:56–67.

Swift J.W., Strogatz S.H., and Wiesenfeld K. (1992) Averaging of globally coupled oscillators. Physica D 55:239–250

Terman D. and Wang D. (1995) Global competition and local cooperation in a network of neural oscillators. Physica D 81:148–176.

Thom R. (1975) Structural Stability and Morphogenesis. W.A. Benjamin: Reading, MA.

Traub R.D. and Miles R. (1991) Neuronal Networks of the Hippocampus. Cambridge University Press, Cambridge.

Troy W. (1978) The bifurcation of periodic solutions in the Hodgkin-Huxley equations. Quarterly of Applied Mathematics, 36:73–83.

Tsodyks M., Mitkov I., and Sompolinsky H. (1993) Pattern of Synchrony in Inhomogeneous Networks of Oscillators with Pulse Interactions. Physical Review Letters 71:1280–1283.

Tsuda I. (1992) Dynamic Link of Memory – chaotic memory map in nonequilibrium neural networks. Neural Networks 5:313–326.

Usher M., Schuster H.G., and Niebur E. (1993) Dynamics of populations of integrate-and-fire neurons, partial synchronization, and memory. Neural Computation 5:570–586.

Wang X.J. and Rinzel J. (1995) Oscillatory and Bursting Properties of Neurons. In Arbib MA (Ed) Brain Theory and Neural Networks, The MIT press, Cambridge, MA.

Wiggins S. (1994) Normally Hyperbolic Invariant Manifolds in Dynamical Systems. Springer-Verlag, New York.

Williams T.L. (1992) Phase Couplings in Simulated Chains of Coupled Oscillators Representing the Lamprey Spinal Cord. Neural Computation 4:548–558.

Wilson H.R. and Cowan J.D. (1972) Excitatory and inhibitory interaction in localized populations of model neurons. Biophys J 12:1–24.

Wilson H.R. and Cowan J.D. (1973) A Mathematical theory of the functional dynamics of cortical and thalamic nervous tissue. Kybernetik 13:55–80.

Wilson M. and Bower J.M. (1991) A computer simulation of oscillatory behavior in primary visual cortex. Neural Computation 3:498–509.

Winfree A, (1967) Biological Rhythms and the behavior of Populations of Coupled Oscillators. Journal Theoretical Biology 16:15–42.

Winfree A. (1974) Patterns of phase compromise in biological cycles. Journal of Mathematical Biology 1:73–95.

Winfree A. (1980) The Geometry of Biological Time. Springer-Verlag, New York.

Woodward D.E. (1990) Synthesis of phase-locking patterns in networks with cyclic group symmetries. SIAM Journal on Applied Mathematics 50:1053–1072.

Zak M. (1989) Weakly connected neural nets. Applied Mathematics Letters 3:131-135.

Index

Applied Mathematical Sciences

(continued from page ii)

(continued on next page)

Applied Mathematical Sciences

(continued from previous page)